Linear Integrated Circuits for Technicians

Linear Integrated Circuits for Technicians

Frank R. Dungan

Gavilan College

BRETON PUBLISHERS
A division of Wadsworth, Inc.
North Scituate, Massachusetts

This book is dedicated to my wife, Ruth Ann, for her patience and encouragement throughout the preparation of this manuscript.

Breton Publishers
A Division of Wadsworth, Inc.

Library of Congress Cataloging in Publication Data

Dungan, Frank R., 1924–
  Linear integrated circuits for technicians.

  Includes index.
  1. Linear integrated circuits.   I. Title.
TK7874.D86   1984   621.381'73   83–15336
ISBN 0–534–03124–2

Printed in the United States of America
   3 4 5 6 7 8 9—88 87 86 85

Linear Integrated Circuits for Technicians was sponsored by George J. Horesta. It was designed and prepared for publication by Ellie Connolly. Marilyn Prudente served as copy editor. The cover designer was Stephen Wm. Snider; the illustrator, George Nichols. The book was set in Caledonia by Modern Graphics, Inc.; printing and binding was by The Maple-Vail Book Manufacturing Group.

Contents

Preface

Industry demands that electronic technicians possess a thorough knowledge of modern linear electronics. Unfortunately, however, available text material on linear integrated circuits has been unequal to the task. Existing books have ranged from highly theoretical, design-based presentations to project-oriented books addressed to the electronics hobbyist. Integrated circuit "handbooks" are available, but these are more useful to the technician working in industry than they are to the student who is encountering the subject for the first time. This book has been written to remedy the situation and provide future technicians with the knowledge of linear integrated circuits that is needed on the job or that is prerequisite for other electronics courses.

Linear Integrated Circuits for Technicians is a descriptive approach to the subject that minimizes mathematical analysis and design considerations while emphasizing applications and troubleshooting techniques. It is intended for use in the first course in linear integrated circuits in an electronics curriculum. Prerequisites include a course in dc/ac fundamentals and a basic understanding of semiconductor devices. The mathematics in the presentation is limited to algebra and trigonometry. To ensure student understanding of the material, a full range of learning aids is presented in the book including chapter objectives, chapter introductions, frequent examples, and extensive end-of-chapter questions and problems. For reference, an appendix containing an extensive series of data sheets is included. A breakdown of the chapters is as follows:

Chapter 1—Linear and digital signals and devices are compared, an overview of linear applications is given, and a simple explanation of fabrication and packaging is provided.

Chapter 2—The simple operational amplifier is introduced as the building block for linear integrated circuits. The 741 is used as a model because it is the most available device that is easily used and extremely versatile. For clarification, this chapter provides a review of the internal stages of the op amp, its operating characteristics, and some safety precautions for its use.

Chapter 3—The basic op amp circuits, many of which are used as the basis for further examination in later chapters, are introduced in Chapter 3. A thorough knowledge of the circuits in this chapter is essential for the reader to easily gain an understanding of the circuits to follow.

Chapter 4—For devices to operate within specified limits, voltage and current regulation is essential. This vital subject is covered in Chapter 4. The additional topics of heat sinking and regulator protection circuits are also included.

Chapter 5—Oscillator circuits are examined in this chapter. Fundamentals of the three basic types of oscillators are covered. Included are the more commonly used oscillators and voltage controlled oscillators.

Chapter 6—The various wave-shaping circuits are taken up in Chapter 6. Here the basic circuits are used as the basis for more complicated circuits. Wave-shaping, an extremely important subject, is given full coverage.

Chapter 7—Active filters, important in all analog systems, are covered at length in Chapter 7. Simple design techniques are demonstrated, and the reader is introduced to the concept of higher-order filters.

Chapter 8—The major signal processing circuits, important circuits for industrial operations, are introduced. Instrumentation devices and amplifiers are examined.

Chapter 9—Important subjects for both digital and analog device study, the digital-to-analog converter and the analog-to-digital converter, are discussed in this chapter. Several applications for these devices are demonstrated.

Chapter 10—An introduction to communications systems is provided in Chapter 10. This chapter covers the basic concepts of modulation and demodulation in AM and FM systems in a straightforward, easily understood manner. Compandors and phase-locked loop devices are also covered.

Chapter 11—One of the more versatile devices, the timer, is covered in Chapter 11. The well-known 555 timer is used as the model, but other devices used in more specialized systems are also covered.

Chapter 12—Many special circuits and devices are introduced and examined. Circuits used in consumer products and other practical circuits are discussed.

Chapter 13—Troubleshooting, always an important subject for the electronics technician, is discussed in Chapter 13. Several practical troubleshooting methods are discussed, as are typical integrated circuit failures.

An appendix at the end of the text includes manufacturers' data sheets that provide additional circuits and applications. Also provided at the end of the text are a glossary of terms and answers to selected end-of-chapter questions and problems. The accompanying laboratory manual, *Experiments in Integrated Circuits for Technicians*, provides correlated lab work and is available from Breton Publishers. An instructor's manual, also available from the publisher, provides answers to the text problems and solutions to lab experiments.

The author extends thanks and appreciation to George Bruce, Spartanburg Technical College, Spartanburg, South Carolina; Jerry D. Mullen, Tarrant County Junior College—South Campus, Fort Worth, Texas; Gerald E. Jensen, Western Iowa Technical Community College, Sioux City; and Lou Gross, Columbus Technical Institute, Columbus, Ohio for their many valuable suggestions during the preparation of the manuscript. Also, thanks are extended to the editorial and production staff of Breton Publishers for their invaluable assistance in the publication of this text.

Linear Integrated Circuits

1.1 Introduction

The purpose of this first chapter on linear integrated circuits (ICs) is to provide an overview of the subject and to introduce basic concepts. You will learn what linear integrated circuits are, how they are used, and why they are used. Finally, you will see how linear circuits are implemented with hardware. This chapter provides the background and base upon which you will build your knowledge of linear IC principles and applications. This foundation will put the concept of linear ICs into perspective so that you can relate them directly to the field of electronics. The information presented here will give you a clear understanding of the need for and uses of linear ICs.

Linear ICs are so widely used today that it is almost impossible to think of electronic equipment without them. Linear ICs are used in virtually every area of electronics, even in digital systems. They have greatly improved the reliability and capability of electronic equipment. At the same time, they have greatly reduced the cost of building and repairing electronic equipment. There is also strong potential for further advances and improvements.

1.2 Objectives

When you complete this chapter, you should be able to:

☐ Distinguish between analog and digital signals.

☐ Define the characteristics of analog signals.

☐ Distinguish between linear and digital applications.

☐ Define the purposes of linear systems.

☐ Describe the fabrication, packaging, and numbering of linear ICs.

1.3 Comparing Linear and Digital Signals and Devices

There are two basic types of electronic signals and techniques used to generate, transfer, and process information. These are digital and analog signals.

Digital Signals

Digital signals are essentially a series of pulses or rapidly changing voltage levels (usually two) that vary in discrete steps or increments. This two-level, rapid-switching characteristic is fundamental to all digital signals.

Figure 1–1 shows several examples of digital signals. Notice how these signals switch sharply between two distinct levels. In Figure 1–1A, the signal switches between 0 V and +5 V, the ideal values for the standard logic signals. A more practical example is shown in Figure 1–1B, in which the transition occurs between some level above 0 V (0.8 V) and some level less than the ideal +5 V (2.8 V). The signal in Figure 1–1C alternates between equal positive (high) and negative (low) values. Finally, Figure 1–1D represents a signal in which the highest value is 0 V and the lowest value is −5 V, ideal values for a negative-going signal. The sharpness of the rise and fall of the signals shown in Figure 1–1 would be ideal, but, practically, the ideal condition cannot be met because of the finite delay time in transition from one level to another inherent in all electronic devices and circuits, although it can be approached. Electronic circuits that generate and process on-off, true-false, and up-down signals are called digital, logic, or pulse circuits.

Analog Signals

The second form of signal is the *analog* signal. We will concentrate on analog signals because they are used in linear systems. An analog signal is an ac or dc

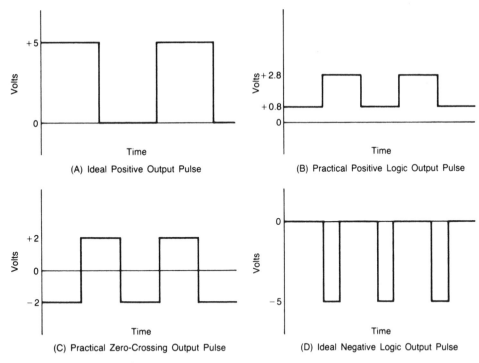

Figure 1-1. Digital Signal Waveforms

voltage or current that varies smoothly or continuously. It is a signal that does not change abruptly or in steps.

Analog signals can exist in a wide variety of forms. Figure 1-2 shows six types of analog signals. The sine wave, shown in Figure 1-2A, is the most common analog signal. Some examples of sinusoidal (sine wave) applications include the audio tones and radio waves. A fixed dc voltage, as shown in Figure 1-2B, is also an analog signal, even though it may appear to be a digital signal. Consider the fact that a battery has a fixed dc output. The output value will gradually decrease over time, even if the battery is unused. Other examples of analog signals are a pulsating dc voltage, shown in Figure 1-2C, and a varying dc voltage or current. Note that these varying signals can assume either positive values, as in Figure 1-2D, or negative values, as in Figure 1-2E. Finally, Figure 1-2F shows a random but continuously varying waveform. There are an infinite number of such analog signals.

Some Comparisons

Now we will examine analog and digital methods by using devices that are familiar to you. Table 1-1 lists a number of such items. The light bulb is an interesting

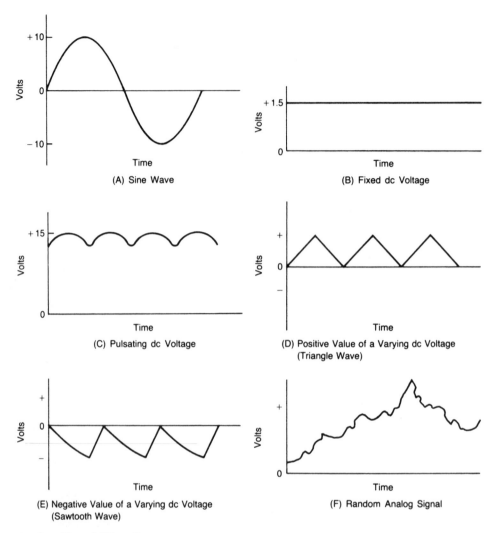

Figure 1–2. Analog Signal Waveforms

example to note, because it can be either a digital or an analog device, depending on how it is used. By using a dimmer in the circuit, the amount of current through the light bulb can be set to any level we choose, up to its maximum rated value. We can vary the current through the bulb continuously and thus vary its brightness over an infinite number of levels. The light bulb is an analog device when used in this manner.

The light bulb can also be used as a digital device, for example, when the current through the bulb and the brightness of the bulb vary in discrete steps. Without the dimmer in the circuit, a switch turns the lamp on or off at a set brightness level. Another example of the light bulb used as a digital device is

Table 1–1	Comparison of Analog and Digital Devices	
	Analog	Digital
	Electric light bulb	Electric light bulb
	Volume control on a radio or TV set	TV channel selector
	Variac or dimmer control	Any type on-off switch or switch with positive detents
	Automobile speedometer	Automobile odometer
	Typical wall clocks and wristwatches	Digital clocks and wristwatches
	Meters: standard voltmeter, ohmmeter, and ammeter	Meters: digital readout multimeters

the three-way lamp, which has three different light levels that are obtained by discrete steps of a switch.

Many of the other devices listed in Table 1–1 that we think of in analog terms also appear in the digital column. For example, the speedometer on your car is an analog device, but some newer models of cars may offer the speed indicator in digital form. But for comparison, consider the analog speedometer and the odometer portion of that speedometer. While the speed is indicated by the needle in a continuously varying and smooth manner, the odometer indicates the miles traveled in a digital manner. Since the odometer records mileage in increments of one mile or one-tenth mile, it is a digital device. You should be able to list many more comparisons of analog and digital devices on your own.

1.4 Linear Systems Defined

A *linear system* can be defined as any unit or assembly of components that generates or processes the analog signals we have discussed. Such systems are everywhere, performing a multitude of tasks. We have but to look in our homes to see examples of their uses. Radio and television receivers are prime examples of applications of these systems. The power supplies of the receivers are regulated with analog devices; the filtering action of the television tuner and radio station selector chooses only specific frequencies, rejecting all others; and the various amplifiers in both radio and television receivers are linear devices.

Today, almost all the units and assemblies mentioned in the preceding paragraph are packaged as linear integrated circuits, the subject of this book.

1.5   An Overview of Linear Applications

The broad capabilities of linear integrated circuits make them an important and vital part of electronic circuitry. Table 1–2 lists some applications of linear ICs. The first column identifies the unit or system that the IC is used in, and the second column identifies the purpose or function of the IC. These applications will be discussed in detail in later chapters.

Table 1–2 lists only a few of the many uses of linear ICs. Some additional applications are analog computers, timers, phase-locked loops, logarithmic amplifiers, and multipliers. You can probably list even more applications.

Table 1–2	Applications of Linear Integrated Circuits	
	System	**Purpose**
	Power supplies	Regulate voltage or current
	Radio and TV receivers	Amplify audio, RF, and IF signals
	Signal generators	Generate and shape desired waveforms such as square, triangle, and sawtooth waves
	Instrumentation amplifiers	Process output signals from instruments and transducers, such as strain gauges and thermocouples
	Communications systems	Generate, modulate, and detect desired frequencies as well as filter and amplify modulated signals
	Converters	Convert voltage to current and current to voltage (Important interface devices are analog-to-digital (ADC) and digital-to-analog (DAC) converters)
	Oscillators	Provide necessary amplification and phase inversion

1.6 Fabrication of Linear Integrated Circuits

You do not need to know how a device is constructed in order to use it, but there are advantages to understanding the construction. By knowing how the IC is fabricated, you will be better prepared to follow handling precautions and to select the right device for the job at hand.

There are four basic ways to manufacture integrated circuits:

1. Monolithic (the most widely used method)
2. Thin-film
3. Thick-film
4. Hybrid

Hundreds of integrated circuits are fabricated side by side on a single silicon wafer that is approximately 0.25 mm thick and two or more inches in diameter. Each of the ICs may contain thousands of elements. The circuitry is built up layer by layer.

Monolithic Construction

A *monolithic* IC is constructed entirely on a single silicon chip. The various junctions that make up components such as diodes, transistors, resistors, and capacitors are formed by diffusing semiconductor materials into the basic semiconductor substrate. The materials to be diffused into the substrate are in gaseous form, deposited on the substrate through masking operations, under high temperature. With this method, the entire circuit is formed on a single base, providing the name monolithic IC.

Thin- and Thick-Film Construction

Thin-film and *thick-film* techniques are used primarily for constructing passive networks such as resistor ladders, filters, attenuators, and phase-shift networks. Such ICs are manufactured by depositing resistive and conductive materials, through a series of masking steps, on a nonconducting base such as ceramic. The networks formed can be made extremely small, and component tolerances can be made closer than equivalent components made by the monolithic method. For this reason, thin-film and thick-film techniques are preferred for high precision circuits.

In the thin-film technique, resistors, capacitors, and conductors are formed from thin layers of metals and oxides that are deposited on an insulating substrate such as glass or ceramic. The resistors are usually formed of either tantalum or nichrome, the resistive value being determined by the length, width, and thickness of the material. The capacitors are constructed by depositing a thin metal layer on the substrate, followed by an oxide coating for a dielectric, and topped by another thin metal film. The capacitor value is determined by the area of the plates and the thickness and type of the dielectric material used. The interconnecting conductors are extremely thin strips of gold, aluminum, or platinum deposited on the substrate. To insure that the desired films are deposited in the desired locations on the substrate, a series of masking steps is used, as in the monolithic process.

The thick-film technique uses a silk-screen process to form resistors, capacitors, and conductors on an insulating substrate. In this process, a very fine wire screen is placed over the substrate and a metalized ink is forced through the screen with a squeegee. The painted surfaces are submitted to temperatures higher than 600 degrees centigrade (°C) and harden to form the interconnecting conductors. Resistors and capacitors are formed in the silk-screen process by forcing materials in paste form through a wire screen onto the substrate, and then submitting the substrate to extremely high temperatures. Resistors with tolerances as low as ±0.5% can be obtained by carefully trimming the resistive material after construction, using a sandblasting or laser trimming technique. Thick-film capacitors are limited to relatively low values.

Hybrid Construction

Hybrid ICs are combinations of monolithic, thin-film, or thick-film circuits. A hybrid IC may consist of two or more monolithic chips interconnected in a single package; a monolithic circuit combined with a thin-film or thick-film passive network; or monolithic circuits, thin-film or thick-film circuits, and individual semiconductor component chips combined in a single package. Because of the different techniques required to construct the various circuits, hybrid ICs are more complex and, therefore, more expensive than other types.

A Fabrication Procedure

A procedure used by Signetics Corporation will serve to demonstrate a typical fabrication of linear ICs. There are three major steps in the Signetics procedure—masking, etching, and diffusion.

Fabrication of a typical IC begins with a polished silicon wafer. The silicon is treated with a *dopant,* an impurity, to give the wafer a positive electrical charge. During this *initial oxidation,* a thin layer of silicon dioxide, SiO_2, a very pure form of glass, is grown on the surface, as shown in Figure 1–3A. The oxidized wafer then is subjected to masking.

Masking

Masking, sometimes called photomasking, is the name given to a series of steps that selectively cuts openings or windows into the oxide, metal, or glass surfaces of a wafer. The masking operation is a photographic process that uses a *photoresist* (film) that is sensitive to ultraviolet (UV) light. When the UV light strikes the photoresist, it hardens. Then, when the photoresist is chemically developed, the hardened area of the resist struck by UV light (clear areas of the mask) will not develop away; the soft areas not struck by the UV light (dark areas of the mask) will develop away. These steps are shown in Figure 1–3B through 1–3E. You should remember that the same pattern is being reproduced on every die (each tiny segment) of the wafer at the same time, as shown in Figure 1–3F.

Etching

After developing, the wafer is submitted to *oxide etching*. During this step, the masking procedure is completed. In the etching process, for example, oxide

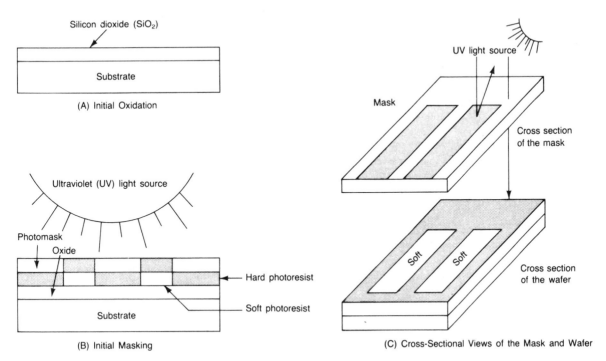

Figure 1–3. Fabrication Steps for an Integrated Circuit (Courtesy of Signetics, a subsidiary of U.S. Philips Corporation, © copyright 1980.) (Continued on p. 10)

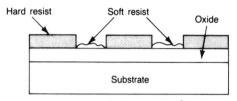

(D) Wafer after First Masking, before Developing

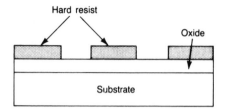

(E) Wafer after Soft Areas (Not Subjected to UV Light) Have Been Developed Away by Chemical Action

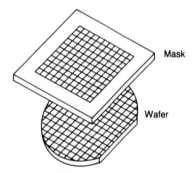

(F) Masking a Full Wafer

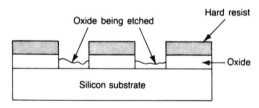

(G) Oxide Being Etched after Soft Resist Is Removed

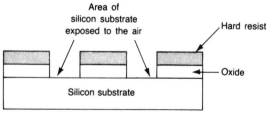

(H) Wafer after Etching

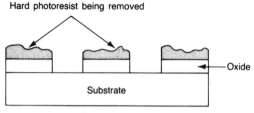

(I) Removal of the Hard Resist, No Longer Needed

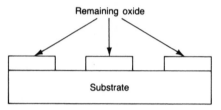

(J) Wafer after Hard Resist Removal, Ready for Dopant Application

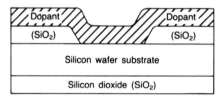

(K) Deposition of Dopants

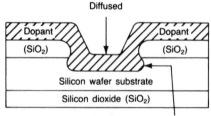

Dopant is made to penetrate the bare material of the wafer surface where the oxide has been removed.

(L) Diffusion of the Dopant into Substrate

Figure 1–3. *Continued*

not protected by hard photoresist will be etched (worn away) (Figure 1–3G) and completely removed down to the bare silicon (Figure 1–3H). Then the wafers are cleaned of the photoresist on the oxide surface because it is no longer needed. The hard layer of photoresist is dissolved completely (Figure 1–3I) and the patterns from the mask remain (Figure 1–3J).

Diffusion

There are two general operations performed in the *diffusion* process:

1. Deposition of dopants onto the wafer's surface
2. Diffusion of the dopant to the desired depth in the wafer

In the diffusion process, electrical characteristics are built into the silicon wafers step by step. Following each masking step (there may be five or more, depending on the product), the wafers are returned to diffusion, where N-type dopants carrying negative (−) charges or P-type dopants carrying positive (+) charges are deposited onto the wafer surfaces (Figure 1–3K). The wafers are then exposed to very high temperatures. The extreme heat supplies the needed energy to diffuse, or drive, the dopants to the desired depths in the wafers (Figure 1–3L). The electrical characteristics of the wafers are thereby changed in predetermined areas—areas that were formed by the mask pattern during the masking operation immediately preceding each diffusion.

The diffusion steps and the corresponding masking steps are used in combination to fabricate individual components of the semiconductors in the silicon wafers.

Once the individual components have been fabricated, they must be connected in order for them to operate as complete circuits. The circuits are completed when additional masking steps are done to perform two functions:

1. Link the individual components together.
2. Provide the necessary contacts that connect the circuit to the leads of the package that covers and protects the finished device.

Thus the fabrication of integrated circuits on a single wafer may require as many as ten or more masking steps plus the correlating diffusion operations that are done after each of the steps.

Final Processing

After fabrication, each circuit is probed by computer-controlled test instruments to distinguish good circuits from bad ones. If a circuit is defective, it is marked with an ink spot to indicate that it should be discarded. A single source of contamination during the fabrication process, such as a scratch, can cause a break in an electrical connection and ruin an entire circuit.

Following probing, the wafers are *scribed* (cut) and *sectioned* into individual dies, each of which is a complete electronic circuit. Individual dies with good circuits are then bonded into packages, with hairlike wires connected from the contacts on the die to the package leads. To make sure that the circuits have not been scratched or otherwise damaged, they are viewed under high- and low-power magnification. The packages are then marked or symboled with the company logo, an identification number, and a product type designation. When these processes are completed, the packages are sealed, put through final testing, and shipped for eventual customer use. The entire production flow, from wafer fabrication to shipment of the final product, requires three to four months.

1.7 Packaging of Linear Integrated Circuits

Linear ICs come in a variety of packages. Package outlines and specifications are shown in Appendix A. Selection of the package type depends upon the application for which the IC is intended. Devices to be used in extreme environmental conditions obviously must have the packaging and high performance that provide the highest level of reliability and integrity. Packaging materials will be discussed later in this section.

Metal Packages

Metal cans, similar to the familiar TO-5 units used for discrete transistors, were used for packaging the first linear ICs and are still in use today. Leads connected to the chip are brought out from the base of the package, just like transistors. These leads can be either pushed through holes in a printed circuit board (PCB) or inserted in special mounts.

The cans are hermetically sealed by forcing inert gas into the cap to remove moisture or other corrosive materials and then welding the cap to the base. The seal prevents any harmful materials from entering and damaging the chip.

Flat-Pack Packages

The *flat pack* was the first package style that was designed specifically for the linear IC. The leads extend from two or four sides of the flat pack, rather than from the base as in the can.

The flat pack is mounted flush on the surface of a PCB with the leads in contact with the copper on the surface rather than pushed through the board. Attaching the flat pack requires welding or resistance soldering. Because of the spread-out configuration, the flat pack requires more board space than the can; however, the reduced height of the flat pack allows more boards to be stacked.

Dual In-Line Packages

The *dual in-line package* (DIP) was designed to improve mounting and soldering capabilities of the linear ICs to PCBs. The DIP is easily inserted in the PCB by pushing the leads through holes drilled in the board. The boards are then either dip or wave soldered. The more useful method of mounting the DIP, particularly in consumer products, is on a mount that is soldered to the PCB. The DIP can be fabricated in many pin styles. Another advantage of the dual in-line package is its ease of fabrication, which results in a lower per unit cost.

Packaging Materials

There are three types of materials used for packaging ICs—plastic, metal, and ceramic. Each possesses certain characteristics that make it suitable for particular uses. It is important to know these characteristics in order to choose the appropriate package for a given application.

The least expensive of the packaging materials is a *plastic*, such as resin, epoxy, silicone, or a combination of these. The plastic is molded around the chip and leads using high temperature and pressure, thus offering good mechanical protection and some small protection from the surrounding environment. It is not hermetically sealed; therefore, it offers no protection against moisture or chemicals in the surrounding air. Because of this, the plastic package is unsuitable for use in damp or corrosive atmospheres. Plastic is used only in the DIP form of packaging. Plastic materials have the lowest thermal dissipation capability of all material types available. This limits the plastic package to use at lower temperatures and power ratings than any of the other materials. The higher thermal resistance of the plastic package results in higher chip temperatures than in other packages because the plastic does not allow as much heat to flow away from the chip.

The *metal-encased* chip provides a better hermetic seal because the joints can be soldered or welded, thus sealing the package. This isolation of the chip allows it to be used in environments that would melt or corrode a plastic-packaged chip. The metal package costs more to manufacture, but manufacturing in volume reduces that cost.

Ceramic packaging offers the highest reliability and integrity against harmful elements in the environment. The chip is hermetically sealed, preventing moisture and corrosive elements from reaching and damaging the chip. It also offers the highest thermal dissipation capability of the packaging materials, which allows its use at much higher temperatures than the other types.

As you have learned, selection of the right IC, the right package, and the right packaging material are equally important. The choices all depend upon the application for which the IC is intended.

1.8 Classification and Numbering of Linear Integrated Circuits

Type classification and numbering of integrated circuits vary with manufacturers. All manufacturers assign code numbers to the devices they produce. In general, standard ICs such as the 741 and 301 are labeled as such by all manufacturers. However, other ICs may have different part numbers assigned, even if they are identical circuits manufactured by different vendors. You also should be aware that even if an IC from one vendor has the same basic circuit and the same part number as an IC from another vendor, the ICs may not be interchangeable.

Most vendors use an alphanumeric code to classify their ICs. The first two or three alphabetic characters identify the vendor. The letters are followed by a numeric designation to classify the type of IC. Some vendors assign a final alphabetic character to designate the package type and material.

Some representative designations from various manufacturers are listed in Table 1–3. Some examples of specific code numbers and products are also shown in the table. Many manufacturers provide a section in their data books and data sheets in which their code numbering system is explained. Two examples of these are shown in Figure 1–4.

Pro-Electron Numbering System

Because of the diversity of numbering systems for linear ICs, an organization called *Pro-Electron*, located in Belgium, has attempted to establish a standard identification system. All vendors would register their IC products by a code of

Table 1–3	Manufacturers' Type Identification Codes		
Manufacturer	**Identification Code**		**Example**
Advanced Micro Devices (AMD)	Am	Am 9614	Dual differential line driver
Fairchild	μA	μA 741	Frequency-compensated op amp
Harris	HA	HA 2000	FET-input op amp
Motorola	MC, MFC	MC 4044P	Frequency phase detector
National Semiconductor	LH, LM	LM 311H	High-performance voltage comparator
RCA	CA, CD	CD 4529	Dual 4-channel analog data selector
Signetics	SE/NE, N/S	NE 544N	Servo amplifier
Texas Instruments	SN	SN 52741	High-performance op amp

six characters, three alphabetic and three numeric. Figure 1–5 shows an example of one such unit. Note that the first alphabetic character establishes the device type. The letter *T* is used for linear ICs, and the letter *U* is used if the device contains both linear and digital circuits. The next two alphabetic and two numeric characters form an identification of the type of device. The final character designates the temperature range of operation, which is shown in Table 1–4. Finally, one additional character may be added to the six-character designation

Fairchild 38510/883A MOS Order Information & Available Products

Fairchild MOS offers standard processing and packaging to cover the majority of requirements. There are only two processing classifications—Unique 38510/883A for HI REL requirements or Matrix VI for commercial/industrial requirements. When a ''garden variety'' plastic or ceramic part is ordered, a Matrix VI, Level 1 (PC) or Level 3 (DC) part is shipped. The order-number formats indicate the appropriate speed-grade, package, temperature range, processing and screening-level combinations. For additional information, contact the nearest Fairchild Sales Office.

Unique 38510/883A Device-Number Format

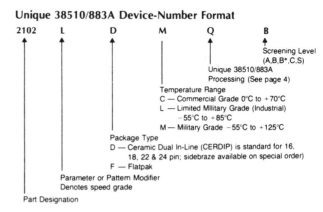

Unique 38510/883A Products

Part Number	Memory Products	Packages		Temperature		
		D	F	C	L	M
2102	1024 × 1-Bit Static RAM	x	x	x	x	x
3341	64 × 4 FIFO	x	s	x	x	x
3342/47	Quad 64/Quad 80-Bit Static Shift Register	x	s	x	x	s
3348/49	Hex 32-Bit Static Shift Register	x	s	x	x	
3351	40 × 9 FIFO	x		x	x	s
3357	Quad 80-Bit Static Shift Register	x	s	x	x	x
3515	512 × 8 4096-Bit ROM	x		x	x	s
3539	256 × 8 2048-Bit Static RAM	x		x	x	
4027	4096 × 1-Bit Dynamic RAM (16 pin)	x	x	x	x	
4096	4096 × 1-Bit Dynamic RAM (16 pin)	x	x	x	x	
F16K	16,384 × 1-Bit Dynamic RAM (16 pin)	x	x	x	x	
General Products						
3262 A/B	TV Sync Generator	x	s	x		
3708	8-Channel Analog Multiplexer	x	x	x	x	x
3815	5-Decade Counter	x		x		
3816	Programmable Divider	x	s	x		
NOTE:	3705 MUX is replaced by 3708					

(A) Numbering System for Fairchild Camera & Instrument Corporation

Figure 1–4. Typical Manufacturer's Numbering System Explanation (Part A: Courtesy of Fairchild Camera & Instrument Corporation, © copyright; Part B: Courtesy of National Semiconductor Corporation, © copyright 1980.) (Continued on p. 16)

PACKAGE

D — Glass/Metal Dual-In-Line Package

F — Glass/Metal Flat Pack

H — TO-5 (TO-99, TO-100, TO-46)

J — Low Temperature Glass Dual-In-Line Package

K — TO-3 (Steel)

KC — TO-3 (Aluminum)

N — Plastic Dual-In-Line Package

P — TO-202 (D-40, Durawatt)

S — "SGS" Type Power Dual-In-Line Package

T — TO-220

W — Low Temperature Glass Flat-Pack

Z — TO-92

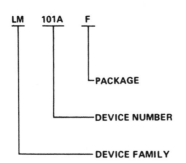

DEVICE NUMBER

3, 4, or 5 Digit Number Suffix Indicators:

A — Improved Electrical Specification

C — Commercial Temperature Range

DEVICE FAMILY

AD — Analog to Digital

AH — Analog Hybrid

AM — Analog Monolithic

CD — CMOS Digital

DA — Digital to Analog

DM — Digital Monolithic

LF — Linear FET

LH — Linear Hybrid

LM — Linear Monolithic

LX — Transducer

MM — MOS Monolithic

TBA — Linear Monolithic

(B) Numbering System for National Semiconductor Corporation

Figure 1–4. *Continued*

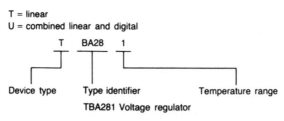

T = linear
U = combined linear and digital

TBA281 Voltage regulator

Figure 1–5. Pro-Electron Numbering System

Table 1–4	Temperature Range Code Used in the Pro-Electron System	
	Code Number	Temperature Range
	0	No designation
	1	$0°$ to $+70°C$
	2	$-55°$ to $+125°C$
	3	$-10°$ to $+85°C$
	4	$+15°$ to $+55°C$
	5	$-25°$ to $+70°C$
	6	$-40°$ to $+85°C$

if there are different versions of the same device. (This is reminiscent of the letter designations following identifiers on the vacuum tube that indicated when some change was made in electrical characteristics or in the tube envelope.)

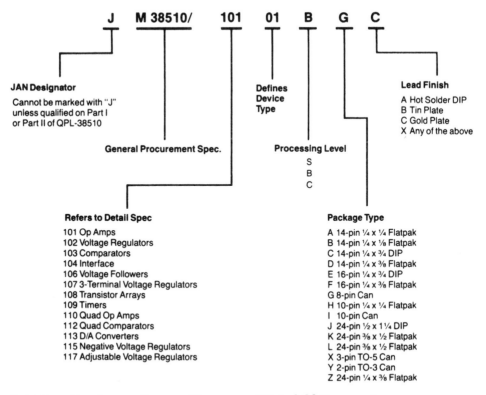

Figure 1–6. JAN Part Numbering System (Courtesy of Fairchild Camera & Instrument Corporation, © copyright.)

Military Numbering System

There is one more important numbering system that you should be aware of—the military system. Devices that meet exacting military standards are assigned a Joint Army-Navy (JAN) part number. The JAN system of numbering involves several groups of characters, as shown in Figure 1–6. The system provides far more information about the device than the standard and Pro-Electron systems show. Most manufacturers who sell their devices to military sources provide a section in their data books that designates the JAN system and the steps taken to satisfy the military standards.

1.9 Summary

  1. Digital and analog signals and techniques are used to generate, transfer, and process information.

  2. Digital signals have only two possible values and are characterized by a fast-switching up-down action.

  3. Analog signals can take any value of voltage or current and are characterized by a continuously changing signal.

  4. The most common type of analog signal is the sine wave.

  5. Some applications of linear systems are power supplies, radio and television receivers, signal generators, and communications systems.

  6. Linear ICs are fabricated in a variety of ways, primarily by diffusion.

  7. Linear ICs are packaged in metal cans, flat-pack packages, and dual in-line packages.

  8. Packaging materials are plastic, metal, and ceramic. These materials offer different environmental tolerances. Ceramic offers the best overall characteristics.

  9. Selection of an IC for a particular application must include environmental considerations.

  10. Classification and numbering of ICs vary with the manufacturer. Reference to the manufacturer's data sheets or data books is necessary when selecting replacement parts.

  11. Of the three methods used in classification and numbering, the JAN method provides the most information.

1.10 Questions and Problems

1.1 Draw a waveform for a digital signal with a positive value of +5 volts and a negative value of −2 volts.

1.2 Draw an analog signal waveform that varies smoothly between +10 volts and −10 volts. What is this type of waveform called?

1.3 Explain the difference between a digital signal and an analog signal.

1.4 Define *linear system*.

1.5   Give three examples of linear systems.

1.6   You are to use a linear system in a tropical coastal region. What type of packaging would you select for the ICs in that system? Why?

1.7   List five applications of linear ICs.

1.8   What factors must you consider when selecting an IC?

1.9   You are to replace an IC in the audio section of a television receiver. You do not have an exact replacement, but you do have similar ICs in stock. What must you do in order to select the proper replacement?

1.10 Which classification and numbering system offers the most information to the user?

The Operational Amplifier— A Building Block

2.1 Introduction

Operational amplifiers, *op amps*, were originally constructed with discrete components within a sealed package. They were used for analog computing circuits, instrumentation, and in control circuits. Basically, the term *op amp* was used to describe certain high-performance dc amplifiers.

Today's op amp is a conveniently packaged, very high gain dc amplifier in the form of an integrated circuit. It is the building block for many circuits, because only a few externally connected components are required to provide feedback, which controls the op amp response.

Circuits based on the op amp as the active device come in many configurations. The most widely used in the electronics industry has a differential input and single output. This configuration will be used throughout this book.

In this chapter you will study the stages contained within the IC op amp, its dc and ac operating characteristics, power supplies, protective circuits, and manufacturers' data sheets.

2.2 Objectives

When you complete this chapter, you should be able to:

- ☐ List the operating characteristics of an op amp.
- ☐ Name the stages of an op amp and state the function of each.
- ☐ List the advantages and disadvantages of an op amp.
- ☐ Define input offset current, input offset voltage, slew rate, gain-bandwidth product, common-mode operation, and common-mode rejection ratio.
- ☐ Determine which op amp configurations allow use of a single power supply.
- ☐ Obtain valuable information from manufacturers' data sheets.

2.3 Op Amp Symbol

Before we examine practical op amp characteristics, let's look at the op amp schematic symbol and some of the identifiers usually found there. The standard symbol is shown in Figure 2–1. This symbol remains the same regardless of the type of op amp, package material, or package configuration being considered.

The triangular shape of the symbol in Figure 2–1 is representative of all amplifiers. The major difference in the op amp symbol is at the input. Note the two input terminals. The minus ($-$) sign indicates an *inverting* input. This means that any ac or dc signal applied to this input will be 180° out of phase at the output. The plus ($+$) sign indicates a *noninverting* input. Any ac or dc signal applied to this input will be in phase at the output.

Power supply terminals are shown above and below the triangle. These terminals may not always be shown on a schematic. The same applies to other terminals, such as those used for null or frequency compensation adjustments.

The manufacturer's part number or type of op amp is generally centered in the triangle. In some general-circuit schematics that do not indicate a specific

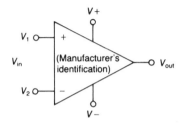

Figure 2–1. Standard Op Amp Schematic Symbol

type of op amp, the symbol may contain other identifiers, such as A_1, A_2, A_3 (representing amplifiers 1, 2, 3).

2.4 Operating Characteristics of the Op Amp

Linear IC op amps provide amplifier characteristics that were unattainable in op amps constructed with discrete components. These unique characteristics are what make op amps so versatile and almost universally used in electronic circuits today.

The ideal op amp would have the following characteristics:

1. Infinite open-loop gain
2. Infinite input impedance
3. Zero output impedance
4. Infinite bandwidth
5. Zero offset, that is, zero output voltage when input voltage is zero
6. Zero drift with temperature change

In practice, there are no ideal op amps. However, the operating characteristics of modern op amps do make them an ideal base upon which to build many practical circuits.

Reading the Data Sheets

Knowing the practical characteristics of linear ICs will provide you with the understanding necessary for you to service or design circuits in which they are used. There are many different linear ICs available, each with its own set of characteristics. Manufacturers provide data sheets that specify typical operating characteristics. It is important that you be able to read data sheets so that you can select the proper device for a specified application.

We will now study a typical data sheet—the μA741C data sheet shown in Figure 2–2. The *C* represents the commercial version of the μA741 op amp, the most commonly used op amp.

The figure lists "absolute maximum ratings." These are electrical limitations that must not be exceeded if the device is to operate properly and if possible damage to the device is to be minimized. These ratings are generally self-explanatory and need not be discussed at this point. Specifics will be examined when necessary as we proceed through this chapter.

Figure 2–2 also lists important parameters for specified electrical characteristics. Some special purpose op amps may have an *input resistance* (impedance) as high as 100 MΩ, but standard op amps are typically 1 MΩ or more. The 741C,

FAIRCHILD

A Schlumberger Company

µA741
Operational Amplifier

Linear Products

Description

The µA741 is a high performance Monolithic Operational Amplifier constructed using the Fairchild Planar epitaxial process. It is intended for a wide range of analog applications. High common mode voltage range and absence of latch-up tendencies make the µA741 ideal for use as a voltage follower The high gain and wide range of operating voltage provides superior performance in integrator, summing amplifier, and general feedback applications.

- **NO FREQUENCY COMPENSATION REQUIRED**
- **SHORT-CIRCUIT PROTECTION**
- **OFFSET VOLTAGE NULL CAPABILITY**
- **LARGE COMMON MODE AND DIFFERENTIAL VOLTAGE RANGES**
- **LOW POWER CONSUMPTION**
- **NO LATCH-UP**

Connection Diagram
8-Pin Metal Package

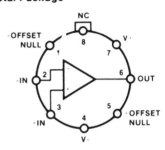

(Top View)

Pin 4 connected to case

Order Information

Type	Package	Code	Part No.
µA741	Metal	5W	µA741HM
µA741A	Metal	5W	µA741AHM
µA741C	Metal	5W	µA741HC
µA741E	Metal	5W	µA741EHC

Connection Diagram
8-Pin DIP

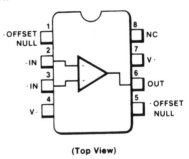

(Top View)

Connection Diagram
10-Pin Flatpak

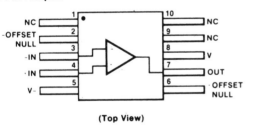

(Top View)

Order Information

Type	Package	Code	Part No.
µA741	Flatpak	3F	µA741FM
µA741A	Flatpak	3F	µA741AFM

Order Information

Type	Package	Code	Part No.
µA741C	Molded DIP	9T	µA741TC
µA741C	Ceramic DIP	6T	µA741RC

Figure 2–2. Typical Data Sheet (Courtesy of Fairchild Camera & Instrument Corporation, Linear Division, © copyright 1982.) (Continued)

μA741 and μA741C

Electrical Characteristics $V_S = \pm 15$ V, $T_A = 25°C$ unless otherwise specified

Characteristic	Condition	μA741			μA741C			Unit
		Min	Typ	Max	Min	Typ	Max	
Input Offset Voltage	$R_S \leq 10$ kΩ		1.0	5.0		2.0	6.0	mV
Input Offset Current			20	200		20	200	nA
Input Bias Current			80	500		80	500	nA
Power Supply Rejection Ratio	$V_S = +10, -20$ $V_S = +20, -10$ V, $R_S = 50$ Ω		30	150		30	150	μV/V
Input Resistance		.3	2.0		.3	2.0		MΩ
Input Capacitance			1.4			1.4		pF
Offset Voltage Adjustment Range			± 15			± 15		mV
Input Voltage Range					± 12	± 13		V
Common Mode Rejection Ratio	$R_S \leq 10$ kΩ				70	90		dB
Output Short Circuit Current			25			25		mA
Large Signal Voltage Gain	$R_L \geq 2$ kΩ, $V_{OUT} = \pm 10$ V	50k	200k		20k	200k		
Output Resistance			75			75		Ω
Output Voltage Swing	$R_L \geq 10$ kΩ				± 12	± 14		V
	$R_L \geq 2$ kΩ				± 10	± 13		V
Supply Current			1.7	2.8		1.7	2.8	mA
Power Consumption			50	85		50	85	mW
Transient Response (Unity Gain)	Rise Time $V_{IN} = 20$ mV, $R_L = 2$ kΩ, $C_L \leq 100$ pF		.3			.3		μs
	Overshoot		5.0			5.0		%
Bandwidth (Note 4)			1.0			1.0		MHz
Slew Rate	$R_L \geq 2$ kΩ		.5			.5		V/μs

Absolute Maximum Ratings

Supply Voltage
μA741A, μA741, μA741E   ± 22 V
μA741C   ± 18 V
Internal Power Dissipation (Note 1)
Metal Package   500 MW
DIP   310 mW
Flatpak   570 mW
Differential Input Voltage   ± 30 V
Input Voltage (Note 2)   ± 15 V
Storage Temperature Range
Metal Package and Flatpak   −65°C to +150°C
DIP   −55°C to +125°C

Operating Temperature Range
Military (μA741A, μA741)   −55°C to +125°C
Commercial (μA741E, μA741C)   0°C to +70°C
Pin Temperature (Soldering 60 s)
Metal Package, Flatpak, and Ceramic DIP   300°C
Molded DIP (10 s)   260°C
Output Short Circuit Duration (Note 3)   Indefinite

Notes

4. Calculated value from BW(MHz) = $\dfrac{0.35}{\text{Rise Time (μs)}}$

5. All $V_{CC} = 15$ V for μA741 and μA741C.

6. Maximum supply current for all devices
   25°C = 2.8 mA
   125°C = 2.5 mA
   −55°C = 3.3 mA

Figure 2–2. *Continued*

for example, has an input resistance of 2 MΩ. The higher the input impedance, the better the op amp will perform. *Input capacitance* may become an important factor when the op amp is to be operated at high frequencies. Typically, this capacitance is less than 2 pF. Our example has an input capacitance of 1.4 pF.

Since all op amps are different, *output resistances* (impedances) may vary from near zero to several thousand ohms. The 741C has an output resistance of 75 Ω. With such low impedence, the output can function as a *voltage source* capable of providing current for a variety of load ranges. Also, with high input impedance and low output impedance, the op amp makes an excellent *impedance-matching* device.

Characteristics Affecting dc Operations

Input bias current, the average of the two input currents (+ and −) with no signal applied, is held to microamperes or less by the high input impedance. This current can cause an imbalance in the op amp, and thus in the output. Using field effect transistors (FETs) at the inputs of the op amp will reduce the amount of input bias current and the imbalance in the op amp.

To obtain zero output voltage, both input currents should be equal. Since this is practically impossible, there must be an *input offset current*, the absolute value of the difference between the two input currents, to maintain a 0-volt output. In other words, one input may require more current than the other. This offset current may go as high as 20 mA in some ICs. The 741C has a typical input offset current of 20 nA.

The output voltage of an op amp should be zero when the voltage at both inputs is zero, but because of the inherent high gain in op amps, any circuit imbalance can cause some output voltage. The output voltage can be maintained at zero by applying a small *input offset voltage*, the absolute value of the voltage between the input terminals, to one of the inputs. This is accomplished by applying a small reference voltage to one of the terminals. The 741C typically requires 2.0 mV.

Op amps, as all other solid-state devices, are susceptible to temperature changes. Alternating current circuits using op amps are less susceptible than are dc circuits. *Drift* is the term used for a change in offset current and offset voltage caused by temperature change. Drift will upset any adjusted imbalance in the op amp and produce output voltage errors.

Characteristics Affecting ac Operations

Frequency compensation capacitors are frequently added to op amps to prevent high frequency output signals from being fed back to the input and causing undesired oscillations. These capacitors may be added externally or internally. The capacitor decreases the gain of the op amp as frequency increases, thus reducing the possibility of such feedback. The 741C uses internal frequency compensation.

Slew rate is the most important characteristic affecting the ac operation of an op amp, because it limits large-signal bandwidth. Slew rate indicates how fast the output voltage of the op amp can change. It can be stated as follows:

$$\text{Slew rate} = \frac{\text{Maximum change in output voltage}}{\text{Change in time}} = \frac{\Delta V_{\text{out(max)}}}{\Delta t} \qquad (2.1)$$

For example, the 741C has a slew rate of 0.5 volts per microsecond (V/μs), which means that output voltage can change no faster than 0.5 volts per microsecond, regardless of how fast the input voltage changes. The frequency compensation capacitor of the op amp, either internal or external, is the most common cause of slew rate limiting. Slew rate is normally specified at unity gain of the op amp. Figure 2–3 illustrates examples of slew rate limiting. The square wave input signal in Figure 2–3A rises and falls rapidly. The output signal shows the limitation of the rapid rise and fall caused by slew rate limiting. The sine wave input of Figure 2–3B results in the distorted output wave shown. As the frequency of the input signal increases, the slew rate limiting becomes more pronounced.

Voltage Gain

The higher the voltage gain of an op amp, the better. Gains of 200,000 are common. This gain is *open-loop gain* (considered as infinite gain), generally at zero hertz or dc, and *without feedback*. Figure 2–4 shows the gain–frequency response curve for the 741C. Most of this gain is sacrificed in practical circuits, because the op amp is normally operated with heavy degenerative or regenerative

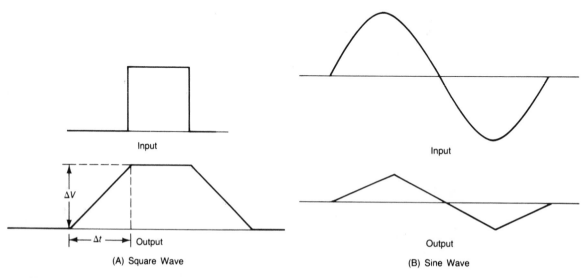

(A) Square Wave

(B) Sine Wave

Figure 2–3. Examples of Slew Rate Limiting

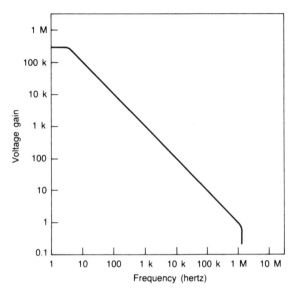

Figure 2–4. Gain Versus Frequency
Response Curve

feedback. Degenerative feedback drastically reduces the gain of the circuit, but it has an advantage, however, in that it increases the bandwidth (*BW*) of the circuit. Notice in Figure 2–4 that as the frequency is increased by one decade (tenfold) the gain is decreased by one decade, and the bandwidth is increased by one decade. For example, the open-loop bandwidth is about 10 Hz with an open-loop gain of 100,000. As gain is reduced to about 100, bandwidth is increased to about 10 kHz. When gain is 10, bandwidth is about 100 kHz. Unity gain, gain of one, occurs at 1 MHz, which is called the *unity-gain frequency* for this example. Many manufacturers use the unity-gain frequency as a reference point for op amps. Different op amps may have different values than the 741.

The *gain-bandwidth product* (*GBP*) is equal to the unity-gain frequency. The *GBP* not only indicates the upper useful frequency of a circuit, but it provides a means of determining the bandwidth for any given gain. Refer again to Figure 2–4. If you multiply the gain and bandwidth of any specific circuit, the product will equal the unity-gain frequency. This relationship can be expressed as follows:

$$GBP = A_v \times BW = \text{Unity-gain frequency} \qquad (2.2)$$

where

GBP = gain-bandwidth product
A_v = amplifier voltage gain
BW = bandwidth

Example 2.1

An op amp circuit has a gain of 100 and a bandwidth of 10 kHz. What is the gain-bandwidth product of the circuit?

Solution

Use Equation 2.2:

$$GBP = A_v \times BW = 100 \times 10,000 \text{ Hz} = 1,000,000 \text{ Hz} = 1 \text{ MHz}$$

Example 2.2

An op amp has a gain of 10 and a bandwidth of 100 kHz. What is the gain-bandwidth product of the circuit?

Solution

Use Equation 2.2:

$$GBP = A_v \times BW = 10 \times 100 \text{ kHz} = 1000 \text{ kHz} = 1 \text{ MHz}$$

By simple algebraic manipulation of Equation 2.2, you can determine the upper frequency limit or bandwidth of any circuit with any given gain by dividing the unity-gain frequency of the op amp used by the given gain, as follows:

$$BW = \frac{\text{Unity-gain frequency}}{A_v} \tag{2.3}$$

Example 2.3

You are using a general purpose op amp that has a unity-gain frequency of 1 MHz. Your circuit is to have a gain of 50. What will be the bandwidth of the circuit?

Solution

Use Equation 2.3:

$$BW = \frac{\text{Unity-gain frequency}}{A_v} = \frac{1,000,000 \text{ Hz}}{50} = 20,000 \text{ Hz} = 20 \text{ kHz}$$

Likewise, the gain of the circuit can be determined by further manipulation of Equation 2.2:

$$A_v = \frac{\text{Unity-gain frequency}}{BW} \tag{2.4}$$

Example
2.4

You are using a 741 op amp, and your circuit has a bandwidth of 3 kHz. What is the gain of the circuit?

Solution

Use Equation 2.4:

$$A_v = \frac{\text{Unity-gain frequency}}{BW} = \frac{1,000,000 \text{ Hz}}{3,000 \text{ Hz}} = 333$$

Noise

Any electronic circuit is susceptible to noise, and op amps are no exception. External noise generated by electrical devices and inherent noise of electronic components can be minimized by proper construction techniques. Internal noise resulting from internal components, bias current, and drift can be minimized by keeping input and feedback resistor values as low in value as possible and still maintain circuit requirements. You should remember that undesired noise will be amplified along with the desired signal.

Common-Mode Rejection

The input circuit of the IC op amp is a differential amplifier stage. Associated with this input stage is a feature called *common-mode rejection*. This means that if the voltage applied to both inputs of the differential amplifier is in phase and has the same amplitude and frequency, the output will be zero. A voltage at the output can be produced only if a difference of potential is introduced at the inputs. The ability of an op amp to reject common-mode signals while amplifying differential signals is called the *common-mode rejection ratio* (CMRR). The CMRR is usually expressed in *decibels* (dB). The higher the rating, the better the common-mode rejection. Common-mode operation will be discussed in more detail later in this chapter.

2.5 The Three Stages in Op Amp Construction

While it is not necessary to know what is inside the op amp in order to use it, knowledge of the op amp and its operation is enhanced by understanding the three stages of its construction. Figure 2–5 is a block diagram of a typical op amp that shows the three stages—a differential amplifier, a high-gain voltage amplifier, and an output amplifier. It also shows the power supply and input and output connections.

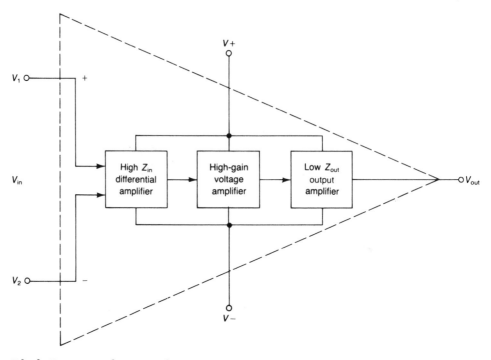

Figure 2–5. Block Diagram of a Typical Op Amp

Review of Differential Amplifiers

The input stage of virtually all op amps is a *differential amplifier*. The differential amplifier is ideally suited to IC fabrication because it has no capacitors; it requires only transistors and resistors.

A simple differential amplifier is shown in Figure 2–6. The amplifier has two input (V_1 and V_2) and two output ($\pm V_{out}$) terminals. The differential output is taken between the two collectors. Resistor R_E assures an equal voltage across the emitter-base junctions of both transistors. With discrete component construction, this circuit would require matched-pair transistors (Q_1 and Q_2) to provide the desired symmetry. However, transistors fabricated in ICs have almost identical characteristics, so the circuit is very nearly symmetrical. Transistor biasing is accomplished with base resistors (R_B) and collector resistors (R_C). The circuit amplifies the difference between the two inputs. Output voltage (V_{out}), taken between the two collectors, is equal to voltage input 1 (V_1) minus voltage input 2 (V_2) times the amplifier gain (A_v). Expressed in equation form,

$$V_{out} = A_v (V_1 - V_2) \tag{2.5}$$

In special cases, the two input signals will be identical except that they will be 180° out of phase. Because the signals are 180° out of phase, the differential

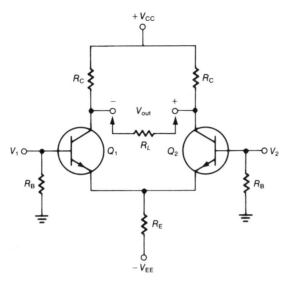

Figure 2–6.   Basic Differential Amplifier
Circuit

(difference) voltage (V_D) between the inputs is twice the amplitude of either input signal. This relationship can be stated as follows:

$$V_D = V_1 - V_2 \qquad\qquad (2.6)$$

**Example
2.5**

If input V_1 is equal to $+10$ mV and input V_2 is equal to -10 mV, what is the amplitude of the difference voltage?

Solution

Use Equation 2.6:

$$V_D = V_1 - V_2 = +10 \text{ mV} - (-10 \text{ mV}) = +10 \text{ mV} + 10 \text{ mV} = 20 \text{ mV}$$

Analyzing the effects of each input separately provides a better understanding of why the circuit responds as it does. Refer to Figure 2–6, and first consider the effects of V_1. A positive-going signal at the base of transistor Q_1 causes an amplified negative-going signal at the collector of Q_1. By emitter-follower action, V_1 also couples a positive-going signal to the emitter of transistor Q_2, which causes an amplified positive-going signal at the collector of Q_2.

V_2 swings negative at the same time V_1 swings positive. This places a negative-going signal on the base of Q_2, which drives the collector of Q_2 more positive. This means that V_1 and V_2 acting together tend to force the collector of Q_2 positive. Also, through emitter-follower action, Q_2 couples a sample of V_2 to the emitter of Q_1. The negative-going signal of V_2 on the emitter of Q_1 causes Q_1 to conduct more and drive its collector more negative. Thus, V_1 and V_2 acting together tend to force the collector of Q_1 more negative.

Acting together, V_1 and V_2 produce an amplified output signal. A load is normally connected between the two output terminals on the collectors. A large output voltage is developed across the load (R_L) because the signals at the two output terminals are 180° out of phase.

If desired, the load could be connected between either output terminal and ground. The voltage available at the remaining terminal would have the same amplitude but the opposite phase.

Constant-Current Source

Most differential amplifiers used in ICs have a *constant-current source* in the emitter circuit. Figure 2–7 shows a practical current source, using a zener diode to maintain a constant voltage (V_Z) across the emitter-base junction. Resistor R_s is a current-limiting series resistor for zener diode protection.

Assume Q_1 is a silicon transistor with an emitter-to-base voltage drop (V_{BE}) of exactly 0.6 V. Then the voltage across the emitter resistor (R_E) is as follows:

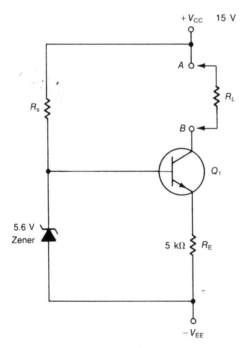

Figure 2–7.   Constant-Current Source

$$V_{R_E} = V_Z - V_{BE}$$
$$= 5.6 \text{ V} - 0.6 \text{ V} = 5.0 \text{ V} \tag{2.7}$$

The emitter current (I_E) is

$$I_E = \frac{V_{R_E}}{R_E}$$
$$= \frac{5 \text{ V}}{5000 \ \Omega} = 0.001 \text{ A, or } 1 \text{ mA} \tag{2.8}$$

Since Q_1 draws a very small amount of base current, for practical purposes, consider the collector current to be 1 mA. Shorting across points A and B will cause 1 mA of current to flow. Connecting a load resistor (R_L) across these same points will have little effect on current flow as long as R_L is not made too large. For proper transistor action, the collector-base junction must be reverse biased, so the collector voltage must not drop lower than the base voltage. Figure 2–7 shows a base voltage of 5.6 V and a V_{CC} of 15 V. Therefore, the collector voltage will be the same as base voltage when

$$V_L = V_{CC} - V_B$$
$$= 15 \text{ V} - 5.6 \text{ V} = 9.4 \text{ V} \tag{2.9}$$

The collector current (I_C) is essentially 1 mA, so R_L will drop 9.4 V when its value is

$$R_L = \frac{V_L}{I_C}$$
$$= \frac{9.4 \text{ V}}{0.001 \text{ A}} = 9400 \ \Omega \tag{2.10}$$

Keeping R_L well below this value will allow the circuit to act as a constant-current source. Figure 2–8 shows the current source circuit connected to the differential amplifier circuit. The current delivered to the differential amplifier is independent of changes in the input signals.

Common-Mode Signal

A look at another characteristic of the differential amplifier will demonstrate the importance of the constant-current source.

As discussed earlier, a common-mode signal is one that is identical at both input terminals. Since one of the major advantages of the differential amplifier is its ability to reject common-mode signals, it should not respond to such signals. This becomes obvious if we remember that

$$V_{out} = A_c (V_1 - V_2) \qquad \text{(repeat of Equation 2.5)}$$

If $V_1 = V_2$, then $V_1 - V_2 = 0$ and V_{out} would be zero for all common-mode signals.

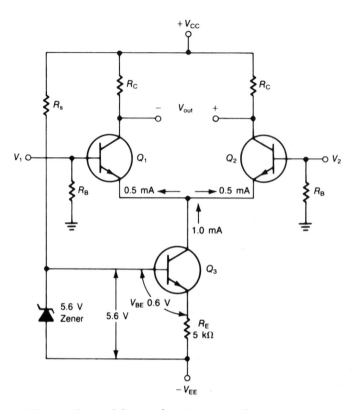

Figure 2–8. Differential Amplifier with a Constant-Current Source

Consider what happens when the two signals shown in Figure 2–9 are applied to the circuit shown in Figure 2–8. The two signals can be classified as common-mode signals because V_1 and V_2 are identical; that is, they are in phase and of equal amplitude.

Both V_1 and V_2 are at zero volts at time T_0. A constant 1 mA current is provided by the current source (see Figure 2–8). This current is split, 0.5 mA to each transistor. Both V_1 and V_2 swing positive to $+100$ mV at time T_1 (see Figure 2–9). Under ordinary circumstances you would expect both transistors to conduct more when the bases swing more positive. But remember that the current source is providing exactly 1 mA, and since both bases swing positive by the same amount, both transistors will still draw exactly the same 0.5 mA current as before. The current does not change and, therefore, the output voltage (V_{out}) does not change.

The resultant rejection of the common-mode signal and amplification of the differential signal is a ratio, called the common-mode rejection ratio (CMRR). It is the ratio of difference gain (A_D) to common-mode gain (A_{CM}), usually expressed in decibels (dB).

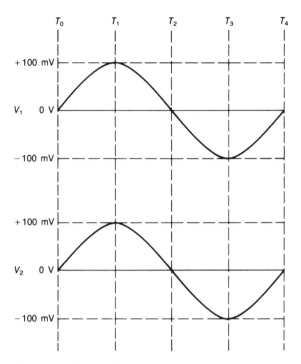

Figure 2–9. Common-Mode Signals

$$\text{CMRR} = 20 \log \frac{A_D}{A_{\text{CM}}} \tag{2.11}$$

Comparing a change in the output voltage (ΔV_{out}) to a change in the differential input voltage (ΔV_D) will provide the difference gain (A_d):

$$A_D = \frac{\Delta V_{\text{out}}}{\Delta V_D} \tag{2.12}$$

Example 2.6

A change of 10 mV at the differential input causes a change of 10 V at the output. What is the difference gain? The common-mode gain? The common-mode rejection ratio?

Solution

Use Equation 2.12:

$$A_D = \frac{\Delta V_{\text{out}}}{\Delta V_D} = \frac{10 \text{ V}}{0.01 \text{ V}} = 1000$$

If a common-mode voltage change (ΔV_{CM}) of 10 mV causes an output voltage change (ΔV_{out}) of 1 mV, the common-mode gain (A_{CM}) is

$$A_{\text{CM}} = \frac{\Delta V_{\text{out}}}{\Delta V_{\text{CM}}}$$

$$= \frac{1 \text{ mV}}{10 \text{ mV}} = 0.1 \qquad (2.13)$$

Recall that CMRR is 20 log times the ratio of A_D to A_{CM}. Therefore, using the previous calculations including Equation 2.11,

$$\text{CMRR} = 20 \log \frac{A_D}{A_{\text{CM}}} = 20 \log \frac{1000}{0.1} = 80 \text{ dB}$$

In practice, the amplifier may not entirely reject the common-mode signal. But a good amplifier will amplify the desired signal by a large amount while attenuating the undesired frequency by a similar amount. Thus, the output may indeed result in the desired signal being 80 dB higher than the undesired signal, as the calculations in Example 2.6 show.

Review of Darlington Pairs

The second stage of the op amp is a very high gain voltage amplifier stage. This amplifier stage is usually constructed with *Darlington pairs* and provides most of the gain of the op amp.

Recall from earlier studies of transistor amplifiers that the dc beta (β_{dc}) of a transistor is known as the dc current gain and is designated on most data sheets as h_{FE}. The higher β_{dc} is, the higher the dc current gain of the transistor.

Recall also that one way to increase beta is to connect two transistors, as illustrated in Figure 2–10. The overall beta of the two transistors is equal to the product of the individual betas, that is,

$$\beta = \beta_1 \beta_2 \qquad (2.14)$$

Example 2.7

If $\beta_1 = \beta_2 = 100$, calculate the effective beta of the Darlington pair.

Solution

Use Equation 2.14:

$$\beta = \beta_1 \beta_2 = 100 \times 100 = 10,000$$

Obviously, then, cascading additional Darlington pairs will result in the high

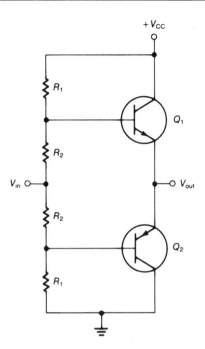

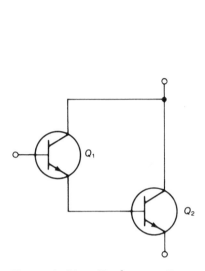

Figure 2–10.   Darlington Pair

Figure 2–11.   Class B Push-Pull Emitter
                   Follower

gain for which op amps are noted. Remember that a typical op amp may have an open-loop voltage gain of 200,000 or more.

Review of Push-Pull Amplifiers

The third, or output, stage of the op amp is usually a class B push-pull emitter follower. Figure 2–11 shows a typical push-pull circuit. Assume biasing near cutoff, as you will recall from earlier studies.

The push-pull circuit action is obtained by combining the two emitter followers, with the output taken from the junction of the two emitters. The NPN transistor, Q_1, handles the positive half cycle of input voltage, while the PNP transistor, Q_2, handles the negative half cycle. The output voltage then combines the opposite halves and produces a complete sine wave.

In the typical op amp, equal positive and negative voltages provided by the split supply will ideally produce a *quiescent* (no input signal) output voltage of zero.

A typical schematic diagram for an op amp, identifying the three stages we have discussed, is shown in Figure 2–12. (This is the 741 op amp described in the data sheet of Figure 2–2.)

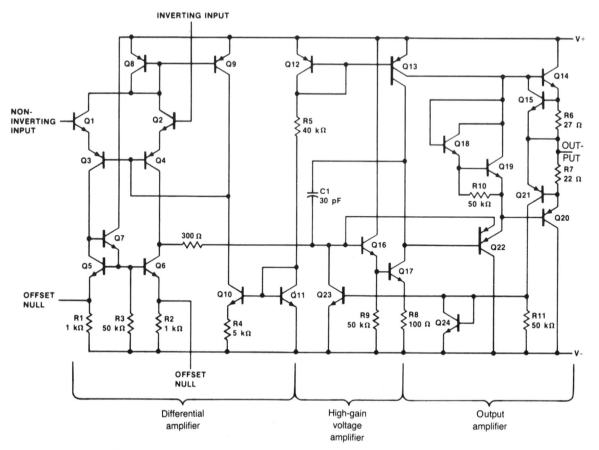

Figure 2–12. Schematic Diagram of a 741 Op Amp (Courtesy of Fairchild Camera & Instrument Corporation, Linear Division, © copyright 1982.)

2.6 **Power Supply Considerations**

Most op amps require a dual power supply (equal positive and negative voltages) for proper operation. Such a power source allows the op amp output to swing positive and negative with respect to ground, a particularly useful feature in dc circuits and audio applications.

Dual power sources can be provided by connecting two batteries in series, as illustrated in Figure 2–13, with the common connection being reference ground. Such a supply is portable, but all batteries must be fresh if the circuit is to operate properly.

Battery-operated power supplies can provide the basic voltages required for op amp operation, but many op amp applications require very stable, noise-free, regulated voltages. For such applications, IC *dual tracking* voltage regulators may be used, as shown in Figure 2–14. Input capacitor C_1 provides steady

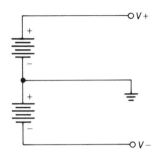

Figure 2–13. Dual ± Battery Power Source

positive and negative voltages to the regulator, and the two 0.1 μF capacitors are bypass capacitors that provide stability. Table 2–1 lists several such ICs.

Some special op amps are designed to operate from single-voltage power supplies. These supplies usually are positive voltage types. In the quiescent state the output voltage should measure approximately halfway between the maximum voltage supply and ground.

In some limited applications, standard op amps use a single power supply. We will study some of those applications later.

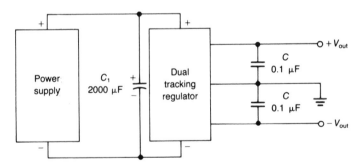

Figure 2–14. Dual Tracking IC Voltage Regulation

Table 2–1	Dual Tracking Voltage Regulators	
	Type	**Voltage Range**
	LM 125	± 15 V
	LM 126	± 12 V
	LM 127	+ 5 V, − 12 V
	NE/SE 5551N	± 5 V
	NE/SE 5552N	± 6 V
	NE/SE 5553N	± 12 V
	NE/SE 5554N	± 15 V
	NE/SE 5555N	+ 5 V, − 12 V
	RM/RC 4195	± 15 V

Heat Effects

Because the IC chip is so small, special care must be taken to keep the heat generated by internal currents within limits specified by the manufacturer. Input and output conditions will vary these currents. With no input signal and no output load, the state of the transistors and resistors of the internal circuits determine the supply current. When input signals are applied and an output load is connected, the power supply current through the device increases. If the generated heat becomes excessive, the device may be destroyed.

Data sheets provide values of parameters related to power requirements. Refer back to Figure 2–2. Note, for example, *supply voltage* of the 741C op amp is limited to ±18 volts. Operating at this maximum power supply value for any length of time will degrade the characteristics of the device to a point where it cannot function properly, and exceeding this maximum value can destroy the device.

Power Consumption with No Load

Remember that any device that carries current will generate heat and, therefore, will dissipate (consume) power. Refer again to the data sheet of Figure 2–2. Note that the *operating temperature range maximum* is +70°C and the *maximum power consumption* is 85 mW. The total no-load power consumption (P_{CC}) of the device is the product of power supply voltage (V_{CC}) and current (I_{CC}). When using a dual power supply, the total no-load power consumption can be determined as follows:

$$P_{CC} = (+V_{CC})(+I_{CC}) + (-V_{CC})(-I_{CC}) \tag{2.15}$$

If a single power supply is used, the power consumption is

$$P_{CC} = (+V_{CC})(+I_{CC}) \tag{2.16}$$

Example
2.8

A typical circuit has $V_{CC} = \pm 15$ V and $I_{CC} = \pm 2.8$ mA. Under no-load conditions, how much power is consumed in the device?

Solution

Use Equation 2.15:

$$\begin{aligned}
P_{CC} &= (+V_{CC})(+I_{CC}) + (-V_{CC})(-I_{CC}) \\
&= (+15 \text{ V})(+2.8 \text{ mA}) + (-15 \text{ V})(-2.8 \text{ mA}) \\
&= 42 \text{ mW} + 42 \text{ mW} = 84 \text{ mW}
\end{aligned}$$

This is less than the maximum 85 mW indicated by the data sheet (Figure 2–2), and the device will not consume more power than this under the no-load condition.

Power Consumption under Load

More current from the power supply flows in the device under load conditions. The extra current includes current received from or delivered to the load. The total power dissipation (P_D) of the device under these conditions is the power delivered from the source (P_s) minus the power consumed in the load (P_L):

$$P_D = P_s - P_L \tag{2.17}$$

P_L is the product of load voltage (V_L) and load current (I_L):

$$P_L = V_L I_L \tag{2.18}$$

Example 2.9

A load drawing 17 mA at +8 V is connected to the circuit of Example 2.8. If the power supplied to the device is 250 mW, what is the internal power dissipation of the device?

Solution

Use Equation 2.18:

$$P_L = V_L I_L = 8 \text{ V} \times 17 \text{ mA} = 136 \text{ mW}$$

Use Equation 2.17:

$$P_D = P_s - P_L = 250 \text{ mW} - 136 \text{ mW} = 114 \text{ mW}$$

Note that *internal power dissipation* for the standard DIP has a maximum rating of 670 mW for ambient temperatures to 70°C (Figure 2–2). Therefore, the device is within specifications.

Circuit Current Requirements

The *rated supply current* is an indication of the maximum current drawn from the power supply under no-load, no-signal conditions. The power supply must provide at least that amount, plus any increase in current requirements of the device caused by the addition of an input signal and a load. Therefore, the *maximum* current required by the circuit with an input signal and a load will determine the *minimum* power supply current requirements. A good practice is to select a power supply that will provide a current value of *at least 10 percent higher* than the maximum circuit requirements. This will ensure sufficient current for the circuit and will not load down the supply voltage.

Example
2.10

A power supply is to be selected for the circuit of Example 2.9. The load draws 17 mA. What is the power supply current requirement? Use the typical I_{CC} of the circuit in Example 2.8.

Solution

The maximum circuit current $(I_{CC(max)})$ requirement must include the load current (I_L) plus the maximum rated no-load device current (I_{CC}):

$$I_{CC(max)} = I_L + I_{CC}$$
$$= 17 \text{ mA} + 2.8 \text{ mA} = 19.8 \text{ mA} \qquad (2.19)$$

The minimum power supply current $(I_{PS(min)})$ should provide at least 10 percent greater current, therefore,

$$I_{PS(min)} = I_{CC(max)} + 0.1(I_{CC(max)})$$
$$= 19.8 \text{ mA} + \approx 2 \text{ mA} = 21.8 \text{ mA} \qquad (2.20)$$

The power supply selected for this circuit, therefore, must be capable of providing a minimum of 22 mA while maintaining the required ± 15 volts.

Other Factors to Consider

Factors other than voltage and current needs must be considered when selecting a power supply to ensure that the supply will not interfere with the proper operation of the device or damage it. Such considerations are *long-term stability*, *ac stability* of the device, and *protective circuits*.

Long-term stability is necessary to provide a constant output voltage over a long period of operation, assuming constant load conditions, ambient temperature, line voltage, and circuit output control adjustments. Such stability can be ensured by regulating the load with line voltage regulators, checking power supply filter circuits for minimum ripple voltage, or using a common ground to eliminate the possibility of ground loops.

The ac stability of the device may be improved by installing bypass capacitors between all power supply connections and ground. The capacitors should be greater than 0.01 μF and located as close to the device terminals as possible. Also, the leads should be kept short to reduce the possibility of resonance with lead inductance. This precautionary measure will prevent undesirable feedback through the internal impedance of the power supply.

The last factor mentioned that should be considered when choosing a power source was protective circuits. Protective circuits are so important that they will be discussed under a separate heading.

Protective Circuits

Protective circuits should be provided for both the device and the power supply. The device, for example, can be destroyed if the power supply terminals are reversed, even momentarily. One method of protection from such a possibility is to place diodes (D_1 and D_2) in the connecting lines, as shown in Figure 2–15. The breakdown voltage of the diodes must be greater than the supply voltage used. Then, if the power supply is inadvertently connected with reversed polarity, the diodes will be reverse biased, act as open circuits, and prevent current from destroying the op amp.

Overvoltage Protection

Most op amps operate with maximum values of about ±20 volts. Therefore, to prevent performance degradation and the possibility of destroying the op amp, *overvoltage protection* should be considered. Figure 2–16 illustrates the use of zener diodes (Z_1 and Z_2) as overvoltage protective devices. The zener breakdown

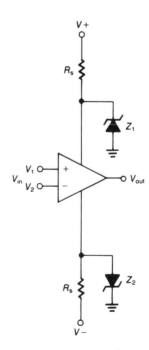

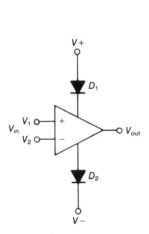

Figure 2–15. Diode Protection for Power Supply Connection Lines

Figure 2–16. Zener Diode Overvoltage Protective Circuit

voltage should be slightly higher than that of the power supply operating value, but lower than the maximum allowable supply voltage.

The series resistors (R_s) are inserted to limit the zener current and prevent excessive current drain from the power supply. The proper resistance value can be determined as follows:

$$R_s = \frac{V - V_Z}{I_Z} \qquad (2.21)$$

where

$V - V_Z$ = the voltage drop across the resistor
I_Z = the desired zener current

Example 2.11

Refer to Figure 2–16. What value of resistance (R_s) should be used to limit op amp voltage to 20 V? You also must limit power supply current drain to a safe level in case the voltage rises to 25 V.

Solution

First, select a zener diode that will satisfy the requirement. A look at a data book would indicate a possible choice of a 1N968, which has V_Z = 20 V and I_Z = 15 mA. Then, use Equation 2.21:

$$R_s = \frac{V - V_Z}{I_Z} = \frac{25\text{ V} - 20\text{ V}}{15\text{ mA}} = 333\ \Omega$$

You should use a standard 330-ohm resistor in the circuit.

Load-Protection Circuits

Most commercial power supplies have built-in *load-protection* circuits that automatically reduce the voltage or limit the current of the supply if something goes wrong with the power supply (PS). These protection circuits are in addition to the standard fuses and circuit breakers.

One such circuit is the *overvoltage crowbar circuit* illustrated in Figure 2–17. If the output voltage exceeds a predetermined value, the comparator circuit generates a signal that causes the silicon-controlled rectifier (SCR) to conduct. The resistance of the SCR and the voltage drop across it are reduced to a low value. Therefore, the load is almost shorted and receives a very low voltage from the supply, thereby preventing damage to the load from overvoltage. The SCR series resistance R is a current-limiting resistor that prevents current from destroying the SCR.

Another common protective method protects both the load device and the power supply if the load current exceeds a predetermined value. Such excess

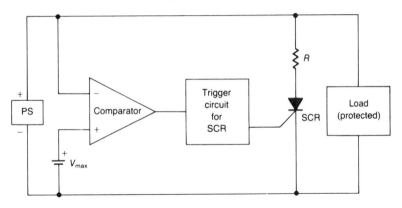

Figure 2–17. Block Diagram of an Overvoltage Crowbar Circuit for Load Protection

current may occur because of too many devices being powered by the source or because of a short of the outputs.

The block diagram in Figure 2–18 shows a simple *foldback current-limiting circuit*. When load impedance goes too low, the power supply voltage decreases, drawing excessive current from the supply. The excessive load current causes an increased voltage drop across R_1 that activates the sensing device. The sensing device causes R_2 to increase in value and thereby drop more of the input voltage, leaving less voltage for the load. The reduced output voltage causes the supply current to drop to a safe level. The power supply output voltage returns to its predetermined value when the conditions that caused the increased current load are removed.

The current-limiting resistance, R_2, of Figure 2–18 is actually a transistor in the op amp—Q_{15} in Figure 2–12. The transistor acts as a variable resistor by varying the base voltage applied to it.

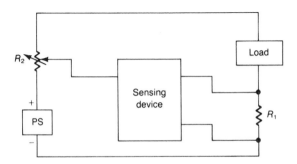

Figure 2–18. Block Diagram of a Foldback Current-Limiting Circuit

2.7 Single Voltage Supply Systems

Op amps are designed specifically to operate using symmetrical dual voltage supply systems. Most other linear ICs operate from a single positive or single negative power supply, although there are special circuits where op amps can also be operated from a single voltage supply. It should be obvious, however, that the single voltage supply must provide the same amount of voltage as did the dual voltage supply it replaces. For example, a single supply to operate a μA741C must provide 36 V maximum to replace the ± 18 V maximum provided by a dual supply normally used. In practice, operating voltages are limited to ± 15 V, so the single supply must provide 30 V. If a single positive supply is used, it is connected to the positive voltage terminal of the op amp, and the negative terminal is grounded. The inverse is true if a single negative supply is used. Figure 2–19 shows the proper methods of connecting the single supply.

When using a single voltage supply, the input signal should be isolated from the input terminals through a capacitor because the input terminals are above ground. Examples of circuits using single supply systems will be studied in later chapters.

Inverter and Converter Circuits

If the single power supply available to you cannot provide enough voltage for correct operation of the active device, inverter or converter circuits may be used. An *inverter* is a circuit that converts a dc input voltage to a higher or lower ac output voltage. The dc input voltage is used as the supply voltage for an oscillator, whose ac voltage output is then stepped up or down through transformer action. Similarly, a *converter* is a circuit that converts a dc input voltage to a higher or lower dc output voltage. The dc input voltage is used to supply voltage to an oscillator, whose ac voltage is stepped up or down through transformer action. However, the secondary of the transformer has a rectifier-filter circuit that converts the ac into the desired dc level.

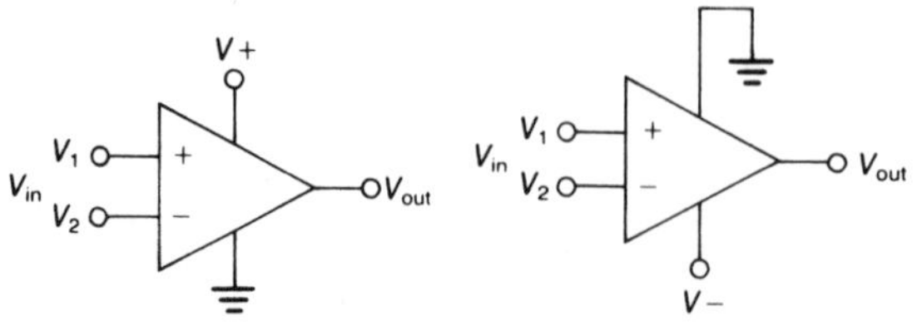

(A) Positive Supply Connection   (B) Negative Supply Connection

Figure 2–19. Block Diagram of Single Voltage Power Supply Connections to Op Amp

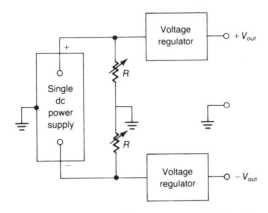

Figure 2–20. Block Diagram of a Single Voltage Power Supply Converted to a Dual Voltage Power Supply

2.8 Dual Power Supply Systems

Power supplies for op amp circuits must provide exactly equal but opposite voltages. These exact values must be maintained during the operation of the device if it is to operate properly; that is, the device should provide true outputs for any given input. Each terminal of the supply must provide equal amounts of regulation if it is to maintain a balanced output.

Single power supplies can be converted to dual power supplies quite easily. One approach is to use a circuit like the one shown in Figure 2–20. Variable resistors in the voltage divider network ensure equal and opposite voltages for balance. While this circuit can be used with most single power supplies, be aware that the output of the power supply must be floating with respect to its case, which is grounded.

2.9 Summary

1. Operational amplifiers (op amps) are very high gain dc amplifiers, conveniently packaged in the form of integrated circuits.

2. Op amps are the basic building blocks for many modern electronic circuits.

3. The most widely used configuration of the op amp has a differential input and single output.

4. The difference between the op amp symbol and the regular amplifier symbol is that the op amp shows two inputs.

5. The minus sign on the symbol indicates an inverting input, and the plus sign indicates a noninverting input.

6. Power supply terminals may not always be shown on the symbol.

7. A manufacturer's data sheet provides much valuable information.

8. Absolute maximum ratings must never be exceeded if damage to the op amp is to be prevented.

9. Operating characteristics of the op amp must be considered in order to select the right device to accomplish the desired result in any given application.

10. The most important characteristic affecting ac operation is the slew rate.

11. Manufacturers use the unity-gain frequency as a reference point for op amps. Unity-gain frequency is equal to the gain-bandwidth product.

12. The three stages in op amps are differential input, high-gain amplifier, and output.

13. The second stage provides most of the amplification of the op amp and is generally constructed with Darlington pairs.

14. Most op amps require dual power supply systems, but some special application devices can operate from single supply systems.

15. Basic op amp circuits can be operated from battery-operated sources, but many op amp applications require noise-free regulated voltage sources.

16. Protective circuits should be provided for both the op amp and the power supply.

17. Protective circuits can be internal or external to the device.

18. Many modern op amps have built-in short circuit protection.

2.10 Questions and Problems

2.1 Explain what the plus and minus signs on the op amp schematic symbol represent.

2.2 Why is it necessary to be able to read a manufacturer's data sheet?

2.3 List the characteristics of an ideal op amp.

2.4 What would you expect to happen if you momentarily applied reversed polarities to power an op amp?

2.5 What is the slew rate of the μA741C? (See Figure 2–2.)

2.6 Calculate the bandwidth of a general purpose op amp with a unity-gain frequency of 1 MHz and a circuit gain of 100.

2.7 Calculate the voltage gain of a circuit with a bandwidth of 5 kHz and a unity-gain frequency of 1.5 MHz.

2.8 What is the amplitude of the difference voltage (V_D) of an op amp that has input V_1 equal to $+2$ mV and input V_2 equal to -2 mV?

2.9 Calculate V_{out} of a circuit with an amplifier gain of 100, $V_1 = +2$ mV, and $V_2 = +1$ mV.

2.10 Given a difference gain (A_D) of 500 and common-mode gain (A_{CM}) of 0.1, calculate the common-mode rejection ratio (CMRR).

2.11 What is the effective gain (beta) of a Darlington pair if the beta of $Q_1 = 80$ and the beta of $Q_2 = 100$?

2.12 A typical op amp circuit has $V_{CC} = \pm 18$ V and $I_{CC} = \pm 2.8$ mA. Calculate the power consumption of the device under no-load conditions.

2.13 A device with no-load power consumption of 100 mW is connected to a load drawing 20 mA at +9 V. If the power supplied to the device is 300 mW, what is the internal power dissipation of the device?

2.14 What is the minimum current requirement of a power supply selected for the circuit described in Problem 13?

2.15 What value of resistance would be required to limit op amp voltage to 18 volts if it is desirable to limit power supply current drain to a safe level in case the voltage rises to 22 volts? The zener diode used would have $V_Z = 18$ V and $I_Z = 12$ mA. Refer to Figure 2–16.

Chapter 3

Basic Operational Amplifier Circuits

3.1 Introduction

In Chapter 2 we defined the op amp as the building block for linear circuit applications. In this chapter we will discuss basic op amp circuits, their performance, and their design. The basic linear applications discussed in this chapter are those in which the output signal is directly proportional to the input signal. Both dual-supply and single-supply circuits will be explored. The concept of virtual ground and degenerative feedback also will be examined.

The comparator is not an amplifier, but it is introduced at this point because its qualities as a sensing device are important to the chapters that follow.

3.2 Objectives

When you complete this chapter, you should be able to:

☐ Explain the difference between the inverting and noninverting inputs to the op amp.

☐ Identify the different types of amplifier circuits by observation.

☐ Design and predict the performance of the various types of amplifiers.

☐ Distinguish between dual-supply and single-supply op amp circuits.

☐ Discuss dc output offset voltage and know how to minimize its effect.

☐ Identify the input and feedback elements in the various types of amplifier circuits.

☐ Determine the effects of feedback upon the op amp circuits and how to alter circuit performance by altering the feedback elements.

☐ Discuss the concept of virtual ground.

☐ Discuss the comparator and its application as a sensing device.

3.3 Noninverting Amplifiers

The circuit in Figure 3–1 shows a common configuration of a *noninverting amplifier*. The name stems from the input signal being applied to the plus (+), or noninverting, input of the op amp. The output signal, then, is simply an amplified version of the input signal; that is, it is not inverted but always in phase. Resistor R_1 is the *input element*. Resistor R_f is the *feedback element,* so called because part of the output voltage is diverted, or feeds back, to the minus (−), or inverting, input of the op amp. The feedback element is also the determining factor for gain and bandwidth, as you will discover later. For most amplifier circuits, a degenerative signal is fed back to the input. Regenerative feedback will be studied in a later chapter.

For reasons that will be explained later in this chapter, there is virtually zero voltage between the (+) and (−) input terminals of the op amp. Therefore, both terminals are at the same potential, that is, V_{in}, and V_{in} dropped across R_1 causes current I to flow. This current I flows through R_f, developing a voltage drop V_{R_f}, determined as follows:

$$V_{R_f} = IR_f = \frac{R_f}{R_1}V_{in} \tag{3.1}$$

The output voltage V_{out} of the noninverting amplifier is determined by adding

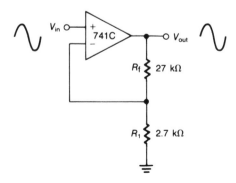

Figure 3–1. Noninverting Amplifier

V_{in}, which is the voltage drop across R_1, to V_{Rf}, which is the voltage drop across R_f, so that

$$V_{out} = V_{in} + \frac{R_f}{R_1}V_{in} \tag{3.2a}$$

Factoring out V_{in}, we have

$$V_{out} = \left(1 + \frac{R_f}{R_1}\right)V_{in} \tag{3.2b}$$

Recall that the voltage gain (A_v) of a circuit is the ratio of the output voltage (V_{out}) to the input voltage (V_{in}):

$$A_v = \frac{V_{out}}{V_{in}} \tag{3.3a}$$

Therefore, by algebraic manipulation of Equations 3.2b and 3.3a, the voltage gain of the noninverting amplifier is

$$A_v = 1 + \frac{R_f}{R_1} \tag{3.3b}$$

In practical terms, then, we can state that the voltage gain of a noninverting amplifier will always be greater than unity (1), regardless of the values assigned to resistors R_1 and R_f, except in the special case where $R_f = 0\ \Omega$ and $R_1 \approx \infty\ \Omega$. Under these conditions, A_v is equal to 1. (The circuit shown in Figure 3–10 is such a special case.)

One of the intrinsic characteristics of an op amp without feedback is the *open-loop gain* (A_{OL}). With feedback present, we have the *closed-loop gain* (A_{CL}) that is equal to A_v. In terms of the noninverting amplifer,

$$A_{CL} = 1 + \frac{R_f}{R_1} \tag{3.3c}$$

The *loop gain* (A_L) is the ratio of the open-loop gain (A_{OL}) to the closed-loop gain (A_{CL}):

$$A_L = \frac{A_{OL}}{A_{CL}} \tag{3.4}$$

The open-loop input impedance Z_{in} of the noninverting amplifier is, for all practical purposes, the intrinsic input resistance Z_{iin} of the op amp itself. For closed-loop operation this impedance becomes

$$Z_{in} = Z_{iin} \times A_L \tag{3.5a}$$

In either case, this impedance is high enough to minimize input circuit loading.

The output impedance Z_{out} is the ratio of the intrinsic output resistance Z_{iout} of the op amp to the loop gain (A_L):

$$Z_{out} = \frac{Z_{iout}}{A_L} \tag{3.5b}$$

Example
3.1

Refer to Figure 3–1 and to the 741C data sheet (Figure 2–2). Determine (a) closed-loop gain, (b) loop gain, (c) input impedance, and (d) output impedance.

Solutions

  a. Using Equation 3.3c, the closed-loop gain is
$$A_{CL} = 1 + \frac{R_f}{R_1} = 1 + \frac{27 \text{ k}\Omega}{2.7 \text{ k}\Omega} = 11$$
  b. The 741C data sheet (Figure 2–2) indicates the typical open-loop gain is 200,000. Using Equation 3.4, the loop gain of the circuit is
$$A_L = \frac{A_{OL}}{A_{CL}} = \frac{200,000}{11} = 18,200$$
  c. The 741C data sheet indicates the typical input resistance is 2 MΩ, so from Equation 3.5a,
$$Z_{in} = Z_{iin} \times A_L$$
$$= 2 \text{ M}\Omega \times 18,200 = 36.4 \text{ M}\Omega$$
  d. The 741C data sheet indicates the typical output resistance is 75 Ω. Using Equation 3.5b, the output impedance is
$$Z_{out} = \frac{Z_{iout}}{A_L} = \frac{75 \text{ }\Omega}{18,200} = 0.004 \text{ }\Omega$$

Example 3.1 demonstrates the results of adding feedback to the circuit. The loop gain is increased and the output impedance is decreased. Since the output impedance in such a circuit is so small, it can, in general, be ignored.

Example
3.2

Refer to Figure 3–1. Assume R_f = 100 kΩ and R_1 = 1 kΩ. Determine (a) A_{CL} and (b) A_L.

Solutions

  a. From Equation 3.3c,
$$A_{CL} = 1 + \frac{R_f}{R_1} = 1 + \frac{100 \text{ k}\Omega}{1 \text{ k}\Omega} = 101$$
  b. From Equation 3.4,
$$A_L = \frac{A_{OL}}{A_{CL}} = \frac{200,000}{101} = 1980$$

3.4 Inverting Amplifiers

An *inverting amplifier* has the input signal applied to the minus (−) terminal of
the op amp, as shown in Figure 3–2. In such an amplifier the output signal is
180° out of phase, or inverted, compared with the input signal. Resistor R_1 is
the input element, and resistor R_f is the feedback element.

The output voltage (V_{out}) of the inverting amplifier is

$$V_{out} = -\left(\frac{R_f}{R_1}\right)V_{in} \qquad (3.6)$$

The − sign in Equation 3.6 indicates only that the circuit being evaluated is an
inverting amplifier. It should not be construed as a negative output voltage,
except in the case of a positive dc input voltage.

Virtual Ground

To better understand the operation of an inverting amplifier, you must first
understand the concept of *virtual ground*. Virtual ground is a point at which
voltage is zero with respect to ground, yet is isolated from ground. In practical
terms, this means that no current can flow into or out of this point.

The inverting input of Figure 3–2 acts as a virtual ground. In this circuit,
a positive-going signal, V_{in}, is applied to the inverting input terminal of the op
amp. An inverted output signal is then fed back to the inverting input terminal
through feedback resistor R_f. The voltage between the inverting and noninverting
input terminals is essentially zero, therefore the inverting terminal is at zero
volts, or ground potential. However, it is not a true ground point because of the
high input impedance of the op amp and should never be used as a true ground
point. This is an important point to remember when analyzing op amp circuits.

In the previous section we learned that A_v is equal to V_{out}/V_{in} and that A_{CL}
is equal to A_v. Therefore, for the inverting amplifier,

$$A_{CL} = -\frac{R_f}{R_1} \qquad (3.7)$$

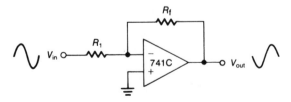

Figure 3–2. Inverting Amplifier

The ratio of R_f to R_1, then, will determine whether the voltage gain is less than, equal to, or greater than unity (1).

Through algebraic manipulation of Equations 3.4 and 3.7, we can now state that the loop gain, A_L, for the inverting amplifier is

$$A_L = \frac{A_{OL}}{A_{CL}} = \frac{A_{OL}}{\left(-\dfrac{R_f}{R_1}\right)} = A_{OL}\left(-\frac{R_1}{R_f}\right) \tag{3.8}$$

The input impedance of the inverting amplifier is the value of the input element. In practice, this value will always be much less than that for a noninverting amplifier. The output impedance of the inverting amplifier is determined in the same manner as for the noninverting amplifier:

$$Z_{out} = \frac{Z_{iout}}{A_L} \qquad \text{(repeat of Equation 3.5b)}$$

Example 3.3

Refer to Figure 3–2. Assume $R_1 = 2\ k\Omega$, $R_f = 22\ k\Omega$, $V_{in} = 0.25\ V$, and the op amp is a 741C. Determine (a) output voltage, (b) closed-loop gain, (c) loop gain, (d) input impedance, and (e) output impedance.

Solutions

a. Using Equation 3.6,
$$V_{out} = -\left(\frac{R_f}{R_1}\right)V_{in} = -\left(\frac{22\ k\Omega}{2\ k\Omega}\right)0.25\ V = -2.75\ V$$

b. Using Equation 3.7,
$$A_{CL} = -\frac{R_f}{R_1} = -\frac{22\ k\Omega}{2\ k\Omega} = -11$$

c. Using Equation 3.8,
$$A_L = A_{OL}\left(-\frac{R_1}{R_f}\right) = 200,000 \times -0.09 = -18,200$$

d. Input impedance is simply the value of R_1, or 2 kΩ.

e. Using Equation 3.5b,
$$Z_{out} = \frac{Z_{iout}}{A_L} = \frac{75\ \Omega}{-18,200} = -0.004\ \Omega$$

The inverting amplifier can be easily and quickly adapted to invert the polarity of an input signal without altering the amplitude of that signal. Simply make R_1 and R_f equal. This results in a *unity-gain inverter*.

Replacing R_f with a variable resistor allows the closed-loop gain of the inverting amplifier to be controlled. Alternatively, switching in different values of feedback resistors, as shown in Figure 3–3, allows gain preselection. Any desired gain can be preselected by carefully selecting the values of the feedback resistors used.

ac Operation with a Single Power Supply

The circuits discussed in Sections 3.3 and 3.4 have been based on the op amp being powered by a dual power supply. However, it is possible to operate the circuits from a single power supply. In order to do that, the *quiescent dc output voltage* of the op amp must be set to one-half the supply voltage. It is necessary that ac coupling capacitors for the input and output signals be used in such applications.

To properly operate from a single-supply voltage, the op amp circuit must be able to produce both negative- and positive-going signals. Properly biased, and with no input signal applied, the dc output voltage of the circuit will be one-half that of the supply voltage. This dc output voltage will be the quiescent, or resting, voltage. When an ac input signal is applied to the circuit, the output signal will vary about the quiescent voltage. The op amp output signal is a combination of the quiescent dc voltage and the amplified ac component superimposed upon it. The dc component can be considered to be an *output offset voltage* that should be removed before applying the desired output signal to another stage. (We will discuss dc output offset voltage more fully in the next section.)

Basic single-supply powered circuits for the inverting ac amplifier and the noninverting ac amplifier are shown in Figure 3–4A and B. Note that the supply connections are included in the diagram to indicate that single-supply voltage is used, and that the V^- connection of the op amp goes directly to ground. The

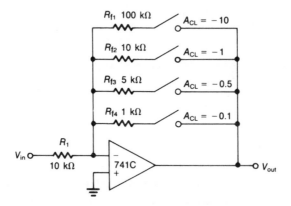

Figure 3–3.   Switchable Feedback
Resistances for Gain
Preselection

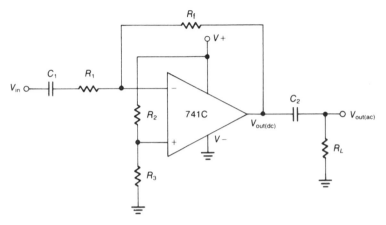

(A) Inverting ac Amplifier with Single-Supply Voltage

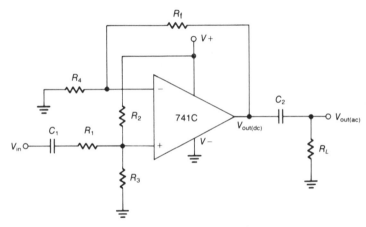

(B) Noninverting ac Amplifier with Single-Supply Voltage

Figure 3–4. Single-Supply Powered Amplifier Circuits

voltage divider network R_2–R_3 applies a voltage to the desired input terminal of the op amp to set the dc output voltage at one-half the supply voltage $(V^+/2)$ with no input signal applied. Capacitor C_1 couples the input signal to the circuit. Capacitor C_2 removes the dc output offset voltage and couples the output signal from the circuit.

The quiescent dc output voltage can be determined as follows:

$$V_{out(dc)} = \frac{R_3}{R_2 + R_3}(V^+)$$ (3.9)

In general, however, resistors R_2 and R_3 are made equal to each other so that $V_{out(dc)} = V^+/2$. This being the case, the maximum peak-to-peak output voltage that can be expected without distortion cannot exceed that same value. Therefore, the maximum peak-to-peak input voltage ($V_{in(max, p-p)}$) that can be applied to the circuit is

$$V_{in(max, p-p)} = \frac{V^+}{A_{CL}} \tag{3.10}$$

The closed-loop gain, A_{CL}, of the single-supply inverting amplifier is determined by using Equation 3.7 (as for the dual-supply circuit), and for the noninverting amplifier by using Equation 3.3C.

The desired low-frequency response and the impedance of either the input or load determines the values for capacitors C_1 and C_2. The value for input capacitor C_1 is determined by

$$C_1 = \frac{1}{2\pi f_c R_1} \tag{3.11}$$

where

f_c = cutoff frequency
R_1 = input resistance

and the value for output capacitor C_2 is determined by

$$C_2 = \frac{1}{2\pi f_c R_L} \tag{3.12}$$

where

f_c = cutoff frequency
R_L = load resistance

The cutoff frequency, f_c, is equal to approximately $0.1 f_{op}$, where f_{op} is the lowest desired operating frequency.

Example 3.4

Refer to Figure 3–4A. Design an inverting ac amplifier circuit with a closed-loop gain of 10, an input impedance of 10 kΩ, an operating frequency of 60 Hz, and a load of 20 kΩ. The single-supply voltage is to be +30 V. Calculate the maximum peak-to-peak input voltage.

Solution

First, evaluate the information given.

Since the input impedance of the inverting circuit is equal to the value of the input resistor, R_1 must equal 10 kΩ. Given a gain of 10, R_f must equal $10R_1$, or 100 kΩ. Resistors R_2 and R_3 can be any value desired, as long as they are equal, but the general guide is to make them equal to $2R_f$. Therefore, R_2 and R_3 are both given values of 200 kΩ, which provides the $V^+/2$ value necessary for $V_{out(dc)}$.

Now calculate the values for the coupling and decoupling capacitors using Equations 3.11 and 3.12:

$$C_1 = \frac{1}{2\pi f_c R_1} = \frac{1}{6.28 \times 6 \text{ Hz} \times 10 \text{ k}\Omega}$$
$$= 2.65 \ \mu\text{F} \qquad \text{(Select a standard 2.7 } \mu\text{F.)}$$
$$C_2 = \frac{1}{2\pi f_c R_L} = \frac{1}{6.28 \times 6 \text{ Hz} \times 20 \text{ k}\Omega}$$
$$= 1.33 \ \mu\text{F} \qquad \text{(Select a standard 1.5 } \mu\text{F.)}$$

Finally, calculate the maximum peak-to-peak input voltage using Equation 3.10:

$$V_{\text{in(max, p-p)}} = \frac{V^+}{A_{\text{CL}}} = \frac{30 \text{ V}}{-10} = -3 \ V_{\text{p-p}}$$

Figure 3–5 shows the completed circuit design.

3.5 Offset Considerations

An ideal op amp will have zero output voltage when the input voltage is zero, but practical commercial op amps are not ideal. Therefore, in practical circuits there will be a small dc output voltage, even when there is zero input. This voltage is called *dc output offset voltage* (V_{os}) and results from the following op amp characteristics:

1. Input bias current
2. Input offset current
3. Input offset voltage
4. Drift

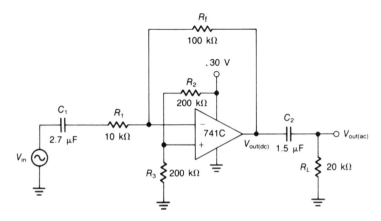

Figure 3–5. Circuit Design for Example 3.4

In op amp ac amplifier circuits, the coupling capacitors remove the dc output offset voltage. Therefore, the discussion in this section pertains to dc amplifier circuits.

Input Bias Current

Since the input impedance of a practical op amp is not infinite, there will be a very small, but finite, amount of current flowing into the (+) and (−) input terminals. The average of these two currents, I_{b+} and I_{b-}, is termed the *input bias current*, I_b:

$$I_b = \frac{I_{b+} + I_{b-}}{2} \tag{3.13}$$

With no signal applied to the circuit, I_b flows through both the input and feedback resistors. The current flowing in the resistors develops a voltage that appears as a dc input voltage. This voltage is then amplifed by the op amp and appears as the dc output offset voltage, V_{os}, which can be calculated by using the following equation:

$$V_{os} = I_b R_f \tag{3.14}$$

Input Offset Current

In practical op amp circuits, the bias currents flowing into the two input terminals will not be equal. The difference between the two currents, I_{b+} and I_{b-}, is termed the *input offset current*, I_{oi}, and is calculated as follows:

$$I_{oi} = I_{b+} - I_{b-} \tag{3.15}$$

The input offset current will cause a small output offset voltage:

$$V_{os} = I_{oi} R_f \tag{3.16}$$

A simple method of correction for V_{os} resulting from input current imbalances is shown in Figure 3–6. Resistor R_2 is a *current-compensating resistor* placed in

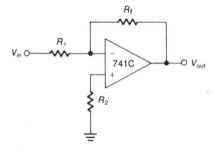

Figure 3–6.   Input Current-Correction Circuit

series with the (+) input terminal of the op amp. The value of R_2 should be equal to the parallel value of R_1 and R_f, or

$$R_2 = \frac{R_1 R_f}{R_1 + R_f} \tag{3.17}$$

The voltage developed across R_2 will be equal to that developed across R_1 and will therefore cancel, resulting in zero input voltage and zero output voltage.

Input Offset Voltage

Imbalances in the input circuits of the practical op amp are caused by mismatches in the internal circuitry. Such imbalances present another error factor, called *input offset voltage*, V_{oi}, which can result in V_{os}. Figure 3–7 shows a model where V_{oi} is represented by the battery. V_{os} can be calculated as follows:

$$V_{os} = V_{oi}\left(1 + \frac{R_f}{R_1}\right) \tag{3.18}$$

A current-compensating resistor (R_2 in Figure 3–6) in series with the (+) input cannot correct this problem.

Example 3.5

Assume an inverting amplifier circuit using a 741C op amp. Input impedance is 10 kΩ and gain is 10. The (+) input terminal is grounded. Using the typical parameters shown, calculate the dc output offset voltage (V_{os}) as a result of (a) the input bias current, (b) the input offset current, and (c) the input offset voltage. Parameters for the 741C are as follows:

	Typical	Maximum
Input bias current (I_b)	80 nA	500 nA
Input offset current (I_{oi})	20 nA	200 nA
Input offset voltage (V_{oi})	2.0 mV	6.0 mV

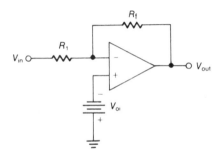

Figure 3–7. Input Offset Voltage Model

Solutions

First, evaluating the facts given should tell you that R_f must be 100 kΩ. To solve (a), use Equation 3.14:

$$V_{os} = I_b R_f = 80 \text{ nA} \times 100 \text{ k}\Omega = (80 \times 10^{-9} \text{ A})(100 \times 10^3 \text{ }\Omega) = 8 \text{ mV}$$

To solve (b), use Equation 3.16:

$$V_{os} = I_{oi} R_f = 20 \text{ nA} \times 100 \text{ k}\Omega = (20 \times 10^{-9} \text{ A}) (100 \times 10^3 \text{ }\Omega)$$
$$= 2 \times 10^{-3} \text{ V} = 2 \text{ mV}$$

To solve (c), use Equation 3.18:

$$V_{os} = V_{oi}\left(1 + \frac{R_f}{R_1}\right) = 2 \text{ mV} \times 11 = (2 \times 10^{-3} \text{ V})11$$
$$= 22 \times 10^{-3} \text{ V} = 22 \text{ mV}$$

Example 3.6

In worst-case conditions, the maximum parameters of the 741C would have to be considered. For comparison purposes, repeat the calculations performed in Example 3.5, this time using the maximum parameters given.

Solutions

To solve (a), use Equation 3.14:

$$V_{os} = I_{oi} R_f = 200 \text{ nA} \times 100 \text{ k}\Omega = (200 \times 10^{-9} \text{ A}) (100 \times 10^3 \text{ }\Omega)$$
$$= 50 \text{ mV}$$

To solve (b), use Equation 3.16:

$$V_{os} = I_{oi} R_f = 200 \text{ nA} \times 100 \text{ k}\Omega = (200 \times 10^{-9} \text{ A})(100 \times 10^3 \text{ }\Omega)$$
$$= 20 \text{ mV}$$

To solve (c), use Equation 3.18:

$$V_{os} = V_{oi}\left(1 + \frac{R_f}{R_1}\right) = 6 \text{ mV} \times 11 = (6 \times 10^{-3} \text{ V})11 = 66 \text{ mV}$$

In each case, V_{os} shows a significant increase over the typical values.

Example 3.7

Calculate the value required for a current-compensating resistor, R_2, to be placed in series with the (+) input terminal of the circuit described in Example 3.5.

Solution

Use Equation 3.17:

$$R_2 = \frac{R_1 R_f}{R_1 + R_f} = \frac{10 \text{ k}\Omega \times 100 \text{ k}\Omega}{10 \text{ k}\Omega + 100 \text{ k}\Omega} = 9.1 \text{ k}\Omega$$

Remember, however, that this resistor will correct only for input current imbalances, not for input offset voltage.

Null and Balance Procedures

While the current-compensating resistor corrects for the dc output offset voltage that results from the input bias current, other methods must be used to correct errors that result from the input offset voltage. Many op amps, including the 741C, provide terminals used specifically for cancelling the dc output offset voltage. The terminals are generally identified on data sheets as *offset null*, or *balance*, terminals. These terminals are internally connected to affect the internal circuitry of the op amp. The offset voltage may be nulled by connecting a 10 kΩ potentiometer across null pins 1 and 5 as shown in Figure 3–8. The wiper arm of the potentiometer is connected to V^-. In general, a high-turn, wire-wound potentiometer, capable of providing the small voltage changes necessary to affect the internal circuitry is used. The potentiometer is adjusted until the output voltage, with zero input, is also zero. This method of cancelling the dc output offset voltage is called *internal nulling*.

External nulling methods may be necessary for op amps that have no provisions for internal nulling. Data sheets usually indicate which method is used. Figure 3–9 shows typical external nulling circuits.

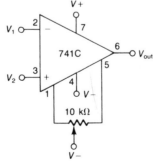

Figure 3–8. Offset Null Circuit for the μA741C Eight-Lead MINIDIP

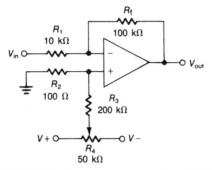

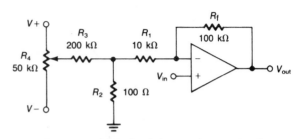

(A) External Nulling Circuit for an Inverting Amplifier

(B) External Nulling Circuit for a Noninverting Amplifier

Figure 3–9. Typical External Nulling Circuits for Op Amps Having No Provisions for Internal Nulling (Courtesy of Signetics, a subsidiary of U.S. Philips Corporation, © copyright 1976.)

Drift

Offset current and offset voltage changes can occur because of aging of the components, supply voltage changes, or temperature changes in the op amp. Changes resulting from temperature fluctuations are called *drift*. Drift cannot be eliminated, but it can be minimized by maintaining constant temperature around the circuit and by selecting op amps with offset characteristics that change minimally with variations in temperature.

3.6 Voltage Follower

The simple op amp circuit shown in Figure 3–10 is called a *voltage follower*, but it has many other names, each related to its application. The circuit is called voltage follower, or *source follower*, because the output voltage follows, or is an exact reproduction of, the input (source) voltage. The circuit is simply a unity-gain noninverting amplifier; therefore, it is sometimes referred to as a *unity-gain amplifier*. The circuit is also called a *buffer amplifier*, or an *isolation amplifier*, because its high input impedance and low output impedance buffers, or isolates, an input signal from its load.

This circuit is that special case for noninverting amplifiers that was men-

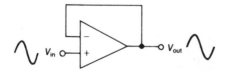

Figure 3–10. Voltage Follower Circuit

tioned earlier in the chapter where $A_v = 1$, since $R_f = 0\ \Omega$ and $R_1 \approx \infty\ \Omega$. This gain of one is demonstrated by using Equation 3.3b:

$$A_v = 1 + \frac{R_f}{R_1} = 1 + \frac{0\ \Omega}{\infty\ \Omega} = 1 + 0 = 1$$

The input voltage is applied directly to the $(+)$ input terminal, and since we consider the voltage between the $(+)$ and $(-)$ input terminals to be zero, it follows that

$$V_{out} = V_{in} \tag{3.19}$$

and

$$A_{CL} = \frac{V_{out}}{V_{in}} = 1 \tag{3.20}$$

With no input or feedback resistances in the circuit, the input impedance is extremely high and, for all practical purposes, is equal to the op amp's intrinsic input impedance. The output impedance is very low, because it is simply the intrinsic output of the op amp divided by the open-loop gain of the op amp.

Example 3.8

Refer to Figure 3–11 and Figure 2–2. Determine (a) input impedance, Z_{iin}; (b) output impedance, Z_{out}; (c) output voltage, V_{out}; and (d) load current, I_L.

Solutions

a. Typical intrinsic input impedance of the 741C is 2 MΩ.
b. The output impedance of the 741C is 75 Ω, and the typical open-loop gain is 200,000, therefore,

$$Z_{out} = \frac{75\ \Omega}{200,000} = 0.000375\ \Omega$$

c. From Equation 3.19,

$$V_{out} = V_{in} = 5\ V_{p\text{-}p}$$

d. From Ohm's Law,

$$I_L = \frac{V_L}{R_L} = \frac{5\ V}{10\ k\Omega} = 0.5\ mA$$

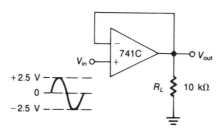

Figure 3–11. Voltage Follower Circuit for Example 3.8

3.7 Summing Amplifiers

A *summing amplifier*, shown in Figure 3–12, is constructed by connecting two or more resistors simultaneously to the $(-)$ input terminal of an op amp. The output voltage of the circuit is the sum of the individual input voltages in terms of individual resistances:

$$V_{out} = -\left(\frac{R_f}{R_1} V_1 + \frac{R_f}{R_2} V_2 + \frac{R_f}{R_3} V_3\right) \tag{3.21a}$$

and simplified,

$$V_{out} = -R_f\left(\frac{V_1}{R_1} + \frac{V_2}{R_2} + \frac{V_3}{R_3}\right) \tag{3.21b}$$

If the feedback resistor is made equal to the value of the input resistors, the output voltage is

$$V_{out} = -(V_1 + V_2 + V_3) \tag{3.22}$$

By making the feedback resistor larger than any of the input resistors, each input voltage is increased by a set gain and the individual results are added. The input impedance for each input is simply the value for the corresponding input resistor, and the gain for each input can be adjusted individually by choosing the desired ratio between each corresponding input resistor and the feedback resistor.

Audio Mixer

A summing amplifier is useful as an audio mixer. The input voltages developed across the input resistors do not interact, since they are common to the inverting input, which is at virtual ground. If the input voltages of Figure 3–12 were microphone inputs, the ac voltages developed by each microphone would be added (mixed) at the output, as demonstrated by Equation 3.22. If volume controls are placed between each of the input resistors and the summing point, the relative volume of the various inputs can be controlled for a more desirable output sound. Figure 3–13 shows this application, where each input affects the output according to the volume control setting, so the gain for each input is

$$A_c = -\frac{R_f}{R_1 + R_2} \tag{3.23}$$

Averaging Amplifier

A variation of the summing amplifier is the *averaging amplifier* that gives an ouput voltage equal to the average of the input voltage (where N = number of inputs).

$$V_{out(av)} = \frac{-(V_1 + V_2 + V_3 + \dots + V_n)}{N} \tag{3.24}$$

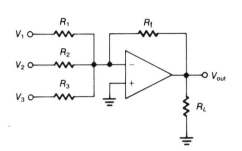

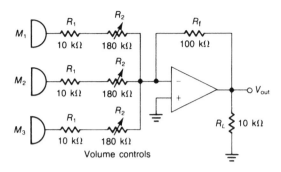

Figure 3–12. Three-Input Summing Amplifier

Figure 3–13. Three-Microphone Input Audio Mixer

For a three-input amplifier as in Figure 3–12, averaging is accomplished by making each of the input resistors equal in value, and the feedback resistor equal to one-third that value.

Example 3.9

Refer to Figure 3–12. Assume $R_1 = 5$ kΩ, $R_2 = 10$ kΩ, $R_3 = 20$ kΩ, and $R_f = 50$ kΩ. Find the output voltage if $V_1 = 0.5$ V, $V_2 = 0.75$ V, and $V_3 = 1$ V.

Solution

Using Equation 3.21b,

$$V_{out} = -R_f\left(\frac{V_1}{R_1} + \frac{V_2}{R_2} + \frac{V_3}{R_3}\right) = -50 \text{ k}\Omega\left(\frac{0.5 \text{ V}}{5 \text{ k}\Omega} + \frac{0.75 \text{ V}}{10 \text{ k}\Omega} + \frac{1 \text{ V}}{20 \text{ k}\Omega}\right)$$

$$= -50 \text{ k}\Omega(0.1 \text{ mA} + 0.075 \text{ mA} + 0.05 \text{ mA}) = -11.25 \text{ V}$$

Example 3.10

Find the output voltage of the five-input averaging amplifier illustrated in Figure 3–14.

Solution

Using Equation 3.24,

$$V_{out(av)} = \frac{-(V_1 + V_2 + V_3 + V_4 + V_5)}{5}$$

$$= \frac{-(2.5 \text{ V} + 1 \text{ V} + 1.5 \text{ V} + 2 \text{ V} + 3 \text{ V})}{5} = -2 \text{ V}$$

Recall that the − sign in front of the equations in this chapter indicates only that the output signal is inverted from the input signal.

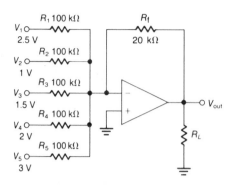

Figure 3–14. Five-Input Averaging Amplifier

Single-Supply Summing Amplifier

A single-supply voltage op amp summing amplifier circuit is shown in Figure 3–15. Note the coupling capacitors, C_1 through C_4, and the grounded power terminal. Resistors R_4 and R_5 are made equal in value, forming a voltage divider network that biases the quiescent dc output voltage at one-half the V^+ supply voltage. The output voltage is determined in the same manner as for the dual-supply amplifier, that is, by using Equation 3.21b. The value for the input coupling capacitors is determined by using Equation 3.11, and for the output coupling capacitor by using Equation 3.12.

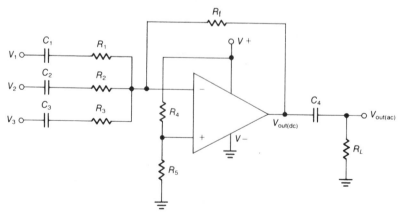

Figure 3–15. Summing Amplifier with Single-Supply Voltage

3.8 Difference Amplifiers

A basic *difference amplifier*, sometimes called a *differential amplifier*, has input signals applied simultaneously to the ($+$) and ($-$) input terminals of the op amp, as shown in Figure 3–16. A short analysis of the figure will simplify circuit understanding.

First, assume that point A is shorted to ground. The resulting circuit would appear as a simple inverting amplifier and, therefore, the output voltage would be calculated by using Equation 3.6, with V_1 substituted for V_{in}:

$$V_{out} = -\left(\frac{R_f}{R_1}\right)V_1$$

Now assume the short at point A is removed and input signal V_1 is shorted to ground. The result is essentially a noninverting amplifier that feels the voltage at point A at the ($+$) terminal input. This voltage is produced by the voltage divider action of resistors R_2 and R_3 and the input voltage V_2:

$$V_{(A)} = \left(\frac{R_3}{R_2 + R_3}\right)V_2 \tag{3.25}$$

where

$$V_{(A)} = \text{voltage at point } A$$

The noninverting output voltage for this circuit, then, is the product of the voltage at point A and the voltage gain of a noninverting amplifier; that is,

$$V_{out} = A_v V_{(A)} = \left(1 + \frac{R_f}{R_1}\right)\left(\frac{R_3}{R_2 + R_3}\right)V_2 \tag{3.26}$$

The output voltage for the difference amplifier is the sum of the inverted output, Equation 3.6, and the noninverted output, Equation 3.26:

$$V_{out} = -\left(\frac{R_f}{R_1}\right)V_1 + \left(1 + \frac{R_f}{R_1}\right)\left(\frac{R_3}{R_2 + R_3}\right)V_2 \tag{3.27}$$

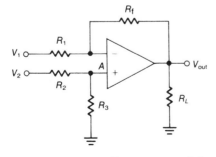

Figure 3–16.   Difference Amplifier

The voltage gain of the difference amplifier (A_{vD}) is set by all four resistors, using R_1 as a base element with relationships where

$$R_2 = R_1 \tag{3.28a}$$

and

$$R_3 = R_f = R_1 A_{vD} \tag{3.28b}$$

A major advantage of the difference amplifier described here is its ability to sense a small differential voltage buried in a larger signal. It can thus measure as well as amplify that small signal.

Subtractor

One variation of the difference amplifier is called a *unity-gain analog subtractor*, a *voltage subtractor*, or an *analog mathematical circuit*. This circuit is constructed simply by making all four resistors equal in value. When used in this application, the output voltage is the difference between V_2 and V_1:

$$V_{\text{out}} = V_2 - V_1 \tag{3.29}$$

Designing a Difference Amplifier

Designing a difference amplifier or a unity-gain analog subtractor is very simple. For the difference amplifier, determine the desired gain of the circuit, then select any standard resistance value for R_1 and R_2, which are equal, and multiply that value times the desired gain, which will provide the resistance value for R_3 and R_f. For a unity-gain analog subtractor, simply set all resistance values equal.

Example 3.11

Design a difference amplifier with a gain of 20.

Solution

First, select a value for R_1, say 5 kΩ. Next, determine the values for R_2, R_3, and R_f using Equations 3.28a and 3.28b.

$$R_2 = R_1 = 5\ \text{k}\Omega$$
$$R_3 = R_f = R_1 A_{vD} = 5\ \text{k}\Omega \times 20 = 100\ \text{k}\Omega$$

The completed design circuit is shown in Figure 3–17. By replacing resistors R_3 and R_f with 5 kΩ resistors, the circuit becomes a unity-gain analog subtractor.

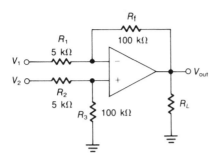

Figure 3–17. Difference Amplifier with Gain of 20

Single-Supply Difference Amplifier

A single-supply voltage op amp difference amplifier circuit is shown in Figure 3–18. The voltage gain is determined in the same manner as for the dual-supply circuit of Figure 3–16, that is, by using Equation 3.27. R_4 is made equal to R_3, forming the voltage divider for proper biasing, but with the added restriction that

$$\frac{R_f}{R_1} = \frac{R_3}{2R_2} \tag{3.30}$$

The value for input coupling capacitors is determined by using Equation 3.11, and for the output coupling capacitor by using Equation 3.12.

The circuit of Figure 3–18 can also be used for the unity-gain analog subtractor. For unity-gain, resistors R_1, R_2, and R_f must all be equal. R_4 is equal to R_3, and the value for R_3 is determined by Equation 3.30.

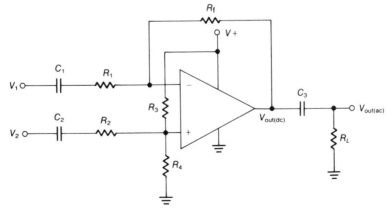

Figure 3–18. Difference Amplifier with Single-Supply Voltage

3.9 Norton Amplifiers

The Norton amplifier is designed for use in ac amplifier circuits that operate from a single power supply. It can, however, be used in many of the dc circuits previously discussed.

The schematic symbol for a single Norton amplifier is shown in Figure 3–19. The circled arrow pointing toward the noninverting (+) input identifies the Norton amplifier as a current-driven device. Power supply connections are not shown because there are usually two or four Norton amplifiers within a single IC package, sharing power supply and ground pins.

The operation of a Norton amplifier is different from that of a standard op amp. A standard op amp is a voltage-driven device that requires no significant input current, using instead negative feedback to keep the two inputs at the same potential (near zero volts). However, a Norton amplifier is a current-driven device whose input potentials are approximately 0.7 volts (the bias voltage of the input transistors of the device). Negative feedback in this device is used to keep the two input currents equal, hence the Norton amplifier is sometimes called a *current-differencing* amplifier.

Single-supply operation of the Norton amplifier requires that the output be biased to a dc level equal to one-half that of the power supply. Standard current-mirror biasing, shown in Figure 3–20, sets the quiescent (resting) dc output voltage at one-half the V^+ supply level, with the power supply connected across pins 7 and 14. Values for R_f and R_b are typically in the megohm range

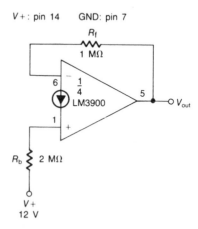

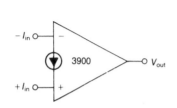

Figure 3–19.   Symbol for the Norton Op Amp

Figure 3–20.   Current-Mirror Biasing for the Norton Op Amp (Courtesy of National Semiconductor Corporation, © copyright 1980.)

because of the very small mirror current requirements of the Norton amplifier. In general practice, the biasing resistor R_b is made twice the value of the feedback resistor R_f.

A commonly used Norton amplifier is the LM3900, a device that consists of four independent, dual input, internally frequency-compensated amplifiers, as illustrated in the package configuration of Figure 3–21. Such a device can be used in many of the applications of a standard op amp. Performance as a dc amplifier that uses only a single supply is not as precise as a standard op amp that operates with split supplies, but it is adequate in many less critical applications.

Precautions

Certain precautions must be observed when using the current-differencing (Norton) amplifier. Since this is a current-driven device, series resistors must *always* be connected to each input. In biasing, resistors R_f and R_b provide the series resistance. Current into the noninverting (+) input is established by V^+, through R_b, so that

$$I^+ = \frac{V^+}{R_b} \tag{3.31}$$

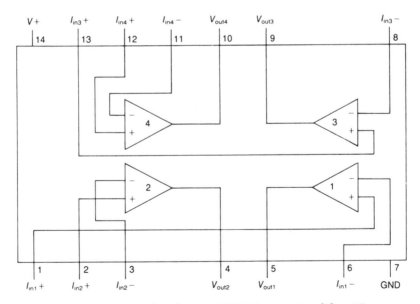

Figure 3–21. Package Configuration for the LM3900 Norton Amplifier IC (Courtesy of National Semiconductor Corporation, © copyright 1980.)

Negative feedback through R_f keeps the two input currents equal, so that

$$I^- = I^+ = \frac{V^+}{R_b} \tag{3.32}$$

Since the inverting $(-)$ input is at ground (virtual ground) level, the dc output voltage $V_{out(dc)}$ is the voltage dropped across R_f by I^-:

$$V_{out(dc)} = R_f I^- \tag{3.33}$$

Substitute Equation 3.32 into Equation 3.33:

$$V_{out(dc)} = \frac{V^+}{R_b}(R_f) \quad \text{or} \quad V_{out(dc)} = \frac{R_f}{R_b}(V^+) \tag{3.34}$$

Thus the output is biased to a dc level one-half that of the supply voltage.

Example 3.12

Refer to Figure 3–20. Calculate the output dc voltage level.

Solution

Using Equation 3.34,

$$V_{out(dc)} = \frac{R_f}{R_b}(V^+) = \frac{1 \text{ M}\Omega}{2 \text{ M}\Omega}(12 \text{ V}) = 6 \text{ V}$$

The power supply for the Norton amplifier IC should never become reversed in polarity, nor should the device be inadvertently installed backwards in a socket. Such actions will result in an unlimited current surge within the device that will fuse the internal conductors and destroy the unit.

Output short circuits either to ground or to the positive power supply should be brief. The device can be destroyed by long-lasting short circuits because of excessive junction temperatures caused by the large increase in IC chip power dissipation.

Unintentional signal coupling from the output to the noninverting input can cause oscillations. High-value biasing resistors used in the noninverting input circuit make this input lead highly susceptible to unintentional ac signal pickup. A quick check for this condition is to bypass the noninverting input to ground with a capacitor. Careful lead dress or locating the noninverting input biasing resistor closer to the IC will help prevent this problem.

Operation of the Norton amplifier can best be understood by remembering that the input currents are differenced at the inverting input terminal and that this difference current flows through the external feedback resistor to produce a large voltage output swing. The maximum output voltage swing of the amplifier is approximately $(V^+ - 1) \text{ V}_{p-p}$.

Inverting Norton Amplifier

An inverting Norton amplifier is shown in Figure 3–22. Capacitors C_1 and C_2 couple the ac signals into and out of the amplifier while blocking dc. The gain of the ac signal is set by resistors R_f and R_1. Biasing is established by V^+ and resistor R_b at the noninverting input. The input impedance Z_{in} is equal to resistance R_1.

While this circuit looks very much like an inverting amplifier constructed with a standard op amp, there are several differences to be observed. The LM3900 has a typical output impedance of 8 kΩ and a typical output current of 5 mA or less. Therefore, it is necessary to take care to prevent loading. The output is loaded by both R_f and load resistance R_L. Keeping both R_f and R_L as high in value as practicable will minimize this loading effect. This technique is opposite from that used for the standard op amp circuit where R_f is to be kept reasonably low in value to minimize output offset voltages. Formulas for the inverting Norton amplifier are given by the following equations. Output ac voltage:

$$V_{out(ac)} = -\frac{R_f}{R_1}(V_{in(ac)}) \tag{3.35a}$$

or

$$V_{out(ac)} = A_{ac} \times V_{in(ac)} \tag{3.35b}$$

where

$$A_{ac} = \text{signal gain}$$

ac gain:

$$A_{ac} = -\frac{V_{out(ac)}}{V_{in(ac)}} = -\frac{R_f}{R_1} \tag{3.36}$$

$V+$: pin 14   GND: pin 7

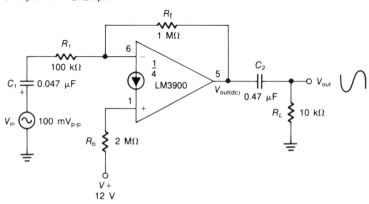

Figure 3–22.   Inverting ac Norton Amplifier Circuit (Courtesy of National Semiconductor Corporation, © copyright 1980.)

Output dc voltage level:

$$V_{out(dc)} = \frac{R_f}{R_b}(V^+) \qquad \text{(repeat of Equation 3.34)}$$

Input impedance:

$$Z_{in} = R_1 \tag{3.37}$$

Cutoff frequency:

$$f_c = \frac{1}{2\pi C_1 R_1} \tag{3.38}$$

Input capacitor C_1:

$$C_1 = \frac{1}{2\pi f_c R_1} \qquad \text{(repeat of Equation 3.11)}$$

Output capacitor C_2:

$$C_2 = \frac{1}{2\pi f_c R_L} \qquad \text{(repeat of Equation 3.12)}$$

Example 3.13

Refer to Figure 3–22. Calculate (a) $V_{out(dc)}$, (b) $V_{out(ac)}$, (c) f_c, (d) Z_{in}, and (e) A_{ac}.

Solutions

a. Using Equation 3.34,
$$V_{out(dc)} = \frac{R_f}{R_b}(V^+) = \frac{1\ M\Omega}{2\ M\Omega}(12\ V) = 6\ V$$

b. Using Equation 3.35a,
$$V_{out(ac)} = -\frac{R_f}{R_1}(V_{in(ac)}) = -\frac{1\ M\Omega}{100\ k\Omega}(100\ mV_{p\text{-}p}) = -\frac{1000\ k\Omega}{100\ k\Omega}(0.1\ V_{p\text{-}p})$$
$$= -1\ V_{p\text{-}p}$$

c. Using Equation 3.38,
$$f_c = \frac{1}{2\pi C_1 R_1} = \frac{1}{6.28 \times 0.047\ \mu F \times 100\ k\Omega}$$
$$= \frac{1}{(6.28)(47 \times 10^{-9}\ F)(100 \times 10^3\ \Omega)} = 34\ Hz$$
(Note that this is cutoff frequency, equal to approximately $0.1 f_{op}$. Therefore the circuit should be operated at 340 Hz or higher to prevent signal attenuation.)

d. Using Equation 3.37,
$$Z_{in} = R_1 = 100\ k\Omega$$

e. Using Equation 3.36,
$$A_{ac} = -\frac{R_f}{R_1} = -\frac{1\ M\Omega}{100\ k\Omega} = -\frac{1000\ k\Omega}{100\ k\Omega} = -10$$

Noninverting Norton Amplifier

A noninverting Norton amplifier is produced by simply moving the input source, coupling capacitor C_1, and series resistor R_1 to the noninverting input terminal, as shown in Figure 3–23. Note that biasing does not change and that negative feedback current still flows through R_f. However, in this circuit the input signal current *adds* to the bias current at the noninverting input terminal.

There are some differences to be observed between the noninverting amplifier constructed with the Norton amplifier and that constructed with the standard op amp. The noninverting Norton circuit is more closely related to the inverting Norton amplifier than to the noninverting op amp circuit. For example, in the noninverting Norton amplifier, the source sees an input impedance Z_{in} equal to R_1. The signal gain A_{ac} is equal to R_f/R_1, which produces the same magnitude as does the inverting amplifier. Therefore, for the noninverting Norton amplifier,

$$Z_{in} = R_1 \tag{3.39}$$

and

$$A_{ac} = \frac{R_f}{R_1} \tag{3.40}$$

All precautions observed for the inverting Norton amplifier must also be observed for the noninverting Norton amplifier.

Other applications for the Norton amplifier will be presented in a later chapter. The data sheets in Appendix B provide many additional applications.

3.10 Wideband Amplifiers

Wideband op amps are capable of amplifying a wide range of frequencies, extending from dc up to several hundred megahertz. In practical terms, this means

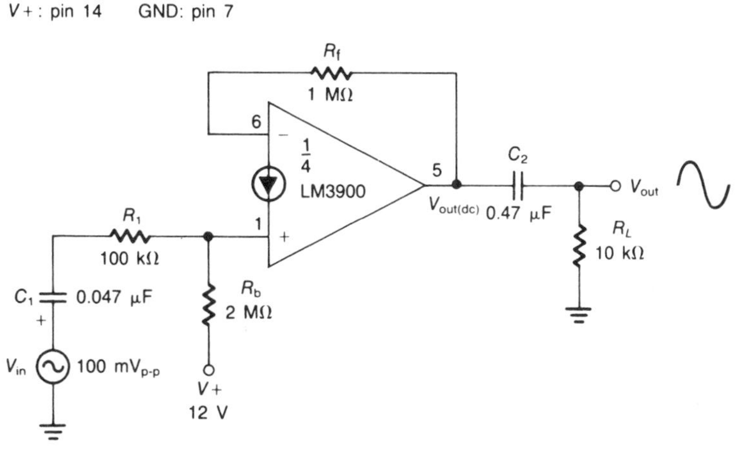

Figure 3–23. Noninverting ac Norton Amplifier Circuit (Courtesy of National Semiconductor Corporation, © copyright 1980.)

that the device can amplify RF (radio frequency), VHF (very high frequency), and UHF (ultrahigh frequency) signals.

The μA733, with a bandwidth (*BW*) of dc to 120 MHz, is used in such typical wideband applications as communications systems, as a video or pulse amplifier, and as a read-head amplifier for magnetic tape, drum, or disc memories. Another wideband device is the HA2620 with a *BW* of dc to 100 MHz. Figure 3–24 shows the HA2620 used as a simple video amplifier.

The LH0061C is a wideband, high speed op amp capable of supplying currents in excess of 0.5 A at voltage levels of ± 12 V. The wide bandwidth (1 MHz), high slew rate (70 V/μs), and high output power capability makes the LH0061C ideal for such applications as ac servo amplifiers, deflection yoke drivers, capstan drivers, and audio amplifiers. Figure 3–25 illustrates the LH0061C used as an ac servo amplifier.

There are many other devices available with lower *BW*s for applications where *BW* requirements are not so wide. Reference to manufacturers' data sheets will provide the necessary information about such devices.

3.11 Comparator as a Sensing Device

An op amp circuit called a *comparator* is not used to amplify signals. Instead, it is used as a *sensing device* to determine if a voltage is greater than or less than

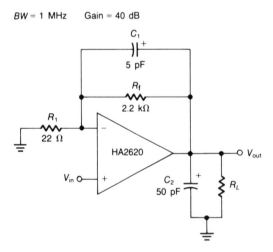

Figure 3–24. Video Amplifier Circuit Using the HA2620 Wideband Op Amp (Courtesy of Harris Semiconductor Products Division, © copyright 1980.)

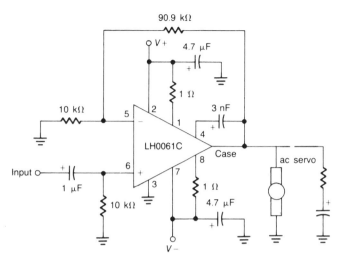

Figure 3–25. ac Servo Amplifier Circuit (Courtesy of National Semiconductor Corporation, © copyright 1980.)

a given reference voltage level. Typically, a comparator operates in the open-loop mode, so that even the smallest difference in input voltages will cause the output of the op amp to saturate.

Level Detector

A simple op amp comparator, called a *level detector*, is illustrated in Figure 3–26A. A reference voltage V_{ref} is connected between the (+) input terminal and ground. The input voltage V_{in} connected to the (−) input terminal is compared

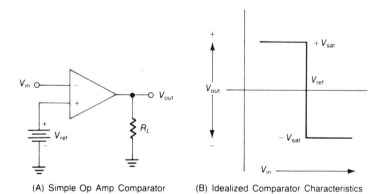

(A) Simple Op Amp Comparator (B) Idealized Comparator Characteristics

Figure 3–26. Simple Sensing Circuit Using a Comparator as a Level Detector

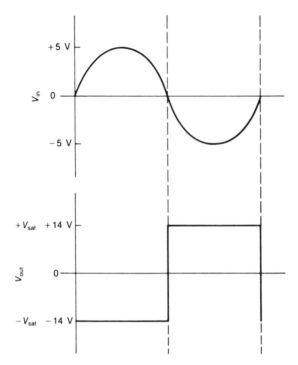

Figure 3–27. Zero-Crossing Detector Input/
Output Relationships

with the reference voltage. If V_{in} is less than V_{ref}, then V_{out} goes positive. If V_{in} becomes greater than V_{ref}, then V_{out} switches to negative. The results are shown in Figure 3–26B, an idealized characteristic curve.

The reference voltage need not be connected to the $(+)$ terminal. If it is connected to the $(-)$ terminal, then the polarity of the output is reversed.

Zero-Crossing Detector

By grounding the $(+)$ input terminal (0 V) of the comparator, we have a *zero-crossing detector*. Figure 3–27 shows the relationship between the input and output voltages. With a small sine wave applied to V_{in}, note that V_{out} is negative when the sine wave is positive, and vice versa. The switch between positive and negative occurs when the input signal crosses the reference voltage, or zero.

Many additional comparator applications will be discussed in later chapters.

3.12 Summary

1. A noninverting amplifier is constructed with the input signal applied to the $(+)$ input terminal of the op amp. The output signal is in phase with the

input signal. The closed-loop gain is always one or greater, since $A_{CL} = 1 + (R_f/R_1)$.

2. An inverting amplifier is constructed with the input signal applied to the $(-)$ input terminal of the op amp. The output signal is 180° out of phase with the input signal. The closed-loop gain can be less than, equal to, or greater than unity, since $A_{CL} = -(R_f/R_1)$, where the $-$ sign indicates only that the output signal is inverted.

3. The input impedance of the inverting amplifier is the value of the input element and will always be less than the input impedance of the noninverting amplifier, which is the intrinsic input impedance of the op amp itself.

4. The inverting amplifier can be used as a unity-gain inverter when the input and feedback elements are equal in value.

5. Error voltage, called dc output offset voltage, is the result of certain op amp characteristics: input bias current, input offset current, input offset voltage, and drift. Null circuits can eliminate all but drift.

6. A voltage follower circuit has no input or feedback resistances. The output signal follows exactly the input, or source, signal. The circuit is also called source follower, unity-gain amplifier, buffer amplifier, and isolation amplifier.

7. A summing amplifier has two or more resistors connected simultaneously to the inverting input terminal of an op amp. The output voltage is the sum of the individual input voltages.

8. An averaging amplifier is a variation of the summing amplifier in which each of the input resistances is made equal in value, and the feedback resistor is proportional to the number of input resistors. The output voltage is the sum of the individual input voltages, divided by the number of the inputs.

9. A difference amplifier has input voltages applied simultaneously to the inverting and noninverting input terminals of an op amp. The output voltage is a combination of the inverting and noninverting output voltages, or, effectively, the difference between the input voltages.

10. A unity-gain analog subtractor, in which all resistances are made equal in value, is a variation of the difference amplifier.

11. Many standard op amp circuits can be constructed with a single-supply power source if certain external components are added.

12. The gain of an amplifier can be controlled by the feedback resistance.

13. Virtual ground is a point at which voltage with respect to true ground is zero. It cannot be used as a ground point because it is isolated from true ground.

14. The Norton op amp is designed specifically to operate from a single power supply. It is used in a multitude of applications. It is sometimes called a current-differencing amplifier, because it operates on the difference of currents flowing into the $(+)$ and $(-)$ input terminals.

15. Wideband amplifiers are those capable of amplifying frequencies over the range from dc to several hundred megahertz.

16. Comparators are used to sense voltage variations, that is, to determine if a voltage is larger or smaller than a reference voltage.

17. Level detectors and zero-crossing detectors are two applications of the comparator.

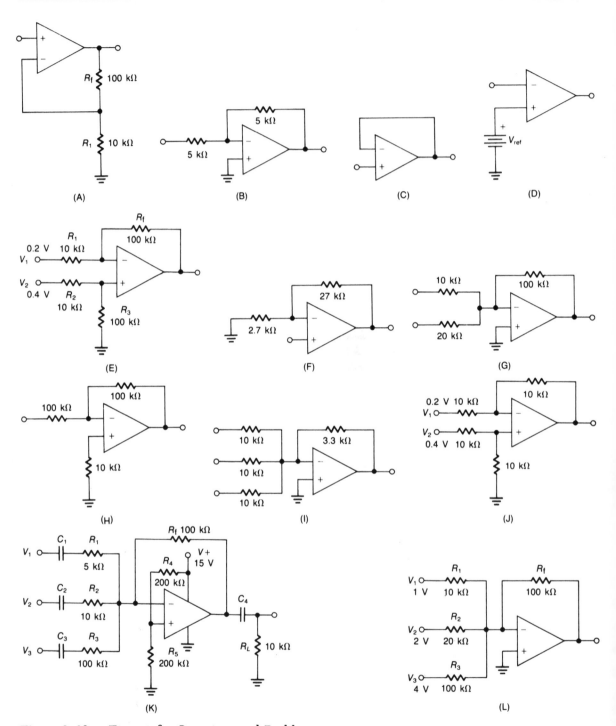

Figure 3–28. Figures for Questions and Problems

3.13 Questions and Problems

3.1 Several op amp amplifiers are shown in Figure 3–28A–L. Identify each amplifier in the figure from the following list of labels (some figure parts may have more than one label):

1. Unity-gain amplifier
2. Averaging amplifier
3. Voltage follower
4. Inverting amplifier
5. Summing amplifier
6. Noninverting amplifier
7. Unity-gain inverter
8. Difference amplifier
9. Unity-gain analog subtractor
10. Comparator

3.2 Calculate the closed-loop gain for the amplifier in Figure 3–28A.

3.3 Calculate the output voltage for Figure 3–28L.

3.4 Calculate the values for each of the capacitors in Figure 3–28K. Assume a frequency response of 30 Hz.

3.5 What is the advantage of a difference amplifier?

3.6 What type of biasing is standard for the Norton amplifier?

3.7 How is output offset voltage that results from drift minimized?

3.8 Explain *offset null* and how it works.

3.9 Design a difference amplifier with a gain of 10. Draw the completed design circuit.

3.10 Design and draw the completed circuit for a noninverting amplifier whose gain is 21. The input element is 5 kΩ.

Regulator Circuits

4.1 Introduction

Several problems can arise in an operating linear IC when variations occur in the supply voltage of a circuit. For example, a change in supply voltage may shift the operating point of an audio amplifier into the nonlinear region, resulting in a distorted output, or it could cause a change in the offset voltage, which would produce an undesirable signal at the output. Such variations in supply voltage are a result of poor regulation.

Current regulation circuits provide a constant current to a load. The current is independent of voltage changes across the load, so such voltage variations will not affect the operation of the circuit.

In this chapter we will discuss methods, linear IC devices, and circuits that can be used for voltage and current regulators.

4.2 Objectives

When you complete this chapter, you should be able to:

☐ Define regulator terms.

☐ Explain basic voltage and current regulator concepts.

☐ Compare series, shunt, and switching regulators.

☐ Explain how op amps are used as comparators in regulator circuits.

☐ Discuss types of available monolithic regulators and their advantages.

☐ Define and discuss hybrid regulators.

☐ Define and discuss regulator protection circuits.

4.3 Fundamentals of Regulators

For proper linear device operation to occur, supply voltage variations, whether caused by temperature, changing load conditions, or power line conditions, must be controlled. Voltage regulator circuits can reduce the amplitude of those output voltage variations to a level that will not interfere with the operation of the device. Voltage regulators can be built into the power supply design, or they can be added externally. If external regulators are added, they may be placed either at the power supply output or at the operating device power terminals (local regulation).

Voltage regulators come in many forms. Some use diodes, transistors, or op amps as active devices. A monolithic voltage regulator is in the form of a linear IC, that is, one single active device that performs the regulating function normally performed by several discrete components.

Types of Regulator Circuits

There are three major types of regulator circuits: series, shunt, and switching. Although each type uses a different approach to regulation, they all have one thing in common. They all use feedback to regulate the output voltage.

Series Regulator

The *series regulator* gets its name from the fact that the variable control element is in series with the load. When a change in load or power supply occurs, the variable element automatically adjusts to offset those changes, effectively maintaining a constant load voltage.

A practical series voltage regulator using discrete components is shown in Figure 4–1. The unregulated output of the power supply is the input to the regulator circuit. Transistor Q_1 provides rapid response to load or power supply changes. Q_1 is operated in its active region and carries the load current. The load current results in a power loss, a major disadvantage of the series regulator circuit. The power dissipated in the transistor can be calculated as follows:

$$P_D = V_{CE}I_C \tag{4.1}$$

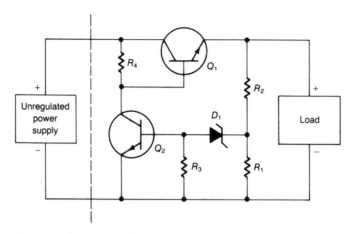

Figure 4–1.   Series Voltage Regulator Using Discrete Components (Courtesy of RCA Corporation, © copyright 1981.)

where

P_D = power dissipated in the transistor
V_{CE} = collector-emitter voltage
I_C = collector current (which is also the load current in this case)

Because of this power loss, the circuit in Figure 4–1 is called a *dissipative voltage regulator* circuit.

Transistor Q_2 senses any change in load voltage and adjusts the voltage at the base of Q_1, which causes Q_1 to change its collector-emitter voltage V_{CE}, returning the load voltage to its original level. Q_2 is biased by the voltage divider network of R_1 and R_2, the zener diode D_1, operated in its breakdown mode, and R_3. There is a constant voltage drop across D_1, so any change across R_1 also is across R_3. This means that the low-input impedance of Q_2 will not load R_1; therefore, R_1 can be a large value, which limits the bleeder current drawn from the load current. Then any change across R_3 changes the bias on Q_2.

Shunt Regulator

In the *shunt regulator* circuit, the variable control element is in parallel with the load. Figure 4–2 shows a shunt regulator circuit using discrete components. As in the series regulator, the unregulated output of the power supply is the input to the regulator circuit. Transistor Q_1 is the shunt element. The amount of current through the circuit is controlled by Q_2, which acts as a variable resistor and controls the amount of base current at Q_1. The voltage divider network of R_1 and zener diode D_1 provides the biasing for Q_2.

Regulator output voltage is determined by the bias circuit of Q_2, one part of which is the zener diode, D_1. To help maintain a constant voltage drop across

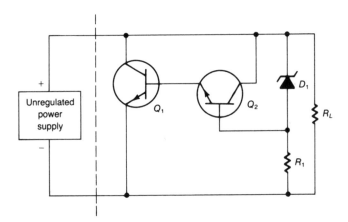

Figure 4–2. Shunt Voltage Regulator Using Discrete Components (Courtesy of RCA Corporation, © copyright 1981.)

the diode with changes in temperature, select a zener diode with a low temperature coefficient and limit diode current flow.

The shunt regulator circuit, just as the series circuit, has the disadvantage of power dissipation in the control transistor. In Figure 4–2 transistor Q_1 is operated in the active region and therefore dissipates power. Such power dissipation in the control transistor prevents maximum power transfer to the load, resulting in low efficiency.

Switching Regulator

The *switching regulator* overcomes the power dissipation disadvantage by switching the control transistor between the cutoff mode and the saturation mode. Recall that at cutoff, the transistor has only a small leakage current flow but high voltage. In the saturation mode, current is high but voltage is very low. In either mode, then, the resultant product of voltage and current is relatively low. This results in low power dissipation in the control transistor and provides high circuit efficiency.

A typical switching regulator circuit using discrete components is shown in Figure 4–3. Again, the unregulated output voltage of the power supply is the input to the regulator circuit. Applying a control signal to the base of the transistor, as shown, will cause it to alternately switch between cutoff and saturation. This alternation allows the input voltage to be periodically applied to the filter circuit, where it is measurable across the diode. The control signal should be high enough in frequency to allow low values to be used for the inductor and capacitor in the filter circuit. Low values of inductance and capacitance result in reduced component size and weight, always a consideration in electronic circuit design.

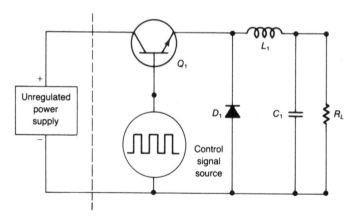

Figure 4–3. Switching Voltage Regulator Using Discrete Components

Duty Cycle

The dc load voltage is determined by the ratio of the amount of time the transistor is in the saturated mode (t_{on}) to the period of the control signal (T). This ratio is called the *duty cycle* and is expressed as a percentage. As an equation,

$$\text{Percent duty cycle} = \frac{t_{on}}{T} \times 100\% \qquad (4.2)$$

The average voltage across the diode is the value of the dc component of the output of the circuit (V_{dc}) and is directly related to the duty cycle. This relationship may be shown as follows:

$$V_{dc} = \frac{V_D(\% \text{ duty cycle})}{100\%} \qquad (4.3)$$

where

V_D = maximum voltage across the diode

Example 4.1

Assume the control signal frequency in Figure 4–3 is 30 kHz and the transistor ON time is 8 μs. Calculate the duty cycle.

Solution

First, recall that the period $T = 1/f$ and solve for T:

$$T = \frac{1}{f}$$

where T is in seconds and f is in hertz

$$T = \frac{1}{30 \times 10^3 \text{ Hz}} = 33.3 \times 10^{-6} \text{ s} = 33.3 \text{ μs}$$

Then solve as follows using Equation 4.2:

$$\% \text{ duty cycle} = \frac{t_{\text{on}}}{T} \times 100\% = \frac{8 \text{ μs}}{30 \text{ μs}} \times 100\%$$

$$= \frac{8 \times 10^{-6} \text{ s}}{33.3 \times 10^{-6} \text{ s}} \times 100\% = 24\%$$

Example 4.2

Assume the maximum voltage across the diode in Figure 4–3 is 50 V. Using the duty cycle calculated in Example 4.1, calculate the dc component of the output voltage of the circuit.

Solution

$$V_{\text{dc}} = \frac{V_D(\% \text{ duty cycle})}{100\%} = \frac{50 \text{ V} \times 24\%}{100\%} = 12 \text{ V}$$

It should now be obvious that if we wish to change the output voltage, the duty cycle must change. The switching regulator accomplishes this by using a sensing device to sample the output voltage. The value of the output voltage is then used to vary the duty cycle of the control signal. In Section 4.5, we will consider how and why an op amp is used for the sensing device.

4.4 Heat Sinking

To fully utilize the various available regulator packages, sufficient attention must be paid to proper heat removal. For efficient thermal management, you must rely on important parameters supplied by the manufacturer. The device temperature depends on the power dissipation level, the means for removing the heat generated by this power dissipation, and the temperature of the body (heat sink) to which this heat is removed.

Remember that any semiconductor device that carries current will dissipate power, generally in the form of heat. This heat is generated at the junction of the device and must flow through the package to ambient (surrounding) air. All packaging material presents some opposition to this flow of heat, designated on data sheets as *thermal resistance* (θ_{JA}). This opposition causes the temperature of the junction (T_J) to rise. If the junction temperature becomes too high, the device will be damaged, or thermal shutdown will occur.

A tabulation by package of various regulators is shown in Figure 4–4. The figure also lists the average and maximum values of thermal resistance for the regulator chip-package combinations and can be used as a guide for selecting a suitable package when designing a regulator circuit.

Determining Heat Sinking Requirements

Thermal characteristics of voltage regulator chips and packages determine that some form of heat sinking is mandatory whenever the power dissipation exceeds the following values at 25°C ambient temperature:

RESISTANCES LISTED AS FOLLOWS
θ_{JC}(TYP) θ_{JC}(MAX) / θ_{JA}(TYP) θ_{JA}(MAX) in C/W

REG TYPE	DEVICE NO./SERIES	IOUT (A)	TO-3 K	4-LEAD TO-3 K	TO-220 U	POWER WATT U1*	TO-39 H	TO-92 W	TO-99 8-LEAD TO-5 H	TO-100 10-LEAD TO-5 H	TO-116 14-PIN PLASTIC D	TO-116 14-PIN CERAMIC D	8 PIN MINIDIP T
POS. 3-TERM.	µA78LXX	0.1					20 40 / 140 190	160 180					
	µA78MXX	0.5			3.0 5.0 / 62 70		18 25 / 120 185						
	µA78CXX	0.5				6 8 / 75 80							
	µA109, µA209, µA309, 5V	1	3.5 5.5 / 40 45										
	µA78XX	1	3.5 5.5 / 40 45		3.0 5.0 / 60 65								
	µA78CB	2	3.5 5.5 / 40 45		3.0 5.0 / 60 65								
	78H05, 5V	5	1.5 2.0 / 37 40										
	µA78HXX	5	2.0 2.5 / 32 .38										
NEG. 3-TERM.	µA79MXX	0.5			3.0 5.0 / 62 70		18 25 / 120 185						
	µA79XX	1	3.5 5.5 / 40 45		3.0 5.0 / 60 65								
POS. ADJ.	µA105/ 305/376	0.012 to 0.045							25 40 / 150 190				160 190
	µA723	0.125								25 50 / 150 190	150 190	125 160	
	µA78MG 4-TERM.	0.5				6 8 / 75 80							
	µA78G 4-TERM.	1		4.0 6.0 / 44 47		6 8 / 75 80							
	µA78HG 4-TERM.	5	2.0 2.5 / 32 38										
NEG. ADJ.	µA104/304	0.020								25 50 / 150 190			
	µA79MG 4-TERM.	0.5				6 8 / 75 80							
	µA79G 4-TERM.	1		4.0 6.0 / 44 47		6 8 / 75 80							

Figure 4–4. Thermal Resistance (θ_{JC}, θ_{JA}) by Device and Package (Courtesy of Fairchild Camera & Instrument Corporation, Linear Division, © copyright 1978.)

0.67 W for the TO-39 package
0.69 W for the TO-92 package
1.56 W for the Power Watt (similar to TO-220 packages)
1.8 W for the TO-220 package
2.8 W for the TO-3 package

At ambient temperatures above 25°C, heat sinking becomes necessary at lower power levels.

To choose a heat sink, you must determine the following regulator parameters:

$P_{D(max)}$—Maximum power dissipation:

$$P_{D(max)} = (V_{in} - V_{out}) I_{out} + V_{in}I_Q$$

where

V_{in} = input voltage
V_{out} = output voltage
I_{out} = output current
I_Q = quiescent current

$T_{A(max)}$—Maximum ambient temperature the regulator will encounter during operation

$T_{J(max)}$—Maximum operating junction temperature, specified by the manufacturer

θ_{JC}, θ_{JA}—Junction-to-case and junction-to-ambient thermal resistance values, specified by the manufacturer

θ_{CS}—Case-to-heat-sink thermal resistance

θ_{SA}—Heat-sink-to-ambient thermal resistance, specified by the heat sink manufacturer

The maximum permissible dissipation without a heat sink is:

$$P_{D(max)} = \frac{T_{J(max)} - T_{A(max)}}{\theta_{JA}} \tag{4.4}$$

If the device dissipation P_D exceeds this amount, a heat sink is required. The total required thermal resistance is then calculated as follows:

$$\theta_{JA(tot)} = \theta_{JC} + \theta_{CS} + \theta_{SA} = \frac{T_{J(max)} - T_{A(max)}}{P_D} \tag{4.5}$$

Case-to-sink and sink-to-ambient thermal resistance information on commercially available heat sinks is normally provided by the heat sink manufacturer. A summary of some commercially available heat sinks is given in Table 4–1.

Table 4–1 | **Heat Sink Selection Guide**

This list is only representative. No attempt has been made to provide a complete list of all heat sink manufacturers. All values are typical as given by manufacturer or as determined from characteristic curves supplied by manufacturer.

θ_{SA} Approx. (°C/W)	Manufacturer and Type
	TO-3 Packages
0.4 (9" length)	Thermalloy (Extruded) 6590 Series
0.4 - 0.5 (6" length)	Thermalloy (Extruded) 6660, 6560 Series
0.56 - 3.0	Wakefield 400 Series
0.6 (7.5" length)	Thermalloy (Extruded) 6470 Series
0.7 - 1.2 (5 - 5.5" length)	Thermalloy (Extruded) 6423, 6443, 6441, 6450 Series
1.0 - 5.4 (3" length)	Thermalloy (Extruded) 6427, 6500, 6123, 6401, 6403, 6421, 6463, 6176, 6129, 6141, 6169, 6135, 6442 Series
1.9	IERC E2 Series (Extruded)
2.1	IERC E1, E3 Series (Extruded)
2.3 - 4.7	Wakefield 600 Series
4.2	IERC HP3 Series
4.5	Staver V3-5-2
4.8 - 7.5	Thermalloy 6001 Series
5 - 6	IERC HP3 Series
5 - 10	Thermalloy 6013 Series
5.6	Staver V3-3-2
5.9 - 10	Wakefield 680 Series
6	Wakefield 390 Series
6.4	Staver V3-7-224
6.5 - 7.5	IERC UP Series
8	Staver V1-5
8.1	Staver V3-5
8.8	Staver V3-7-96
9.5	Staver V3-3
9.5 - 10.5	IERC LA Series
9.8 - 13.9	Wakefield 630 Series
10	Staver V1-3
11	Thermalloy 6103, 6117 Series
	TO-220 Packages (See Note 1)
4.2	IERC HP3 Series
5 - 6	IERC HP1 Series
6.4	Staver V3-7-225
6.5 - 7.5	IERC VP Series
7.1	Thermalloy 6070 Series
8.1	Staver V3-5
8.8	Staver V3-7-96
9.5	Staver V3-3
10	Thermalloy 6032, 6034 Series
12.5 - 14.2	Staver V4-3-192
13	Staver V5-1
15	Thermalloy 6030 Series
15.1 - 17.2	Staver V4-3-128
16	Thermalloy 6072, 6106 Series
18	Thermalloy 6038, 6107 Series
19	IERC PB Series
20	Staver V6-2
20	Thermalloy 6025 Series
25	IERC PA Series

1. Most TO-3 heat sinks can also be used with TO-220 packages with appropriate hole patterns.
2. Most TO-220 heat sinks can be used with the Power Watt package.

IERC 135 W. Magnolia Blvd., Burbank, CA 91502
Staver Co. Inc.: 41-51 N. Saxon Ave., Bay Shore, N.Y. 11706

θ_{SA} Approx. (°C/W)	Manufacturer and Type
	TO-92 Packages
30	Staver F2-7
46	Staver F5-7A, F5-8-1
50	IERC RUR Series
57	Staver F5-7D
65	IERC RU Series
72	Staver F1-7
85	Thermalloy 2224 Series
	Mini Batwing
10	Thermalloy 6069 Series
10.6	Thermalloy 6068 Series
11.7	Thermalloy 6067 Series
13	Thermalloy 6066 Series
20	Thermalloy 6062 Series
26	Thermalloy 6064 Series
	TO-5 and TO-39 Packages
12	Thermalloy 1101, 1103 Series
12 - 16	Wakefield 260-5 Series
15	Staver V3A-5
22	Thermalloy 1116, 1121, 1123 Series
22	Thermalloy 1130, 1131, 1132 Series
24	Staver F5-5C
25	Thermalloy 2227 Series
26 - 30	IERC Thermal Links
27 - 83	Wakefield 200 Series
28	Staver F5-5B
34	Thermalloy 2228 Series
35	IERC Clip Mount Thermal Link
39	Thermalloy 2215 Series
41	Thermalloy 2205 Series
42	Staver F5-5A
42 - 65	Wakefield 296 Series
46	Staver F6-5, F6-5L
50	Thermalloy 2225 Series
50 - 55	IERC Fan Tops
53	Thermalloy 2211 Series
55	Thermalloy 2210 Series
56	Thermalloy 1129 Series
58	Thermalloy 2230, 2235 Series
60	Thermalloy 2226 Series
68	Staver F1-5
72	Thermalloy 1115 Series
	Power Watt (similar to TO-202) Packages (See Note 2)
12.5 - 14.2	Staver V4-3-192
13	Thermalloy 6063 Series
13	Staver V5-1
15.1 - 17.2	Staver V4-3-128
19	Thermalloy 6106 Series
20	Staver V6-2
24	Thermalloy 6047 Series
25	Thermalloy 6107 Series
37	IERC PA1-7CB with PVC-1B Clip
40 - 42	Staver F7-3
40 - 43	Staver F7-2
42	IERC PA2-7CB with PVC-1B Clip
42 - 44	Staver F7-1

Thermalloy Inc.: 2021 W. Valley View Lane, Dallas, TX 75234
Wakefield Engineering, Inc.: Audubon Rd., Wakefield, MA 01880

Source: Courtesy of Fairchild Camera & Instrument Corporation, Linear Division, © copyright 1978.

Example 4.3

Determine the heat sink required for a μA7806C regulator to be used in a circuit that has the following system requirements:

Operating ambient temperature range: 0°C–60°C   (T_A)
Maximum junction temperature: 125°C   (T_J)
Maximum output current: 800 mA   (I_{out})
Maximum input to output differential: 10 V

Solution

A review of the data sheet for the μA7800 series (Appendix C) shows that the device is available in either TO-3 or TO-220 packages. The TO-220 package has a lower cost and better thermal resistance, therefore, it should be selected for this regulator requirement.

θ_{JC} = 5°C/W maximum (from data sheet or Figure 4–4, μA78xx)

$$\theta_{JA(tot)} = \theta_{JC} + \theta_{CS} + \theta_{SA} = \frac{T_J - T_A}{P_D}$$

$$\theta_{CS} + \theta_{SA} = \frac{125°C - 60°C}{0.8\text{ A} \times 10\text{ V}} - 5°C = 3.13°C/W$$

Assuming θ_{CS} = 0.13°C/W, then θ_{SA} = 3°C/W. This thermal resistance value can be achieved with any of the TO-220 package heat sinks selected from Table 4–1.

Tips for Heat Sinking

Here are some tips for better regulator heat sinking:

1. Avoid placing heat-dissipating components such as power resistors next to regulators.

2. When using low dissipation packages, such as TO-5, TO-39, and TO-92, keep lead lengths to a minimum and use the largest possible area of the printed circuit board traces or mounting hardware to provide a heat dissipation path for the regulator.

3. When using larger packages, be sure the heat sink surface is flat and free from ridges and high spots. Check the regulator package for burrs or peened-over corners. Regardless of the smoothness and flatness of the contact between the package and heat sink, air pockets between them are unavoidable unless a lubricant is used. Therefore, for good thermal conduction, use a thin layer of thermal lubricant.

4. In some applications, especially with negative regulators, it is desirable to insulate electrically the regulator case from the heat sink. Hardware kits for

this purpose are commercially available for such packages as the TO-3 and TO-220. The kits generally consist of a thin piece of mica or bonded fiberglass that electrically isolates the two surfaces, yet provides a thermal path between them. The thermal resistance will increase, but some improvement can be realized by using thermal lubricant on each side of the insulator material.

5. If the regulator is mounted on a heat sink with fins, the most efficient heat transfer takes place when the fins are in a vertical plane, because this type of mounting forces the heat transfer from fin to air in a combination of radiation and convection.

6. If it is necessary to bend any of the regulator leads, handle them carefully to avoid straining the package. Furthermore, lead bending should be restricted since bending will fatigue and eventually break the leads.

4.5 Voltage Regulators Using Op Amps

In Section 4.3 we learned that a voltage regulator circuit uses a transistor as a sensing device to detect any change in the output voltage and to drive a second transistor that, in turn, controls the voltage or current of the load. The types of regulator circuits discussed in Section 4.3 offer some good regulation features, but they also have some problems. One problem is the low sensitivity caused by the low gain of the sensing transistor. This low sensitivity prevents slight changes in the load voltage from providing large enough changes in the output of the sensing transistor to affect the operation of the control transistor. Another problem is the slow response time of the sensing transistor. Another is the loading effect that may occur in the sensing divider circuit because of the low input impedance of the sensing transistor. These are problems that cannot be tolerated in circuits that require a high degree of regulation.

Replacing the sensing transistor with an op amp solves these problems for us. The op amp has the high gain necessary to provide a high level of sensitivity and a rapid response to changes in the load voltage. The high input impedance of the op amp prevents loading down devices that are connected to it. The op amp quickly senses slight changes in load voltage, then drives the control transistor that quickly returns the load voltage to its desired value.

Op Amp Series Regulator

Figure 4–5 shows a series voltage regulator with an op amp as the sensing element. The output of the op amp drives the base of the control transistor, Q_1. The op amp is connected as a *voltage level detector*. The amplitude and polarity of the output voltage of the op amp are determined by the more positive of the two input voltages. Recall from Chapter 2 that

$$V_{out} = A_v(V_1 - V_2) \qquad \text{(repeat of Equation 2.5)}$$

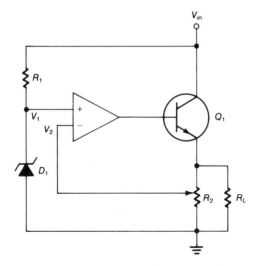

Figure 4–5. Series Voltage Regulator with an Op Amp as the Sensing Element

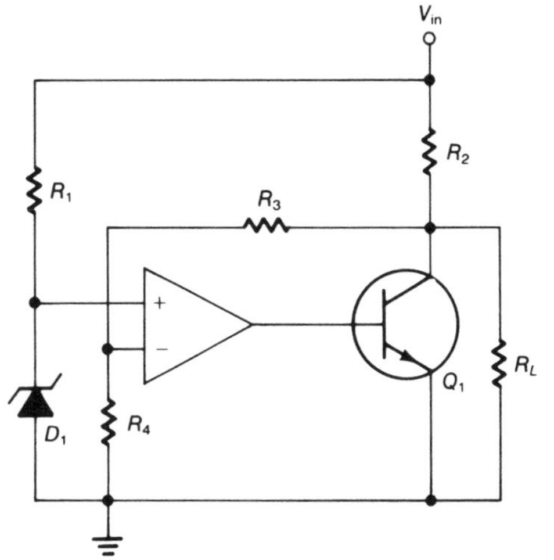

Figure 4–6. Shunt Voltage Regulator with an Op Amp as the Sensing Element

Input signal V_1 is the reference voltage across the zener diode. Input signal V_2 is that portion of the load voltage developed across the potentiometer, R_2. Any change in load voltage will cause a change in the differential voltage, which will cause a change in the output of the detector. Emitter-follower action of Q_1 then causes the load voltage to readjust to its desired value.

The resistance of potentiometer R_2 should be only of high enough value compared with R_L to limit the amount of current and not load down the output. Resistor R_1 limits the current through the zener diode, reducing the temperature rise of the diode and thereby providing increased stability in the circuit. Reducing the gain of the amplifier will also increase the stability of the circuit.

Op Amp Shunt Regulator

A shunt regulator using an op amp as the sensing element is shown in Figure 4–6. The differential voltage between the reference voltage of the zener diode and that portion of output voltage across R_4 determines the output of the comparator, which drives the base of the control transistor, Q_1. The current amplification process of the transistor causes the collector current to change, causing the transistor to act as a variable resistance. The current in Q_1 is thus automatically adjusted inversely to compensate for any load current change. The effect is to keep the supply current constant, which in turn keeps the load voltage constant.

The divider network formed by resistors R_3 and R_4 senses load voltage changes. Current-limiting resistor R_1 helps maintain a steady temperature in the zener diode. Resistor R_2 offers circuit protection by limiting the maximum current drawn from the supply if the load is shorted.

The shunt regulator offers an advantage over the series regulator because the power supply current output is kept constant. This means that there is a constant power drain on the supply and the ac lines feeding it. With little or no change in current in the ac source lines feeding the supply, the lines themselves are better regulated.

Some Considerations

Although op amps do eliminate certain problems, as discussed earlier, there are other possible problems that can arise. To prevent or reduce these new problems, some simple factors must be considered. An unstable circuit can result from the high gain of the op amp and the feedback through the control transistor. Choosing the proper transistor characteristics overcomes this problem. Also, the emitter-follower circuit of the series regulator is susceptible to oscillations. Reducing the lead length reduces the inductance and removes possible unwanted feedback paths.

Another consideration is the stability of the zener reference voltage. Changes in both load and temperature affect zener voltage. To overcome the problem of temperature changes, place a forward-biased zener in series with the reverse-biased reference zener. This works because reverse-biased zeners operating in the breakdown mode have positive temperature coefficients, while forward-biased zeners have negative coefficients. Therefore, temperature compensation occurs, providing a stable reference.

Another important consideration is the response time of the circuit, which determines how rapidly the regulator can adjust the output back to its original value. The slew rate of the op amp affects the response time, so selecting an op amp with a high slew rate will offer the best response time, resulting in best regulation.

Op Amp Switching Regulator

Figure 4–7 shows an op amp comparator that is used in a switching regulator. The comparator replaces the control signal source shown in Figure 4–3. The value of load voltage compared to the zener reference voltage determines the output polarity of the comparator and, therefore, the duty cycle of the transistor. Zener diode D_1 provides a reference voltage (V_Z) to the noninverting input, and the load voltage (V_L) supplies an input to the inverting terminal. When V_L increases to a value greater than V_Z, the output of the comparator swings negative and transistor Q_1 is cut off. This results in a decrease in current through coil L_1 and a decrease in voltage across capacitor C_1 discharging through forward-biased D_2. With C_1 in parallel with the load, V_L is also reduced, thereby holding the load current constant.

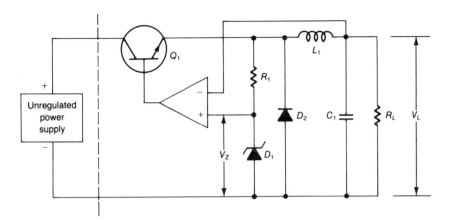

Figure 4–7. Switching Voltage Regulator with Op Amp Comparator
(Courtesy of John Wiley & Sons, Inc., © copyright 1981.)

If V_L decreases to a value less than V_Z, the output polarity of the comparator swings positive, driving Q_1 into saturation. This increases current through L_1, charging C_1 to a higher voltage, and thus increasing V_L, again stabilizing the load current. Diode D_2 is reverse biased at this time to allow charging of C_1.

Dual Voltage Regulator

Since op amps require two supply voltages of equal value but opposite polarity, it is necessary to regulate both negative and positive outputs of the supply. A dual voltage regulator using op amps is illustrated in Figure 4–8.

The two op amps, A_1 and A_2, drive complementary transistors Q_1 and Q_2. A reference voltage is supplied by zener diode D_1 and fed into the (+) inputs of both op amps from the junction of the divider network of R_3 and R_4. Any change in the regulated output is fed back through feedback resistor R_f to the (−) input of the respective op amp.

The regulated negative voltage, $-V_2$, is developed across capacitor C_1, and the regulated positive voltage, $+V_2$, is developed across capacitor C_2.

The zener diode also applies equal and opposite voltages to the (−) inputs of the op amps, and the op amp outputs in turn place equal and opposite voltages at the bases of the transistors. The feedback resistors determine the value of these voltages, and these voltages determine the emitter voltages and, therefore, the load voltage.

Monolithic Regulators

By incorporating all the necessary components on a single chip by the monolithic construction method, voltage regulators can be greatly reduced in size. This

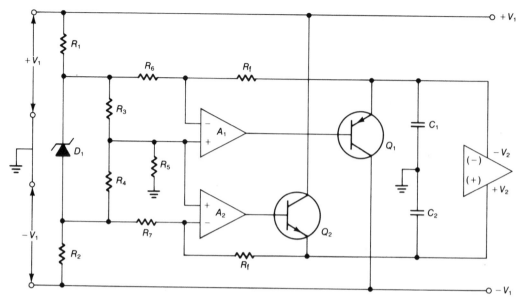

Figure 4–8. Dual Voltage Regulator Constructed with Op Amps (Courtesy of John Wiley & Sons, Inc., © copyright 1981.)

smaller regulator, in the form of a single IC, can then be used for *local regulation;* that is, it can be located physically at any point in the system where it is needed. This local on-card regulation capability eliminates the requirement to regulate the entire power supply, thereby reducing greatly the cost of regulation.

Most monolithic regulators require that some external components be added to provide overload protection, to channel desired currents and voltages to selected loads, and possibly to convert to a switching regulator.

There are three main categories of monolithic regulators:

1. Three-terminal devices with fixed positive or negative outputs
2. Adjustable regulators with positive or negative outputs
3. Dual polarity tracking regulators

Within each category there are a great many regulators available with a variety of parameters.

Fixed Voltage Regulators

Fixed voltage regulators are excellent devices for use at points in a system that requires a single fixed, well regulated voltage. The devices are small, require no external components, and have only three terminals. These characteristics permit these devices to be located directly on the printed circuit board that requires the specified regulated voltage. Fixed voltage regulators are ideal for use in limited space.

The LM340 series are typical three-terminal fixed positive voltage regulators. The 340 series offer output voltages of 5, 6, 8, 10, 12, 15, 18, and 24 volts. A built-in reference voltage drives the (+) terminal of an amplifier. An internal voltage divider, preset to one of the output voltages listed earlier, provides feedback voltage to the (−) terminal of the amplifier. The chip also has a control, *pass*, transistor capable of handling more than 1.5 A of load current if properly heat sinked, and includes *current limiting* and *thermal shutdown*. Figure 4–9 shows the block diagram of the LM340 series.

Thermal Shutdown

When the internal temperature of the device reaches 175°C, the regulator automatically turns off and prevents any further increase in internal temperature. This thermal shutdown is a precaution against excessive power dissipation in the device.

A fixed voltage regulator in its simplest configuration is illustrated in Figure 4–10. The unregulated voltage of the power supply is fed to pin 1, the regulated output is taken from pin 2, and pin 3 is ground. In addition to regulating voltage, this type of device also attenuates ripple. Typical ripple rejection ranges from 55 dB for the 340-5 to 44 dB for the 340-24.

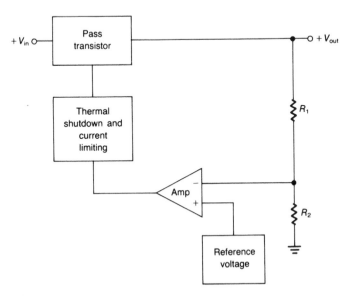

Figure 4–9. Block Diagram of a Typical Three-Terminal Fixed Positive Voltage Regulator (Courtesy of National Semiconductor Corporation, © copyright 1980.)

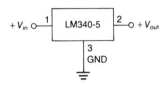

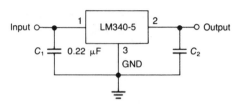

Figure 4–10. Simple Configuration of a Fixed Positive Voltage Regulator (Courtesy of National Semiconductor Corporation, © copyright 1980.)

Figure 4–11. Typical Bypassed Regulator Circuit (Courtesy of National Semiconductor Corporation, © copyright 1973.)

Bypass Capacitors

Power supply feedback may produce oscillations within the IC through lead inductance if the device is more than a few inches from the power supply filter circuit. Figure 4–11 shows a typical bypassed regulator circuit. Bypass capacitor C_1 provides circuit stability. Although no output bypass capacitor (C_2) is needed for stability, one would improve the transient response. If used, C_2 would typically be a 0.1 μF ceramic disc.

If the junction temperature is maintained at 25°C, then with an input range of 7–20 volts, the 340-5 will provide a regulated output of 5 volts. An input range of 27–38 volts will result in a regulated output of 24 volts from the 340-24. Any device in the 340 series requires an input voltage of 2–3 volts greater than the regulated output.

The LM320 series are three-terminal fixed negative voltage regulators, offering output voltages ranging from −5 to −24 volts. The 320 series are similar to the 340 series, including the same features of current limiting, thermal shutdown, and good ripple rejection. Figure 4–12 illustrates the major difference between the series—the pin numbering. Note that pin 1 is ground, pin 2 is the regulated output, and pin 3 is the unregulated input.

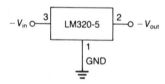

Figure 4–12. Simple Configuration of a Fixed Negative Voltage Regulator (Courtesy of National Semiconductor Corporation, © copyright 1980.)

Adjustable Regulators

Adjustable output voltages can be obtained by adding a few external components to the fixed regulator. Figure 4–13A shows a 340-5 arranged as an *adjustable positive regulator*. Note that pin 3 of the 340 is not grounded, but instead is tied to the junction of R_1 and R_2. This places the regulated output V_{reg} across R_1. Quiescent current I_Q flows through R_2. The output voltage is from pin 2 to ground, and

$$V_{out} = V_{reg} + \left(\frac{V_{reg}}{R_1} + I_Q\right) R_2 \qquad (4.6)$$

Typical I_Q of the 340-5 is 7 mA, with a maximum change ΔI_Q of 1.5 mA over line and load changes.

Figure 4–13B shows a typical *adjustable negative regulator* constructed with the LM320-5 and external components. Capacitor C_2 is optional, but it does improve transient response and ripple rejection. Output is taken between pin 2 and ground, and

$$V_{out} = V_{reg} \left(\frac{R_1 + R_2}{R_2}\right) \qquad (4.7)$$

Dual Polarity Tracking Regulators

Dual polarity tracking regulators must provide both plus and minus regulated voltages, and these voltages must also track. That is, if the plus voltage line changes, the minus voltage line must also change in the opposite direction and by the same amount. For example, if the plus line of a dual tracking regulator suddenly decreases from $+24$ V to $+23$ V, the minus line will simultaneously

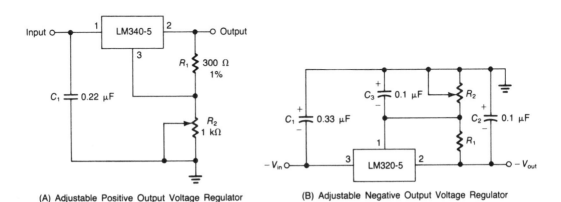

(A) Adjustable Positive Output Voltage Regulator (B) Adjustable Negative Output Voltage Regulator

Figure 4–13. Three-Terminal Voltage Regulators (Part A: Courtesy of National Semiconductor Corporation, © copyright 1973; Part B: Courtesy of National Semiconductor Corporation, © copyright 1980.)

increase from -24 V to -23 V. This tracking ensures that the magnitude of both of the line voltages to ground is constant.

Several methods may be used to provide tracking for dual voltages. One method is to use op amps and external components, as discussed earlier (see Figure 4–8). Another method is to use two single-voltage regulators, as demonstrated in Figure 4–14. This circuit uses a 7815 ($+15$ V output) and a 7915 (-15 V output), combined with the proper external circuitry, to provide dual regulated voltages of ±15 V at 1.0 A. The use of selected single-voltage regulators in this manner can supply a wide range of voltages at high currents.

A typical dual polarity tracking regulator circuit is shown in Figure 4–15. This 4195 regulator IC is designed for local regulation and intended for ease of application. It provides balanced positive and negative 15 V output voltages at currents to 100 mA. Only two external components, two 10 μF bypass capacitors, are required for operation.

Conversion to a Switching Regulator

Through the use of proper external circuitry, many of the standard regulators available today may be converted to switching regulators. Figure 4–16 presents one such conversion using the μA723 as a positive switching regulator. The 723 is a monolithic regulator that consists of a temperature-compensated reference amplifier, error amplifier, power series-pass transistor, and current-limit circuitry. The device features low standby current drain, low temperature drift, and high ripple rejection.

In Figure 4–16, V_{in} provides the $+V$ voltage for the device (pin 12) and the V_{CC} voltage for the series-pass transistor (pin 11). The V_{in} also provides biasing

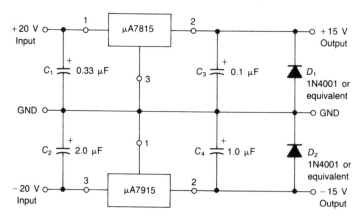

Figure 4–14. Dual-Supply Voltage Regulator Using Two Single-Voltage Regulators (Courtesy of Fairchild Camera & Instrument Corporation, © copyright 1978.)

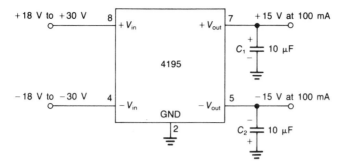

Figure 4–15. Dual Polarity Tracking Regulator (Courtesy of Raytheon Company, Semiconductor Division, © copyright 1978.)

through resistors R_5 and R_7, for external transistors Q_1 and Q_2, which are PNP devices configured as a Darlington pair. Capacitor C_1 is a bypass capacitor to provide stability. A reference voltage from pin 6 is provided through a voltage divider, resistors R_1 and R_2, where the collector output voltage from the Darlington pair is compared with the reference voltage at pin 5, the noninverting input to the error amplifier. Frequency compensation pin 13 is not used in this application. Output voltage pin 10, current limit pin 2, current sense pin 3, and the inverting input to the error amplifier pin 4, are connected together through current-limiting resistor R_6. The regulated output positive voltage is taken across the filter network of inductor L_1 and capacitor C_2.

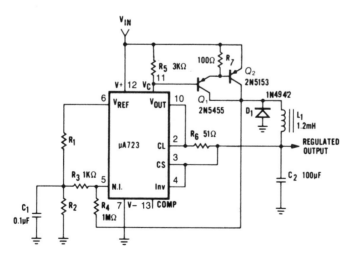

Figure 4–16. Switching Regulator Circuit Constructed with a Standard Regulator and External Components (Courtesy of Fairchild Camera & Instrument Corporation, © copyright 1978.)

Hybrid Regulators

Hybrid regulators use both the monolithic chip and discrete components in chip form (thin-film or thick-film construction methods), all mounted on a ceramic material substrate. Gold wires connect the chips together and the ceramic material insulates the chips from each other. Packaged as a single device, this form of regulator offers matched components that provide a desired output. The package usually consists of a monolithic regulator chip and several transistor chips, with resistors and capacitors formed on the substrate. Because of this multilayered construction, hybrid devices are not as small as monolithic regulators, but they can handle larger currents and dissipate more power.

One example of the hybrid technique is demonstrated by the Fairchild μA78PO5, shown in block diagram form in Figure 4–17. This is a three-terminal positive 5 V hybrid voltage regulator capable of delivering 10 A. It is virtually blowout proof and contains all the protection features inherent in monolithic regulators, such as internal short circuit current limiting through resistor R_{sc}, thermal overload, and safe area protection. This hybrid consists of a monolithic control chip that drives a rugged Mesa transistor and it is contained in a hermetically sealed steel package that provides 70-watt power dissipation.

Step-Down Switching Regulator

The SH1605 is a high-efficiency step-down switching regulator capable of supplying 5 A of regulated output current over an adjustable range from 3.0 V to 30.0 V of output voltage, with up to 150 W of output power. The device incorporates a temperature-compensated voltage reference, a duty-cycle-con-

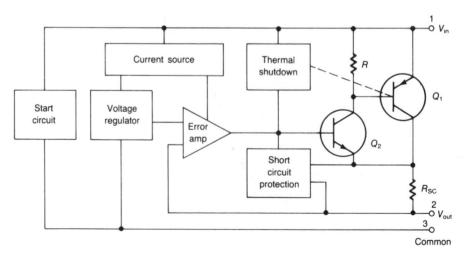

Figure 4–17.   Three-Terminal Hybrid Voltage Regulator (Courtesy of Fairchild Camera & Instrument Corporation, © copyright 1978.)

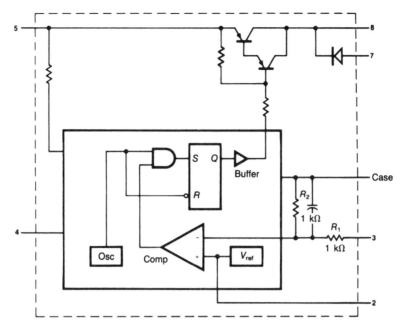

Figure 4–18. Block Diagram of a Hybrid Switching Regulator (Courtesy of Fairchild Camera & Instrument Corporation, © copyright 1978.)

trollable oscillator, error amplifier, high-current high-voltage output switch, and a power diode. A simplified block diagram is shown in Figure 4–18, and a design example is shown in Figure 4–19.

In this circuit (Figures 4–18 and 4–19), when power is first applied, the output voltage V_{out} is low, thus forcing the comparator output into a HIGH state.

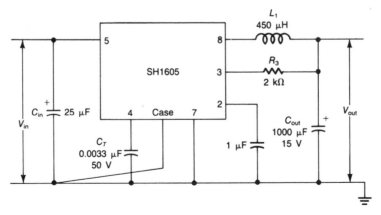

Figure 4–19. Typical Application of a Hybrid Switching Regulator (Courtesy of Fairchild Camera & Instrument Corporation, © copyright 1978.)

As a result, the oscillator freely toggles the output switch (set-reset (SR) flip-flop) on and off at a rate determined by the charge and discharge rate of the timing capacitor C_T. This is a temporary condition that continues until V_{out} has exceeded the reference voltage (V_{ref}) level times the factor set by R_1, R_2, and R_3. The output voltage is expressed as

$$V_{out} = V_{ref} \frac{(R_1 + R_2 + R_3)}{R_1} \tag{4.8}$$

Since the values of R_1 and R_2 (1 kΩ each) inside the SH1605 are established, R_3 can be determined as follows:

$$R_3 = \frac{(R_1 + R_2)(V_{out} - V_{ref})}{V_{ref}} \tag{4.9}$$

where

$$V_{ref} = 2.5 \text{ V}$$

Equilibrium is reached at the completion of the ON cycle when the comparator input has exceeded the reference level. When the comparator output goes LOW, the oscillator output is disabled and Q_1 switches OFF. V_{out} then begins to fall at a rate determined by the ratio of the output voltage to the inductor value:

$$\frac{\Delta I_{L_1}}{t_{off}} = \frac{V_{out} + V_{in}}{L_1} \tag{4.10}$$

where

I_{L_1} = inductor current
L_1 = inductor
t_{off} = time Q_1 is in the OFF state

Whenever V_{out} falls to the level specified by Equation 4.8, the comparator changes state and the output switches ON. It remains in this state until the voltage across the timing capacitor, C_T, reaches a positive threshold level. The rate of C_T charge is determined by the size of the timing capacitor and the magnitude of the constant-current source inside the oscillator. Charging current is typically 25 μA and discharging current is 225 μA. From the following formulas describing ON and OFF duration, the frequency of oscillation can be determined as follows:

$$t_{on} = \frac{C_T \Delta V}{I_C} \tag{4.11}$$

$$t_{off} = \frac{\Delta I_{L_1} \times L_1}{V_{out} + V_{in}} \tag{4.12}$$

where

C_T = timing capacitor
I_C = oscillator charging current
$V = 0.5$ V

Then nominal frequency is calculated by the following equation:

$$\frac{1}{\dfrac{C_T \Delta V}{I_C} + \dfrac{\Delta I_{L_{1(nom)}} \times L_1}{V_{out} + V_{in}}} = \frac{1}{t_{on} + t_{off}} = \frac{1}{T} \tag{4.13}$$

For improved system efficiency, the operating period should always be many times longer than the device transition times. A nominal value is eight times longer. A trade-off must be sought between inductor size and efficiency when selecting the frequency of operation.

Example 4.4

A typical regulator design follows. Refer to Figure 4–19, a design circuit for a step-down switching regulator. Assume the nominal design objectives to be:

$V_{out} = +5$ V	Line regulation = 2%
$I_{out(max)} = 5.0$ A	Load regulation = 2%
$I_{out(min)} = 1.0$ A	Ripple (max) = 0.1 $V_{p\text{-}p}$
$V_{in} = 12$ to 18 V	Efficiency = 70%

Solution

For the solutions that follow, use information from the discussion in this section. First calculate R_3 from Equation 4.9:

$$R_3 = \frac{(R_1 + R_2)(V_{out} - V_{ref})}{R_1} = \frac{(2 \times 10^3 \ \Omega)(5 \text{ V} - 2.5 \text{ V})}{2.5 \text{ V}} = 2 \text{ k}\Omega$$

Since the required $I_{out(min)}$ is 1 A to maintain continuous operation, the peak-to-peak current excursion (twice the value of $I_{out(min)}$) must be equal to 2 A or less, so that

$$\Delta I_{L_1} = 2(I_{out(min)})$$

To calculate the value of L_1, assume the nominal ON time of the system as 60 μs. This value is chosen keeping the efficiency/component size trade-off in mind. Therefore,

$$L_1 = \frac{V_{in(nom)} - V_{out}}{\Delta I_{L_1}} (t_{on})$$

$$= \frac{10 \text{ V}}{2 \text{ A}} (60 \times 10^{-6} \text{ s}) = 300 \ \mu\text{H} \tag{4.14}$$

where

$$V_{in(nom)} = 15 \text{ V}$$
$$t_{on} = 60 \ \mu\text{s}$$
$$\Delta I_{L_1} = 2 \text{ A}$$

An important element in achieving the optimum performance in a switching regulator is to keep the inductor below the specified saturation limits.

Since the timing capacitor controls the 60 μs ON time, C_T can be determined by manipulation of Equation 4.11:

$$C_T = \frac{t_{on}I_C}{\Delta V} = \frac{(60 \times 10^{-6} \text{ s})(25 \times 10^{-6} \text{ A})}{0.5 \text{ V}} = 3000 \text{ pF}$$

where

$I_C = 25$ μA(nominal) from the data sheet

The final step is to determine the requirements for the output capacitor, C_{out}, to obtain the desired value of ripple voltage (V_{ripple}). Consideration must be given to the absolute value of C_{out} as well as to the internal effective series resistance (ESR). Since capacitor size is inversely proportional to the operating frequency, the lowest frequency of operation must be calculated. Minimum operating frequency can be determined by substituting $\Delta I_{L1(max)}$ for $\Delta I_{L1(nom)}$ in Equation 4.13, so that minimum frequency is calculated as follows:

$$f_{min} = \frac{1}{\dfrac{C_T \Delta V}{I_C} + \dfrac{\Delta I_{L1(max)} L_1}{V_{out} + V_{in}}}$$

where

$\Delta I_{L1(max)} = I_{out(max)} - I_{out(min)}$

Inserting values from our circuit, we find that the minimum frequency is

$$\frac{1}{\dfrac{3 \times 10^{-9} \text{ F} \times 0.5 \text{ V}}{25 \times 10^{-6} \text{ A}} + \dfrac{4 \text{ A} \times 3 \times 10^{-4} \text{ H}}{5 \text{ V} + 12 \text{ V}}} \approx 7.7 \text{ kHz}$$

The output capacitor minimum value can now be determined using the following equation:

$$C_{out(min)} = \frac{\Delta I_{L1}}{8 f_{(min)} V_{ripple(max)}}$$

$$= \frac{2 \text{ A}}{8 \times 7.7 \times 10^3 \text{ Hz} \times 0.1 \text{ V}} = 325 \text{ μF} \tag{4.15}$$

The maximum acceptable ESR is, therefore,

$$\text{ESR}_{(max)} = \frac{V_{ripple(max)}}{\Delta I_{L1(max)}} = \frac{0.1 \text{ V}}{4 \text{ A}} = 0.025 \text{ } \Omega$$

Normally the minimum capacitance value for C_{out} should be increased considerably if a low ESR capacitor is not used. As a final step for minimizing switching transients at the input of the device, a low ESR capacitor C_{in} must be used for decoupling purposes between the input terminal and ground.

Universal Switching Regulator

The μA78S40 Universal Switching Regulator Subsystem shown in Appendix D, while not a hybrid circuit, is an excellent example of the available switching regulators. It offers step-up, step-down, and inverting options as well as use as a series pass regulator. The data sheets show the design formulas for the various options and typical designs and operational performance results.

4.6 Current Regulation

Current regulator circuits maintain a constant current in a load independent of changes in that load. This means that variations in load impedances or voltage changes across the load impedances do not affect the circuit operation. An op amp current regulator circuit is shown in Figure 4–20.

Assuming an ideal device, V_1 and V_2 are equal to V_Z, and the current in R_1 is

$$I_1 = \frac{V_Z}{R_1} \tag{4.16}$$

Since an ideal device has zero current entering or leaving the $(-)$ terminal, I_1 must equal I_L. From Equation 4.16 we note that I_1 is determined by two constant terms, V_Z and R_1. Therefore, I_1 and I_L must also be constant. Also note that R_L is not considered in the equation. This means that I_L is independent of R_L. Any change in R_L will change the gain of the circuit, causing V_{out} to vary, which maintains a constant value of I_L. R_2 is a current-limiting resistor for the zener diode.

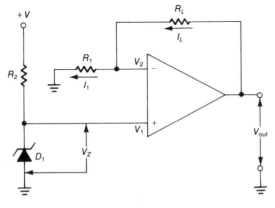

Figure 4–20. Op Amp Current Regulator (Courtesy of John Wiley & Sons, Inc., © copyright 1981.)

Constant-Current Regulator

Any three-terminal voltage regulator can be used as a *constant-current regulator*, as illustrated in Figure 4–21. The output current (I_{out}) dictates the regulator type to be used and is determined by the following equation:

$$I_{out} = \frac{V_{out}}{R_1} + I_Q \tag{4.17}$$

where

V_{out} = the regulator output voltage
I_Q = the quiescent current

The input voltage (V_{in}) must be high enough to accommodate the dropout voltage at the low end, but must not exceed the maximum input voltage rating at the high end.

4.7 Constant-Current Source

An op amp can be used as a *constant-current source* when arranged as in Figure 4–22. The input voltage can be a battery or some other stable reference voltage that delivers a constant current through the input resistor, R_1. This current also

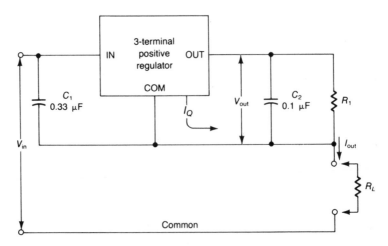

Figure 4–21.   Three-Terminal Voltage Regulator Used as a Constant-Current Regulator (Courtesy of Fairchild Camera & Instrument Corporation, © copyright 1978.)

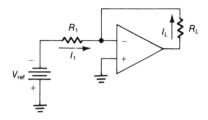

Figure 4–22. Op Amp Constant-Current
Source

flows through the feedback resistor, which functions as the load, R_L. The current
through R_1 and R_L is

$$I_1 = \frac{V_{\text{ref}}}{R_1} \qquad\qquad (4.18)$$

Since V_{ref} and R_1 are constant, I_1 must remain constant. Changes in R_L cannot
alter this fact.

4.8 Voltage-to-Current Converters

Voltage-to-current converters, also called *transmittance amplifiers*, are frequently used for applications such as driving relays and analog meters. Such
converters can drive either floating or grounded loads, depending upon the
specific application.

Floating Loads

For floating loads, either the inverting converter (Figure 4–23A) or the noninverting converter (Figure 4–23B) may be used. These circuits are similar in form
to the inverting and noninverting amplifiers. The major difference is that in the
converter, the load element, R_L, is a relay coil or an analog meter with an internal
resistance representing R_L.

Also, note the similarity between the inverting converter of Figure 4–23A
and the constant-current source of Figure 4–22. The current flowing through
the floating load is

$$I_L = \frac{V_{\text{in}}}{R_1} \qquad\qquad (4.19)$$

and is independent of the value of R_L. The load current for the noninverting
converter of Figure 4–23B is determined in the same manner as that for the
inverting converter of Figure 4–23A.

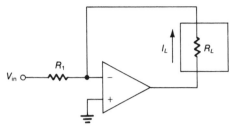
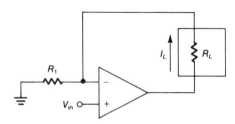

(A) Inverting Voltage-to-Current Converter for Floating Loads

(B) Noninverting Voltage-to-Current Converter for Floating Loads

Figure 4–23.   Voltage-to-Current Converters

Grounded Loads

A different configuration must be used for loads that are grounded on one side. Figure 4–24 shows such a circuit. The load current is controlled by the input voltage (V_{in}) and is determined as follows:

$$I_L = \frac{V_{in}}{R_3} \tag{4.20}$$

when

$$\frac{R_4}{R_3} = \frac{R_2}{R_1} \tag{4.21}$$

4.9 Current-to-Voltage Converters

Figure 4–25 shows a basic *current-to-voltage converter*. Essentially, the circuit is an inverting amplifier without an input resistor. The input current (I_{in}) is applied

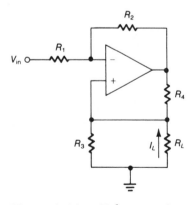

Figure 4–24.   Voltage-to-Current Converter for Grounded Loads

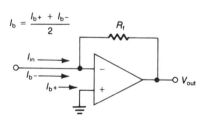

Figure 4–25.   Current-to-Voltage Converter

directly to the inverting input of the op amp. This input current also flows through the feedback resistor, therefore, the output voltage (V_{out}) is

$$V_{out} = I_{in}R_f \tag{4.22}$$

However, care should be taken with such a circuit, because the op amp's bias current (I_b) is added to the input current (I_{in}) and V_{out} becomes

$$V_{out} = (I_{in} + I_b)R_f \tag{4.23}$$

Therefore, the input current should be kept high compared to the input bias current.

4.10 Regulator Protection Circuits

Protective circuits are added on chips to improve reliability and to make regulators immune to certain types of overloads. They protect the regulators against short circuit conditions (current limit), against excessive input-output differential conditions (safe operating area limit), and against excessive junction temperatures (thermal limit). Figure 4–26 shows a basic regulator protection circuit.

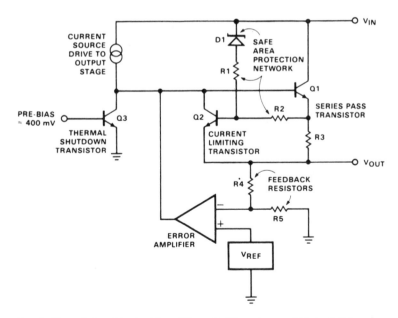

Figure 4–26. Basic Regulator Protection Circuit (Courtesy of Fairchild Camera & Instrument Corporation, © copyright 1978.)

Current-Limiting Protection

The most commonly used protection scheme is *current limiting* for guarding the output series-pass transistor against excessive output currents or short circuit conditions. With low input-output conditions, zener diode D_1 is not conducting, and there is no current flowing through R_1. The current-limiting transistor, Q_2, senses the voltage drop across the current-limiting resistor, R_3. As the output current increases, the drop across R_3 increases. As a result, the Q_2 base-emitter voltage increases until Q_2 begins to conduct, thereby removing the base drive of the series-pass transistor, Q_1. No additional output current can be pulled out since any increase in the output current will cause Q_2 to conduct harder. However, this current-limiting circuit has a slight disadvantage—the voltage developed across R_3 adds to the regulator dropout voltage and degrades the load regulation and output impedance.

A simple way of getting around this problem is to pre-bias the base of Q_2 with a fraction of the base-emitter voltage of Q_1 through R_6 and R_7, as illustrated in Figures 4–27A and B. In these current-limiting schemes, the output current decreases with increasing temperature since the base-emitter voltage of Q_2 is the threshold for preventing the flow of additional regulator output current.

Temperature-Independent Current Protection

Temperature-independent short circuit current protection can be achieved with a slight increase in circuit complexity. Figure 4–27C shows the circuit used in the μa79M00 series of negative regulators to obtain temperature-independent peak output current. At low to medium output current levels, Q_4 is on and Q_5 and Q_6 are off. When the voltage drop across R_3 reaches a predetermined level, Q_4 begins to turn off and current is diverted to Q_5, which in turn causes Q_6 to turn on. Q_1 and, consequently, Q_2 base currents are thus diverted and the output current is prevented from increasing further.

Safe Operating Area Protection

Safe operating area (SOA) protection is included in IC regulators to protect the series-pass transistor against excessive power dissipation by reducing the collector current as the collector-emitter voltage is increased (Figure 4–26). When the input-output voltage differential exceeds the breakdown voltage differential of D_1, current flows through D_1, R_1, R_2, and R_3. The voltage drop across R_3 and, therefore, the Q_2 base voltage both become a function of not only the output current but also the input-output voltage differential. Hence, maximum output current is available when the input-output voltage differential is less than the D_1 breakdown voltage. The safe area protection network reduces the available

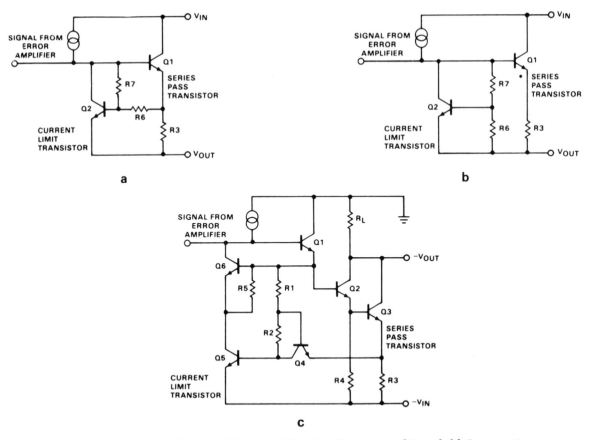

Figure 4–27. Alternate Current-Limiting Circuits (Courtesy of Fairchild Camera & Instrument Corporation, © copyright 1978.)

output current (I_{out}) as the input-output differential $\Delta(V_{in} - V_{out})$ increases at a rate determined by the following equation:

$$\frac{\Delta I_{out}}{\Delta(V_{in} - V_{out})} = -\frac{R_2}{R_1 R_3} \tag{4.24}$$

The safe area protection network thus reduces the available output current as the input-output differential increases, and limits the regulator operation to within the safe operating area of the series-pass transistor. The SOA network is lumped in with the short circuit protection network, and consequently both have the same temperature characteristics.

When selecting a regulator to operate with high input voltage or with high input-output voltage differential, it must be remembered that output current decreases with increased input-output voltage differential. Under heavy load and high input-output conditions, the SOA protection circuit may cause a high output voltage device to *latch up* (lock on to the high voltage-high current condition) after a momentary short, since the input voltage becomes the input-output differential during the short. The regulator may not be able to supply as much current after the fault condition as before. *Latching* will not damage the regulator. Interrupting the power, reducing the load current, or reducing the input voltage momentarily will restore normal operation.

A discrete regulator usually relies on current limiting for overload protection since there is no practical way to sense junction temperature in a separate series-pass transistor. The dominant failure mechanism of this type of regulator, then, is excessive heating of the series-pass transistor. In a monolithic regulator, the series-pass transistor is contained within the thermal overload protection circuit where its maximum junction temperature is limited, independent of input voltage, type of overload, or degree of package heat sinking. It is, therefore, considerably more effective than current limiting by itself. An added bonus to combined thermal and current overload protection is that a higher output current level under normal conditions can be considered, since there is no excessive regulator heating when a load fault occurs.

Thermal Limit Protection

The base-emitter junction of a transistor placed as close as practical to the series-pass transistor is used to sense the chip temperature. The thermal shutdown transistor, Q_3 in Figure 4–26, is normally biased below its activation threshold so that it does not affect normal operation of the circuit. However, if the chip temperature rises above its maximum limit due to an overload, inadequate heat sinking, or other condition, the thermal shutdown transistor turns on, removes the base drive to the output transistor Q_1, and shuts down the regulator to prevent any further chip heating.

4.11 Summary

1. Voltage regulators control supply voltage variations so that proper linear device operation can occur.

2. There are three major types of voltage regulators: series, shunt, and switching.

3. In a series regulator, the variable control element is in series with the load.

4. In the shunt regulator, the variable control element is in parallel with the load.

5. In the switching regulator, the control transistor is switched between the cutoff and saturation modes.

6. The series and shunt regulators have the disadvantage of high power dissipation in the control transistor. The switching regulator has low power dissipation in the control transistor, thereby providing higher circuit efficiency.

7. Heat sinking is required for many applications to prevent damage to devices and circuits due to excessive temperatures.

8. Using op amps as sensing elements in regulator circuits helps to overcome the problems of low sensitivity, slow response time, and loading effects.

9. Monolithic regulators are constructed with all components on a single chip. This allows the regulator to be used for local "on-card" regulation.

10. Most monolithic regulators require that some external components be added in practical applications.

11. The three main categories of monolithic regulators are fixed, adjustable, and dual tracking.

12. Adjustable output voltages may be obtained by adding external components to fixed regulators.

13. Fixed and adjustable regulators can be obtained with either positive or negative outputs.

14. Dual tracking regulators provide both positive and negative voltages that track (follow) each other.

15. Many standard regulators may be converted to switching regulators through the use of proper external circuitry.

16. Hybrid regulators are a combination of monolithic and thin-film or thick-film construction methods that are packaged as a single device.

17. Hybrid devices are larger than monolithic devices, but they are capable of handling larger currents and they dissipate more power.

18. Any three-terminal voltage regulator can be used as a constant-current regulator.

19. An op amp can be used as a constant-current source when properly arranged.

20. Voltage-to-current converters can be used to drive relays or analog meters, with either floating or grounded loads.

21. Current-to-voltage converters operate in a manner similar to constant-current sources.

22. Improved reliability and immunity from certain types of overloads are provided by adding protective circuits to regulators.

23. Current limiting is protection against short circuit conditions and is the most commonly used protection scheme.

24. Safe operating area (SOA) limit is protection against excessive input-output differential conditions.

25. Thermal limit is protection against excessive junction temperatures.

4.12 Questions and Problems

4.1 What might be the result in an audio amplifier if there were a sudden shift in the supply voltage?

4.2 What causes variations in supply voltages?

4.3 Define a monolithic voltage regulator.

4.4 List the three major types of regulator circuits.

4.5 What is the feature that all three types of regulators have in common?

4.6 Explain why the circuit in Figure 4–1 is called a dissipative voltage regulator circuit.

4.7 What distinguishes the shunt regulator circuit from the series regulator circuit?

4.8 What disadvantage do the series and shunt regulator circuits have in common?

4.9 How does the switching regulator overcome the disadvantage described in Problem 4.8?

4.10 What are the advantages of the switching regulator over the series and shunt regulators?

4.11 Refer to Figure 4–3. Assume the control signal frequency to be 50 kHz and the transistor ON time to be 5 μs. Calculate the duty cycle.

4.12 Assume that the maximum voltage across the diode in Figure 4–3 is 60 V. Using the duty cycle calculated in Problem 4.11, calculate the dc component of the output voltage (V_{dc}) of the circuit.

4.13 List three problems that arise in regulators that use transistors as sensing devices.

4.14 What is one simple solution to the problems described in Problem 4.13?

4.15 Discuss some possible problems that may arise from the solution stated in Problem 4.14 and describe some solutions to those problems.

4.16 What is meant by local regulation?

4.17 Describe the three main categories of monolithic regulators.

4.18 What is the major advantage of the fixed voltage regulator?

4.19 Refer to Figure 4–13A. Assume $V_{reg} = 5$ V, $I_Q = 7$ mA, $R_1 = 100$ Ω, and $R_2 = 175.4$ Ω. Calculate V_{out}.

4.20 Refer to Figure 4–13B. Assume $V_{reg} = 5$ V, $R_1 = 3$ kΩ, and $R_2 = 1$ kΩ. Calculate V_{out}.

4.21 Specify two methods of providing tracking for dual voltages.

4.22 State the advantages of the 4195 regulator.

4.23 Explain how hybrid regulators are constructed.

4.24 Explain how current regulators work.

4.25 List the three major protective circuits used in regulators and state what kind of protection each provides.

Basic Oscillator Circuits

5.1 Introduction

Oscillators are as important in electronics as are amplifiers. Oscillators can be found in almost every field of electronics. They are used in computers, communications systems, industrial control and process handling, and radio and television sets. For example, the high-frequency oscillator in the television tuner selects the channel to be viewed. This is perhaps the most common use of the high-frequency oscillator.

Basically, an oscillator generates a continuously repetitive output signal that can be used for timing or synchronizing operations. The output signals can be either sinusoidal (sine) waves or nonsinusoidal (square, triangle, sawtooth) waves. The oscillator is an electronic generator that operates from a dc power supply, has no moving parts, and can produce ac signal frequencies ranging into millions of hertz. A good oscillator will have a uniform output, varying in neither frequency nor amplitude.

In this chapter we will discuss oscillator fundamentals, various types of oscillators, and the design of simple sine wave oscillators using op amps. (Nonsinusoidal wave oscillators will be discussed in Chapter 6.)

5.2 **Objectives**

When you complete this chapter, you should be able to:

☐ Recognize the names of oscillators by observation.

☐ Determine the effects on frequency when reactive elements are varied.

☐ Determine if a crystal is operating in the series-resonant or parallel-resonant mode.

☐ Identify the three basic types of oscillators.

☐ Identify the frequency-determining components of various oscillators.

☐ Explain how each of the oscillators operates.

5.3 **Fundamentals of Oscillators**

You will recall from your study of resonant circuits that excitation of a tank circuit by a dc source tends to cause oscillations in the circuit. These oscillations, or back-and-forth oscillatory motions, are the result of circulating current flow inside the tank circuit. If there were no resistance within the circuit, the oscillations would continue indefinitely. However, there is resistance, and this resistance dissipates energy and damps the oscillations.

For continued oscillation, the energy lost due to resistance must be replaced. In a simple amplifier circuit, this is accomplished through regenerative (positive) feedback obtained by simply feeding back to the input a portion of the amplifier output. However, the feedback signal must be in phase with the input signal in order for the circuit to oscillate. A *phase-shifting network*, as illustrated in Figure 5–1, may be necessary, depending upon the type of amplifier circuit being considered.

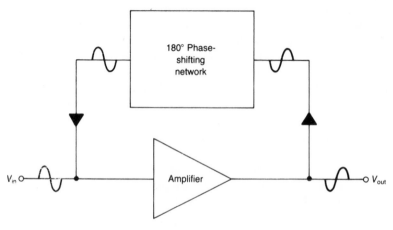

Figure 5–1.   Feedback Loop with 180° Phase-Shifting Network

Once the amplifier begins to operate, the input signal is no longer required. The circuit will continue to oscillate even if the input signal is removed, because the gain of the amplifier replaces the lost energy in the circuit through positive feedback, thus sustaining oscillations. There is a matter for concern, however, because the circuit may be triggered by external noise pulses that cause undesired changes in the oscillating frequency.

For an oscillator to be useful, it must have a constant output. This stability is accomplished through proper selection of components in the *frequency-determining network*. One such network is shown in Figure 5–2. The parallel *LC* (inductor-capacitor) network in the positive feedback loop resonates at a frequency determined by the values of the components. The desired 180° phase shift is produced by the reactance of the components, and the amplifier gain replaces the energy lost in the tank circuit.

Another frequency-determining network is shown in Figure 5–3. The values of the components are selected so that the *RC* (resistor-capacitor) time constants determine the frequency of oscillation, and the desired phase shift is determined by the number of *RC* sections contained in the feedback loop. The amplifier gain replaces the energy lost across the *RC* sections.

In practice, oscillators are naturally *self-starting*. When an amplifier circuit is first turned on, noise pulses are generated. These pulses are fed back to the input of the amplifier through a frequency-determining network. The amplifier action amplifies the pulses, which are again fed back to the input, thereby creating oscillations.

Barkhausen Criterion

An evaluation of the basic oscillator circuit illustrated in Figure 5–4 will show that certain conditions must be met for the oscillator to be self-starting and self-

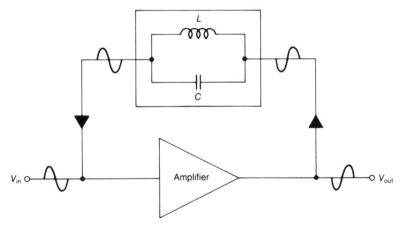

Figure 5–2. *LC* Frequency-Determining Network

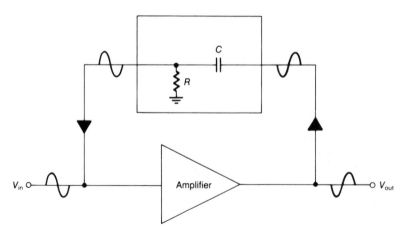

Figure 5–3. *RC* Frequency-Determining Network

sustaining. For the oscillator to produce its own input signal continuously, the product of the amplifier gain (A_v) and the fractional feedback factor (B_v), a small fraction of the output signal that is fed back to the input, must equal one. This condition is called the *Barkhausen Criterion*. Mathematically, oscillator stage gain (A_{osc}) is expressed as follows:

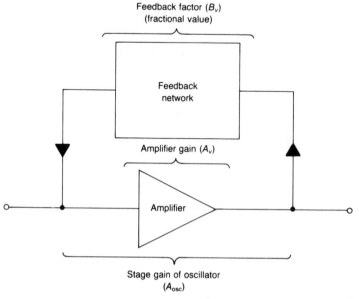

Figure 5–4. Concept of Basic Oscillator Stage Gain

$$A_{\text{osc}} = \frac{A_v}{1 - A_v B_v} \qquad (5.1)$$

When the condition of $A_v B_v = 1$ is met, the oscillator stage gain is infinite, which satisfies one of the requirements for an oscillator—an output signal must be present without an input signal.

When the oscillator first starts, the output increases to a point where $A_v B_v$ is slightly higher than unity. After a few oscillations, however, $A_v B_v$ is reduced to unity. This is accomplished in one of two ways: by amplifier saturation, which damps the signal to its required level, or by placing a nonlinear resistor in the feedback loop, which reduces the feedback factor.

5.4 Classifications of Oscillators

Oscillators are generally classified according to the components used in the frequency-determining networks. The three basic classifications are *LC oscillators*, *RC oscillators*, and *crystal oscillators*. These classifications will be discussed individually.

Briefly summarized, the basic fundamentals common to all oscillators are as follows:

1. Amplification is required to replace circuit losses.
2. A frequency-determining network is required to set the desired frequency of oscillation.
3. A regenerative (positive) feedback signal is required to sustain oscillation.
4. The oscillator is required to be self-starting, with no input signal.

LC Oscillators

The frequency-determining network in the *LC* oscillator is a tuned circuit that consists of capacitors and inductors, connected either in series or in parallel. Figure 5–5A shows a block diagram of a basic *LC* oscillator using an op amp. The feedback network is formed by Z_1, Z_2, and Z_3. Their values determine the oscillating frequency. One or more of the impedances can be made variable if a range of frequencies is desired. The two basic *LC* oscillators are the *Hartley* and the *Colpitts*. If Z_1 and Z_2 are inductors, the circuit is a Hartley oscillator. If Z_1 and Z_2 are capacitors, the circuit is a Colpitts oscillator.

Hartley Oscillator

A Hartley oscillator is illustrated in Figure 5–5B. The inductor is tapped to form two coils, L_{1A} and L_{1B}, corresponding to Z_1 and Z_2 in Figure 5–5A. A

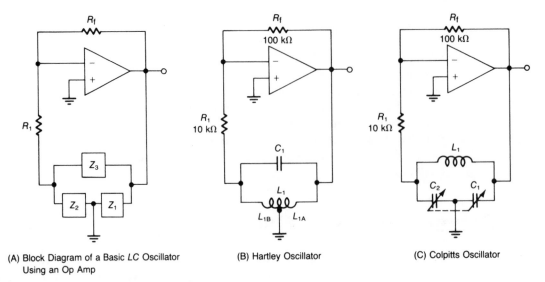

(A) Block Diagram of a Basic *LC* Oscillator
Using an Op Amp

(B) Hartley Oscillator

(C) Colpitts Oscillator

Figure 5–5. *LC* Oscillators

capacitor, C_1, is placed across coil L_1, making the entire coil part of a tuned circuit. Current flow through L_{1A} replaces energy lost in the tank, thus providing the positive feedback necessary for oscillation. The amount of feedback can be controlled by adjusting the position of the coil tap. The *tapped coil* is the identifying feature of the Hartley oscillator.

Oscillating frequency, f_o, is approximated by the following equation:

$$f_o = \frac{1}{2\pi\sqrt{L_{(eq)}C}}$$ (5.2a)

where

$$L_{(eq)} = L_{1A} + L_{1B} + 2L_M$$
$$L_M = \text{mutual inductance}$$

and

$$L_M = K\sqrt{L_{1A} \times L_{1B}}$$ (5.2b)

where

$$K = \text{coefficient of coupling}$$

For L_{1A} and L_{1B} on paper or plastic form, $K = 0.1$; for L_{1A} wound over L_{1B}, $K = 0.3$; and for L_{1A} and L_{1B} on the same iron core, $K = 1$.

Two disadvantages of the Hartley configuration are that (1) the coils tend to be mutually coupled, which makes the frequency of oscillation differ slightly from the calculated frequency; and (2) the oscillating frequency cannot easily be varied over a wide range. The second disadvantage stems from the difficulty of changing the value of an inductor. Some limited frequency variation can be made in the Hartley oscillator by making capacitor C_1 variable.

Colpitts Oscillator

A Colpitts oscillator is shown in Figure 5–5C. Note its similarity to the Hartley oscillator. The identifying feature of the Colpitts oscillator is the *tapped capacitor* arrangement. Frequency is determined by inductor L_1 and the series combination of capacitors C_1 and C_2, so that

$$f_o = \frac{1}{2\pi\sqrt{LC_{(eq)}}} \tag{5.3}$$

where

$$C_{(eq)} = \frac{C_1 C_2}{C_1 + C_2}$$

The disadvantages of the Hartley oscillator are overcome in the Colpitts oscillator. Variable capacitors are readily available. By providing a wide range of capacitance, a wide range of frequencies can be obtained. Note that both capacitors in the Colpitts oscillator in Figure 5–5C are variable, allowing the user to vary the frequency over a wide range.

The Colpitts oscillator is used extensively in AM (amplitude modulation) and FM (frequency modulation) radio receivers. The variable capacitors are ganged so that the tuning dial selects both the mixer/oscillator frequency and the resonant frequencies of the RF amplifier stages. For TV reception, the two capacitors are fixed, and different values of inductance are switched into the circuit. This is accomplished by mounting a fixed-value inductor for each channel on a shaft that is rotated. Each inductor is put into the oscillator tank circuit one at a time as the shaft is rotated, thereby changing the frequency of the circuit to the desired channel frequency.

The disadvantages of the LC oscillator circuits are that relatively pure inductance is required to obtain the desired calculated frequency, and that the input impedance of the amplifier must be infinite so that the output is not loaded. The op amp, with its high input impedance and large open loop gain, overcomes these disadvantages.

RC Oscillators

The *RC* oscillator uses resistance-capacitance networks to determine oscillator frequency. Resistors and capacitors are in great supply and relatively inexpensive, consequently the *RC* oscillator is an inexpensive, easily constructed, relatively stable circuit design for use in low- and audio-frequency ranges. There are basically two types of sine wave-producing *RC* oscillators: the *phase-shift* oscillator and the *Wien-bridge* oscillator. The nonsinusoidal output-producing *RC* oscillators will be discussed in the next chapter.

Phase-Shift Oscillator

As the name implies, the phase-shift oscillator shown in Figure 5–6 employs a phase-shifting RC feedback network that provides the 180° shift necessary to produce the required regenerative feedback. This type of oscillator is typically used in fixed-frequency applications.

The phase difference in an RC circuit is a function of the capacitive reactance, X_C, and the resistance, R, of the network. Resistance does not change with frequency, so the capacitor is the frequency-sensitive component. By careful selection of the components, the amount of phase shift across an RC section can be controlled. Stability can be improved by increasing the number of RC sections, thereby reducing the phase shift across each network. Typically, three RC sections are used.

The frequency of oscillation, f_o, can be obtained as follows:

$$f_o = \frac{1}{2\pi RC\sqrt{6}} \text{ (Hz)} \tag{5.4}$$

where

R's are identical in each of the three sections
C's are identical in each of the three sections
$B_v = 1/29$
Total phase shift = 180°.

For oscillator action to start, A_v must be greater than 29 (Barkhausen Criterion). This gain is set by the ratio of resistors R_f and R_4.

Example 5.1

Refer to Figure 5–6. Assume each RC section has values of $R = 1$ kΩ and $C = 0.1$ μF. Calculate the oscillating frequency, f_o.

Solution

Use Equation 5.4:

$$f_o = \frac{1}{2\pi RC\sqrt{6}} \text{ (Hz)} = \frac{1}{2(3.14) \times (1 \text{ k}\Omega) \times (0.1 \text{ }\mu\text{F}) \times (2.45)}$$

$$= \frac{1}{6.28 \times 1000 \text{ }\Omega \times (0.1 \times 10^{-6}\text{F}) \times \sqrt{2.45}} = 650 \text{ Hz}$$

Designing any oscillator is quite simple. Capacitors are available in many fixed values, so the key to the design is to determine what frequency you wish to

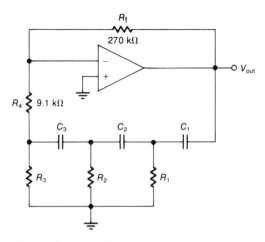

Figure 5–6. Phase-Shift Oscillator

operate with, then select a standard fixed-value capacitor. By algebraic manipulation of Equation 5.4, we can solve for R:

$$R = \frac{1}{2\pi f_o C \sqrt{6}} \ (\Omega) \tag{5.5}$$

The value of R can be found by substituting into Equation 5.5 the values of the frequency and the capacitor you have chosen. The reasoning here is that resistance values are more easily varied than are capacitance values; that is, a wider range of resistor values are available than are capacitor values.

Example 5.2

Design a phase-shift oscillator with three RC sections and an oscillating frequency of 1 kHz.

Solution

Select standard $0.02 \ \mu F$ capacitors for the RC sections, then calculate R. From Equation 5.5,

$$R = \frac{1}{2\pi f_o C \sqrt{6}} \ (\Omega) = \frac{1}{2(3.14) \times (1 \text{ kHz}) \times (0.02 \ \mu F) \times (2.45)}$$

$$= \frac{1}{6.28 \times 1000 \text{ Hz} \times (0.02 \times 10^{-6}F) \times \sqrt{2.45}} = 3250 \ \Omega$$

You would use a standard 20 percent 3300 Ω resistor, measuring low on an ohmmeter.

Wien-Bridge Oscillator

Like the phase-shift oscillator, the Wien-bridge oscillator uses RC networks. However, in the Wien-bridge oscillator, the RC networks are part of a *bridge circuit* that provides both regenerative and degenerative feedback. The networks select the frequency at which the feedback occurs, but they do not shift the phase of the feedback signal.

Feedback is applied to both inputs of the op amp, as shown in Figure 5–7. The frequency-selective network, sometimes called the *lead-lag* network, consists of $C_1 - R_1$ and $C_2 - R_2$ and provides regenerative feedback to the noninverting input terminal. Degenerative feedback is developed across R_3 and R_4 and is applied to the inverting input terminal. R_4 is made variable so that negative feedback can be reduced, because the positive feedback must be greater than the negative feedback in order for the circuit to sustain oscillations. The setting of R_4 is such that the circuit will start oscillating. The ratio of R_4 to R_3 must be two to one for proper operation.

Since resistance values in the degenerative feedback path do not change with frequency, degenerative feedback remains constant. However, regenerative feedback depends on the frequency response of the frequency-selective network, which is frequency sensitive. If oscillator frequency begins to increase, the reactance of C_2 will decrease and shunt some positive feedback to ground, decreasing regenerative feedback. Likewise, if the frequency begins to decrease, the reactance of C_1 becomes greater, causing less voltage to be developed across the $C_2–R_2$ network, and thus reducing regenerative feedback. Therefore, this network forces the oscillator to stay on its operating frequency.

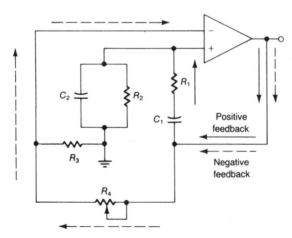

Figure 5–7.   Wien-Bridge Oscillator

The circuit output frequency is determined by the values of C_1, C_2, R_1, and R_2 and can be calculated as follows:

$$f_o = \frac{1}{2\pi\sqrt{R_1 R_2 C_1 C_2}} \text{ (Hz)} \tag{5.6}$$

However, if $R_1 = R_2$ and $C_1 = C_2$, then

$$f_o = \frac{1}{2\pi R_1 C_1} \text{ (Hz)} \tag{5.7}$$

Example 5.3

Refer to Figure 5–7. Assume $R_1 = R_2 = 20$ kΩ and $C_1 = C_2 = 1$ nF. Calculate f_o.

Solution

From Equation 5.7,

$$f_o = \frac{1}{2\pi R_1 C_1} \text{ (Hz)} = \frac{1}{2(3.14) \times (20 \text{ k}\Omega) \times (1 \text{ nF})}$$

$$= \frac{1}{6.28 \times (20 \times 10^3 \ \Omega) \times (1 \times 10^{-9} \text{ F})} = 8 \text{ kHz}$$

Example 5.4

Refer to Figure 5–7. Assume $R_1 = 10$ kΩ, $R_2 = 20$ kΩ, $C_1 = 0.5$ nF, $C_2 = 1$ nF. Calculate f_o.

Solution

From Equation 5.6,

$$f_o = \frac{1}{2\pi\sqrt{R_1 R_2 C_1 C_2}} \text{ (Hz)} = \frac{1}{2(3.14)\times\sqrt{(10 \text{ k}\Omega)\times(20 \text{ k}\Omega)\times(0.5 \text{ nF})\times(1 \text{ nF})}}$$

$$= \frac{1}{6.28\times\sqrt{(10\times10^3 \ \Omega)\times(20\times10^3 \ \Omega)\times(0.5\times10^{-9} \text{ F})\times(1\times10^{-9} \text{ F})}}$$

$$= 1600 \text{ Hz} = 1.6 \text{ kHz}$$

Sine-Cosine Oscillator

It is sometimes desirable to have two sine waves, 90° out of phase, in an electronic system. A circuit that provides the two waves is known as a *sine-cosine* oscillator, or a *quadrature* oscillator. Basically, the circuit consists of two op amps connected as integrators (discussed in Chapter 6) with positive feedback. In Figure 5–8 a dual op amp is used to construct the oscillator. The output of the first op amp is the sine wave, while the output of the second op amp is the cosine wave, 90° out of phase with the sine wave.

The value of resistor R_1 is made slightly less than that of R_2 to ensure that the oscillator circuit starts oscillating. If R_1 has too low a value, the outputs may be clipped. Using a potentiometer for R_1 would allow adjustment for minimum output distortion.

Another possible problem is saturation of the op amps. If this occurs, two zener diodes connected anode-to-anode across C_2 will limit the cosine output at the zener voltage. This limiting circuit is illustrated in Figure 5–8 by Z_1 and Z_2.

The output frequency of the sine-cosine oscillator is

$$f_o = \frac{1}{2\pi R_2 C_2} \text{ (Hz)} \tag{5.8}$$

where

$$R_1 < R_2$$
$$R_2 = R_3$$
$$C_1 = C_2 = C_3$$

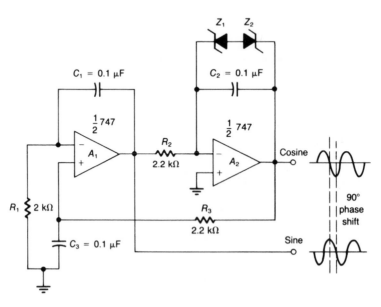

Figure 5–8.　Sine-Cosine (Quadrature) Oscillator

Example 5.5

Calculate the output frequency of the sine-cosine oscillator in Figure 5–8.

Solution

From Equation 5.8,

$$f_0 = \frac{1}{2\pi R_2 C_2} \text{ (Hz)} = \frac{1}{2(3.14) \times (2.2 \text{ k}\Omega) \times (0.1 \text{ }\mu\text{F})}$$

$$= \frac{1}{6.28 \times (2.2 \times 10^3 \text{ }\Omega) \times (1 \times 10^{-7} \text{ F})} = 720 \text{ Hz}$$

Example 5.6

Design a sine-cosine oscillator with an output frequency of 1.5 kHz.

Solution

First, select a standard capacitor, say 0.047 μF. Then, adapt Equation 5.8 to the requirements of the example:

$$R_2 = \frac{1}{2\pi f_0 C_2} \text{ (}\Omega\text{)} = \frac{1}{2(3.14) \times (1.5 \text{ kHz}) \times (0.047 \text{ }\mu\text{F})}$$

$$= \frac{1}{6.28 \times (1.5 \times 10^3 \text{ Hz}) \times (47 \times 10^{-9} \text{ F})} = 2260 \text{ }\Omega$$

You would use a standard 20 percent 2.2 kΩ resistor measuring high on an ohmmeter.

Crystal Oscillators

Oscillator instability is a problem common to all the oscillators we have discussed. This problem stems from several sources: temperature changes, aging of components, quality (Q) of the circuits, and circuit design. The use of crystals in the oscillator circuits provides the desired stability.

A crystal used in oscillator circuits must have the property of *piezoelectricity;* that is, the qualities of (1) generating a difference of potential across its faces when subjected to mechanical pressure, and (2) compressing when a difference of potential is applied across its faces. A crystal has a *natural frequency of vibration,* f_n, that provides an electrical signal from the crystal. This natural frequency of vibration is extremely constant, which makes the crystal ideal for oscillator circuits.

The natural frequency of a crystal is normally determined by its thickness. The thinner the crystal, the higher its natural frequency. Conversely, the thicker the crystal, the lower its natural frequency. There are practical limits, however, on how thin a crystal can be cut without becoming so fragile that it is easily fractured. In general, a crystal will have an upper limit on its natural frequency of around 50 MHz.

Series-Resonant Crystal Circuit

An unmounted crystal, as in Figure 5–9A, appears electrically to be a series-resonant circuit with minimum impedance. In the equivalent circuit, illustrated in Figure 5–9B, the crystal mass that causes vibration is represented by inductor L, crystal stiffness is represented by capacitor C, and the electrical equivalent of internal resistance caused by friction is represented by resistor R.

Parallel-Resonant Circuit

Normally a crystal is mounted between two metal plates that secure the crystal and provide electrical contact. The normal schematic symbol for the crystal is shown in Figure 5–10A. The symbol is derived from the manner in which the crystal is mounted and represents the crystal wafer held between two plates. When mounted in this manner, the crystal appears electrically to be a parallel-resonant circuit with maximum impedance. In the equivalent circuit, illustrated in Figure 5–10B, capacitor C_p represents the metal plates in parallel with the series-resonant circuit of the crystal wafer. The value of this equivalent capacitance is relatively high and at low frequencies has little effect on the series-

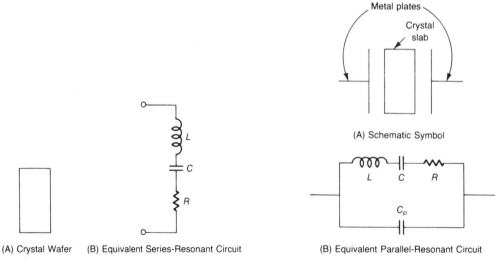

(A) Crystal Wafer   (B) Equivalent Series-Resonant Circuit

Figure 5–9.   Unmounted Crystal

(A) Schematic Symbol

(B) Equivalent Parallel-Resonant Circuit

Figure 5–10.   Mounted Crystal

resonant circuit of the crystal. However, at frequencies above the series-resonant frequency of the crystal, the inductive reactance of the crystal increases, the capacitive reactance decreases, and the crystal appears to be inductive. At a point where the frequency is slightly higher than the series-resonant frequency of the crystal, inductive reactance of the crystal equals the capacitive reactance of the plates, and the circuit is parallel-resonant.

Hartley Circuit

A Hartley crystal-controlled oscillator is shown in Figure 5–11A. The crystal is connected in series with the feedback path. The *LC* network is tuned to the series-resonant frequency of the crystal; therefore, the crystal operates at its series-resonant frequency.

When the oscillator is operating at the crystal frequency, the equivalent circuit offers minimum impedance. Therefore, there is minimum opposition to current flow, and the feedback is maximum. If the oscillator drifts away from the crystal frequency, the impedance increases, reducing feedback, and thereby forcing the oscillator to return to the crystal frequency. The conclusion can be drawn that when the crystal is series-connected, it controls the amount of feedback.

Colpitts Circuit

A Colpitts crystal-controlled oscillator is shown in Figure 5–11B. In this circuit the crystal is again connected in series with the feedback path. The *LC* network is tuned to the crystal frequency, and the series-connected crystal controls the amount of feedback, just as in the Hartley crystal-controlled oscillator.

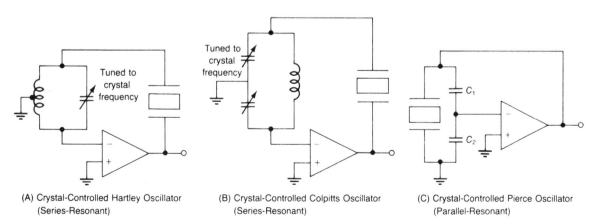

(A) Crystal-Controlled Hartley Oscillator (Series-Resonant)

(B) Crystal-Controlled Colpitts Oscillator (Series-Resonant)

(C) Crystal-Controlled Pierce Oscillator (Parallel-Resonant)

Figure 5–11. Crystal-Controlled Oscillators

Pierce Oscillator

A variation of the Colpitts crystal-controlled oscillator, sometimes called a Pierce oscillator, is illustrated in Figure 5–11C. In this circuit the crystal replaces the inductor of a standard Colpitts oscillator circuit. The crystal operates at its parallel-resonant frequency, which is slightly higher than the series-resonant frequency of the crystal, and appears as an inductance.

Positive feedback results through the 180° phase shift provided by the voltage divider arrangement of capacitors C_1 and C_2. The ratio of these two capacitors also determines the feedback ratio and the crystal excitation voltage. The crystal will provide a very stable output, since its response is extremely sharp, vibrating over a very narrow range of frequencies.

Since the crystal operates in its parallel-resonant mode, it controls the impedance of the tuned circuit. At resonance, circuit impedance is maximum and a large feedback voltage is developed across capacitor C_1. If frequency drifts away from resonance, crystal impedance decreases, and feedback decreases. Thus, by controlling tuned circuit impedance, the crystal effectively determines the amount of feedback and provides a highly stable oscillator.

A final word on crystals: in some applications, such as transmitters, crystals may be mounted inside ovens in which temperature is tightly controlled to insure the high stability and tight frequency control required.

Voltage-Controlled Oscillators

In the oscillators discussed up to this point, changing the frequency of oscillation requires changing the value of one or more of the frequency-determining components. Such an operation requires some manual manipulation by the operator, with a resultant loss of operating time. A device that makes changes in the frequency-determining components automatically and quickly is the *voltage-controlled oscillator* (VCO).

The frequency of the VCO is controlled by a signal voltage; the amount of change in frequency is directly proportional to the input voltage level. The basic circuit in the operation of the VCO is a sine wave oscillator in which the feedback is varied electronically, thus changing the output frequency. Another technique is the generation of a square wave rather than a sine wave, in which case a low-pass filter is used to remove all unwanted harmonic frequencies, leaving the desired sine wave.

One version of a VCO that is shown in Figure 5–12 uses part of a μA3403, a quad op amp device. Amplifier A_1 is an integrator (Chapter 6). It has a triangle wave output whose slope is determined by the input terminal with the more positive input voltage. Amplifier A_2 is a comparator whose output is a constant positive or negative voltage, determined by its input terminal with the more positive input voltage.

If the output of the comparator is initially HIGH, the transistor will be driven to saturation. If $V_{CE(sat)}$ is ignored, then the voltage at the inverting input terminal of A_1 will be equal to $1/3V_c$ (control voltage) due to the voltage divider arrangement of R_1–R_2. The 51 kΩ resistor divider sets the noninverting input voltage

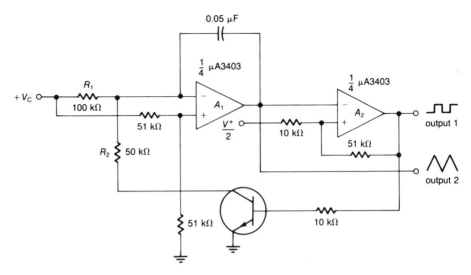

Figure 5–12. Voltage-Controlled Oscillator

to $1/2V_c$. This generates a negative-going ramp at output 2. When the integrator output voltage goes below the voltage at the $(+)$ terminal of A_2, the output of A_2 goes HIGH and the cycle repeats.

The $(+)$ terminal of the comparator has an input voltage determined by $V^+/2$ and a feedback voltage from its output. As the output changes between two values, so does the $(+)$ terminal voltage of A_2. It can therefore be deduced that the period of the triangle wave and, therefore, its frequency, and the frequency of the comparator output, are determined by the value of V_c. If V_c varies, then the output frequency will change. V_c can vary from 0 V to $2(V^+ - 1.5 \text{ V})$ and will provide a wide range of frequencies.

The resulting square wave at output 1 can be filtered through a low-pass filter that removes the unwanted frequency components, and thus provides a desired sine wave. (Filters are discussed at length in Chapter 7.)

There are many forms and uses for the VCO. VCOs are used in FM systems to control the carrier frequency, in phase-locked loops (PLLs), and in many wave-shaping circuits. These applications will be discussed in the following chapters.

5.5 Summary

1. Oscillators are as important to electronics as are amplifiers.
2. An oscillator can provide either sinusoidal or nonsinusoidal output waveforms.
3. Oscillators have no moving parts and produce ac signals.
4. Good oscillators produce uniform output signals with constant amplitude and frequency.
5. Frequency of operation of an oscillator is determined by the components selected for the feedback path.

6. All oscillator circuits must (a) have amplification, (b) have a frequency-determining network, (c) have positive feedback, and (d) be self-starting.

7. An oscillator must meet the requirements of the Barkhausen Criterion in order to be self-starting and self-sustaining; that is, $A_v B_v = 1$.

8. *LC* oscillators use capacitors and inductors, connected either in series or parallel, for the frequency-determining network.

9. A Hartley oscillator is identified by its tapped inductor arrangement.

10. The Hartley oscillator has two disadvantages: inductors are mutually coupled, and changing frequencies is difficult.

11. A Colpitts oscillator is identified by its tapped capacitor arrangement.

12. The operating frequency of the Colpitts oscillator can be easily changed by the use of variable capacitors in the frequency-determining network.

13. The high input impedance and large open-loop gain of the op amp overcome the disadvantages of the standard *LC* oscillator circuits.

14. Resistance-capacitance networks are used for frequency-determining in *RC* oscillators.

15. *RC* oscillators are used primarily in low- and audio-frequency ranges.

16. The two basic types of sine wave-producing *RC* oscillators are the phase-shift oscillator and the Wien-bridge oscillator.

17. The phase-shift oscillator is used principally in fixed frequency applications.

18. Stability in the *RC* phase-shift oscillator is improved by increasing the number of *RC* sections used to provide the necessary 180° phase shift in the feedback network. Three sections are typically used.

19. The Wien-bridge oscillator uses both positive and negative feedback. A lead-lag network determines frequency and applies the positive feedback to the noninverting input, while the negative feedback is applied to the inverting input.

20. A sine-cosine oscillator, sometimes called a quadrature oscillator, consists of two op amps connected as integrators with positive feedback. The output signals from the two op amps are 90° out of phase.

21. Voltage-controlled oscillators (VCOs) are used in a variety of applications. The frequency of the VCO is directly proportional to the input voltage level.

22. Using crystals in oscillator circuits provides a high degree of frequency stability. This stability occurs because of the crystal's natural frequency of vibration.

5.6 **Questions and Problems**

5.1   What types of waveforms can oscillators produce?

5.2   What is the Barkhausen Criterion?

5.3   What determines the stability of the output of an oscillator?

5.4   State the basic fundamental requirements of all oscillators.

5.5   Match each oscillator in the following list with its proper schematic diagram in Figure 5–13:

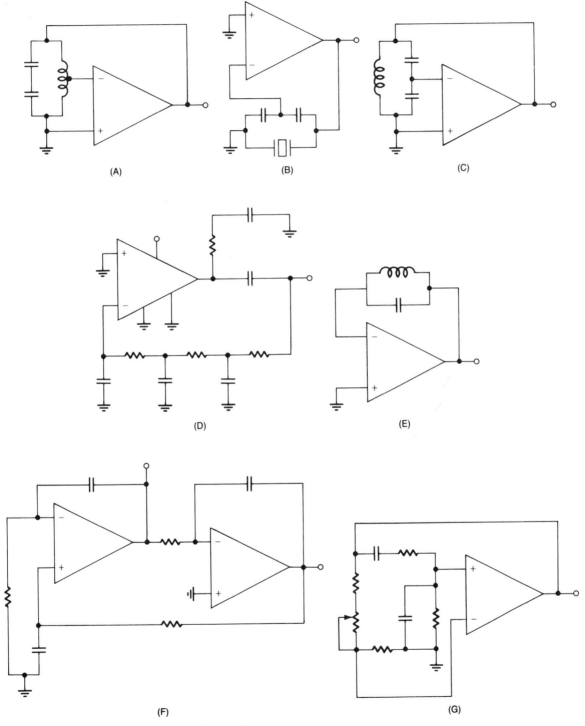

(A) (B) (C)

(D) (E)

(F) (G)

Figure 5–13. Circuit Diagrams for Problem 5.5

1. *LC* oscillator
2. *RC* oscillator
3. Crystal oscillator
4. Hartley oscillator
5. Colpitts oscillator
6. Sine-cosine oscillator
7. Phase-shift oscillator
8. Wien-bridge oscillator
9. Colpitts crystal oscillator

5.6 State the identifying features of (a) the Hartley oscillator and (b) the Colpitts oscillator.

5.7 What are the disadvantages of the Hartley oscillator?

5.8 What determines the frequency of operation of an oscillator?

5.9 What advantages do op amp circuits offer over standard *LC* oscillator circuits?

5.10 Explain the feedback networks of the Wien-bridge oscillator.

5.11 How may the stability of the phase-shift oscillator be improved?

5.12 What oscillator uses both positive and negative feedback?

5.13 Why are crystals used in oscillator circuits? What property of the crystal gives it an advantage?

5.14 Design an op amp phase-shift oscillator that will oscillate at 600 Hz. Draw the schematic diagram and label all components with their values.

5.15 Design an op amp quadrature oscillator that will oscillate at 3 kHz. Draw the schematic diagram and label all components with their values. Show the output waveforms.

5.16 Design an op amp Wien-bridge oscillator that will oscillate at 10 kHz. Draw the schematic diagram and label all components with their values.

Chapter 6

Wave-Shaping Circuits

6.1 Introduction

The basic oscillator circuits described in Chapter 5 were a small sampling of the many wave-shaping circuits available. In this chapter, waveforms other than the sine wave will be discussed. The descriptions of the waveforms are based on the shape of the waves, and the signal generators used to generate the waveforms are classified by the wave shapes they produce. We will look at single-form generators and multiform generators, and we will examine the design of generators using op amps.

6.2 Objectives

When you complete this chapter, you should be able to:

- [ ] Identify and discuss the two types of multivibrators used in analog systems.
- [ ] Identify and discuss the differences between differentiators and integrators.
- [ ] Identify the various types of generators by observation.
- [ ] Design several types of wave-shaping generators using op amps and external components.

□ State the differences between the various types of generators.

□ Evaluate the function generator and explain its operation.

6.3 Multivibrators

There are two types of multivibrators used in analog systems. They are the free-running, or *astable*, multivibrator, and the one-shot, or *monostable*, multivibrator. Another multivibrator, the bistable, is used primarily in digital systems and will not be discussed here.

Astable Multivibrator

A free-running multivibrator has a square wave output whose frequency is determined by the values of the externally connected components. Figure 6–1 shows a 100 kHz free-running multivibrator with an LM311 op amp voltage comparator as the active device. This is a self-starting circuit and requires no external input voltage. The input voltage is replaced with capacitor C_1. Resistors R_4 and R_2 form a voltage divider to feed back a portion of the output to the $(+)$ input. If V_{out} is positive, the voltage on the $(+)$ input is also positive, and capacitor C_1 is charged through the feedback path of resistor R_3. When C_1 charges to a

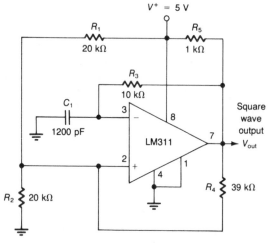

Figure 6–1. 100 kHz Free-Running Multivibrator (Courtesy of National Semiconductor Corporation, © copyright 1980.)

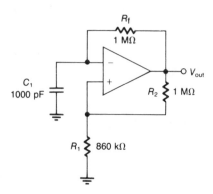

Figure 6–2. Simplified Astable Multivibrator

voltage slightly higher than the voltage at the (+) terminal, V_{out} switches to a negative voltage, and the voltage at the (+) terminal goes negative. Capacitor C_1 now discharges to zero volts and recharges to a negative value. When the voltage charge of C_1 exceeds the negative voltage at the (+) terminal, V_{out} switches back to a positive value. The process continues in this manner. Resistors R_1 and R_2 form a voltage divider to set the quiescent dc output voltage at one-half the supply voltage. Resistor R_5 is a pullup resistor to make the circuit compatible with transistor-transistor logic (TTL) and diode-transistor logic (DTL) circuits.

Simplified Circuit

A simplified astable multivibrator is shown in Figure 6–2. Evaluation of this circuit is the same as for the more complicated circuit of Figure 6–1. The advantage of the simplified version is the ease of design.

The op amp is used as a comparator in this circuit. To determine oscillating frequency, capacitor charge and discharge times are needed. These times are determined by the values of R_f and C_1. If we let the charge time equal t_1 and the discharge time equal t_2, then $t_1 + t_2 = T$, the period of one complete cycle. If we set the value of $R_f = R_2$ and $R_1 = 0.86R_2$, the period, T, is found as follows:

$$T = 2R_fC_1 \tag{6.1a}$$

Since frequency, f, is the reciprocal of the period, T, then

$$f = \frac{1}{T} = \frac{1}{2R_fC_1} \tag{6.1b}$$

where

T is expressed in seconds
f is expressed in hertz
R_f is expressed in ohms
C_1 is expressed in farads

Example 6.1

Refer to Figure 6–2. Calculate (a) the period and (b) the frequency of oscillation.

Solution

a. Use Equation (6.1a):
$$T = 2R_fC_1 = (2)(1 \text{ M}\Omega)(1000 \text{ pF}) = 2(1 \times 10^6 \text{ }\Omega) \times (1 \times 10^{-9} \text{ F})$$
$$= 2 \times 10^{-3} \text{ s} = 2 \text{ ms}$$

b. Use Equation (6.1b):
$$f = \frac{1}{T} = \frac{1}{2 \times 10^{-3} \text{ s}} = 500 \text{ Hz}$$

Monostable Multivibrator

A one-shot multivibrator generates a single output pulse whose pulse width (*PW*) is determined by the values of external components. A trigger signal at one of the input terminals of the op amp is required to generate the output pulse. Figure 6–3 shows a monostable multivibrator with a reference voltage (*V⁺*) using one-half of an LM193 op amp comparator.

The output voltage is compared with the input trigger voltage at the inputs. The output state of the multivibrator is determined by the larger of the input voltages. The trigger pulse is applied to the (−) input terminal through an *RC* high-pass filter circuit. Initially, the output is at zero volts, the normal state for this multivibrator. When a negative-going triggering pulse is applied to the (−) terminal, the output pulse rises to the V^+ reference voltage. This condition is known as the *unstable*, or *timing*, state; that is, it is not the normal state for the circuit. The output pulse will stay in this unstable state for 1 ms, the period of time determined by one time constant of the 0.001 μF capacitor and the 1 MΩ resistor:

$$PW = RC \text{ time constant} \tag{6.2}$$

The output will then return to its stable state until another input pulse is applied.

The 1N914 diode, D_1, prevents the device from triggering on the positive-going edge of the input pulse. Diode D_2 prevents the output signal from swinging negative.

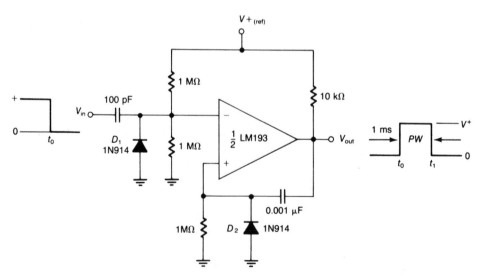

Figure 6–3. One-Shot Multivibrator with *V*+ Reference Voltage (Courtesy of National Semiconductor Corporation, © copyright 1980.)

Example 6.2

Refer to Figure 6–3. Assume that $R = 100\ \text{k}\Omega$ and $C = 100\ \mu\text{F}$ in the feedback line to the $(+)$ terminal. What will be the pulse width of the output pulse?

Solution

Use Equation 6.2:

$$PW = RC = 100\ \text{k}\Omega \times 100\ \mu\text{F} = (1 \times 10^5\ \Omega) \times (1 \times 10^{-4}\ \text{F}) = 10\ \text{s}$$

Example 6.2 should clarify the relationship between component values and pulse duration.

Circuit without Reference Voltage

A monostable multivibrator that requires no V^+ reference voltage is shown in Figure 6–4. In this circuit the negative input pulse is applied to the $(+)$ terminal, resulting in a negative output pulse. The op amp comparator of this circuit has normal output approximately equal to the value of the power supply V^+ voltage. When a negative trigger pulse is applied, diode D_1 conducts, making the $(+)$ terminal less positive than the $(-)$ terminal, and the output switches to the V^- supply voltage level. Diode D_2 is cut off and C_1 is charged toward V^- through feedback resistor R_f. Resistors R_1 and R_2 hold the $(+)$ terminal at the negative voltage level until the voltage at the $(-)$ terminal becomes less negative than the voltage at the $(+)$ terminal. At that point when the $(-)$ terminal voltage

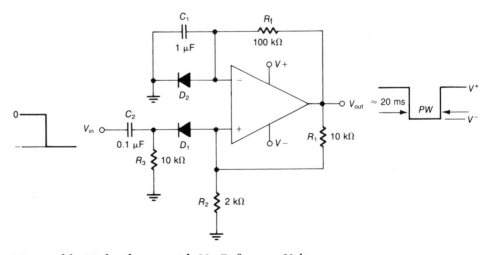

Figure 6–4.   Monostable Multivibrator with No Reference Voltage

drops below the (+) terminal voltage, the output switches back to the V^+ voltage level and completes the pulse duration cycle.

The pulse width of this circuit is determined as follows:

$$PW = R_f C_1 \ln\left(\frac{R_1 + R_2}{R_1}\right) \tag{6.3}$$

where

ln = natural logarithm

A close approximation of the pulse duration can be obtained by the following equation:

$$PW = \frac{R_f C_1}{x} \tag{6.4a}$$

where

$$x = \frac{R_1}{R_2}$$

so that

$$PW = \frac{R_f C_1 R_2}{R_1} \tag{6.4b}$$

Example 6.3

Refer to Figure 6–4. Calculate the pulse duration using (a) Equation 6.3 and (b) Equation 6.4b.

Solution

a.

$$PW = R_f C_1 \ln\left(\frac{R_1 + R_2}{R_1}\right) = 100 \text{ k}\Omega \times 1 \text{ μF} \times \ln\left(\frac{10 \text{ k}\Omega + 2 \text{ k}\Omega}{10 \text{ k}\Omega}\right)$$

$$= (1 \times 10^5 \text{ }\Omega) \times (1 \times 10^{-6} \text{ F}) \times \ln\left(\frac{(1 \times 10^4 \text{ }\Omega) + (2 \times 10^3 \text{ }\Omega)}{1 \times 10^4 \text{ }\Omega}\right)$$

$$= 1.82 \times 10^{-2} \text{ s} = 18.2 \text{ ms}$$

b.

$$PW = \frac{R_f C_1 R_2}{R_1} = \frac{100 \text{ k}\Omega \times 1 \text{ μF} \times 2 \text{ k}\Omega}{10 \text{ k}\Omega}$$

$$= \frac{(1 \times 10^5 \text{ }\Omega) \times (1 \times 10^{-6} \text{ F}) \times (2 \times 10^3 \text{ }\Omega)}{1 \times 10^4 \text{ }\Omega} = 2 \times 10^{-2} \text{ s}$$

$$= 20 \text{ ms}$$

Example 6.3 shows that the approximation formula (Equation 6.4b) provides accuracy to within 10 percent.

Reversing the diodes in the circuit of Figure 6–4 will produce a positive output pulse when a positive input pulse is applied to the (+) input terminal.

6.4 Differentiators

A *differentiator* is a circuit whose output signal is the derivative of its input waveform. The circuit is used in analog computations and wave shaping, but is not limited to those applications.

A simple RC differentiator circuit is illustrated in Figure 6–5A. If V_{out} is much less than V_{in}, ($V_{out} < V_{in}$), then the voltage across the resistance will be much less than the voltage across the capacitance, and

$$V_{out} = RC\frac{dV_{in}}{dt} \qquad (6.5a)$$

where

$$\frac{dV_{in}}{dt} = \text{the change in input voltage over a specified time interval}$$

To ensure that $V_{out} < V_{in}$, either or both R and C must be made very small to cause very small values for V_{out}. Using an op amp in the circuit, as shown in Figure 6–5B, overcomes this problem. The desired result that $V_{out} < V_{in}$ is obtained in this circuit because the input impedance of the op amp with feedback resistor R_f is a resistance equal to $R/(1 + A_r)$. This resistance occurs because the inverting input is at virtual ground, so the capacitor is isolated from R_f. The result is a very small effective resistance comparable to R in Figure 6–5A, but the output voltage is no longer small because of the gain of the op amp.

The op amp differentiator circuit is similar to the inverting amplifier, but the input element is a capacitor. Since the input is applied to the (−) terminal, the output voltage is

$$V_{out} = -R_fC_1\frac{dV_{in}}{dt} \qquad (6.5b)$$

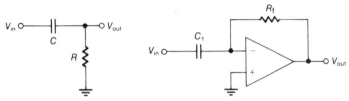

(A) Simple *RC* Differentiator Circuit　　(B) Basic Op Amp Differentiator Circuit

Figure 6–5.　Differentiator Circuits

Reducing High-Frequency Gain Disadvantages

There are two major disadvantages associated with the basic circuit illustrated in Figure 6–5B. These occur because the reactance of the capacitor varies inversely with frequency. Therefore, as frequency increases, capacitive reactance decreases and stage gain increases, creating a tendency for the circuit to go into undesirable oscillations at the high-frequency end of operation. The increase in gain of high frequencies also allows amplification of high-frequency noise signals that may override the desired output signal. Adding external components will reduce the effect of these disadvantages.

Placing a resistor in series with the capacitor, as shown in Figure 6–6A, limits the high-frequency gain to the ratio of R_f/R_1. The output voltage is still calculated by Equation 6.5b. This circuit only acts as a differentiator, however, for input frequencies less than

$$f_{op} = \frac{1}{2\pi R_1 C_1} \tag{6.6}$$

For input frequencies *greater* than those determined by Equation 6.6, the circuit begins to take on the characteristics of an inverting amplifier with the following output voltage gain:

$$A_v = -\frac{V_{out}}{V_{in}} = -\frac{R_f}{R_1} \tag{6.7}$$

In practice, R_1 is usually around 100 Ω. With reference to Equation 6.5b, the time constant of R_fC_1 should be approximately equal to the period of the input signal to be differentiated.

Figure 6–6B shows the addition of a capacitor across the feedback resistor. This capacitor causes the gain of the differentiator to drop off rapidly above a predetermined frequency, thus further reducing the possibility of high-frequency noise signals and undesirable oscillations. The sharp drop in gain also reduces the opportunity for the circuit to take on the characteristics of an inverting amplifier.

For this circuit, there are two calculations to be made: the input frequency, f_{op}, and the upper cutoff frequency, f_c. They may be calculated as follows:

$$f_{op} = \frac{1}{2\pi R_1 C_1} \qquad \text{(repeat of Equation 6.6)}$$

and

$$f_c = \frac{1}{2\pi R_f C_2} \tag{6.8}$$

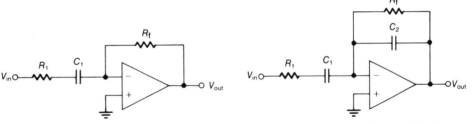

(A) Differentiator with Limited High-Frequency Gain   (B) Differentiator with High-Frequency Cutoff

Figure 6–6. Differentiator Circuits for Limiting Gain Disadvantages

Example 6.4

Design an op amp circuit that will differentiate a 60 Hz input signal. Limit the high-frequency gain to 20.

Solution

With $f_{op} = 60$ Hz, $T = 1/60 = 16.7$ ms; therefore, the time constant $R_f C_1$ from Equation 6.5b must $= 16.7$ ms. Selecting $R_1 = 100$ Ω, then R_f must $= 2$ kΩ, and

$$C_1 = \frac{TC}{R_f}$$
$$= \frac{16.7 \text{ ms}}{2 \text{ k}\Omega} = \frac{16.7 \times 10^{-3} \text{ s}}{2 \times 10^3 \text{ }\Omega} = 8.35 \times 10^{-6} \text{ F} = 8.35 \text{ }\mu\text{F} \qquad \textbf{(6.9)}$$

The final practical circuit is shown in Figure 6–7.

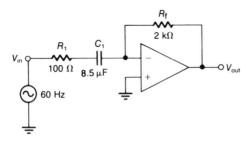

Figure 6–7. 60 Hz Differentiator Circuit with High-Frequency Gain Limit of 20

Example 6.5

Using the practical circuit of Figure 6–7, determine what value capacitor would be required across R_f to cut off frequencies at 400 Hz.

Solution

Use Equation 6.8 and substitute known values from Figure 6–7:

$$C_2 = \frac{1}{2\pi R_f f_c}$$

$$= \frac{1}{6.28 \times 2 \text{ k}\Omega \times 400 \text{ Hz}} = \frac{1}{6.28 \times (2 \times 10^3 \ \Omega) \times (4 \times 10^2 \text{ Hz})}$$

$$= 1.99 \times 10^{-7} \text{ F} \approx 0.2 \ \mu\text{F}$$

Sine Wave Differentiator

For a sine wave input signal, the differentiator output voltage is a function of time and can be calculated as follows:

$$V_{out} = -\omega R_f C V_m \cos(\omega t) \tag{6.10}$$

where

ω = input frequency in radians/second (rad/s) = $2\pi f$
V_m = peak voltage of the input wave

The output signal is a cosine wave, shifted 90° from the input sine wave, but also inverted because the input signal is applied to the (−) input terminal. Therefore, the total phase shift is 270° from the sine wave input signal. The peak output voltage is

$$V_{out(peak)} = \omega R_f C V_m \tag{6.11}$$

The input and output waveforms for the sine wave differentiator are shown in Figure 6–8.

Example 6.6

Refer to Figure 6–7. If the 60 Hz input signal is a 1 V peak sine wave, what will be the peak output voltage?

Solution

Use Equation 6.11:

$$V_{out(peak)} = \omega R_f C V_m = 6.28 \times 60 \text{ Hz} \times 2 \text{ k}\Omega \times 1 \ \mu\text{F} \times 1 \text{ V}$$

$$= 6.28 \times 60 \text{ Hz} \times (2 \times 10^3 \ \Omega) \times (1 \times 10^{-6} \text{ F}) \times 1 \text{ V}$$

$$= 0.75 \text{ V}$$

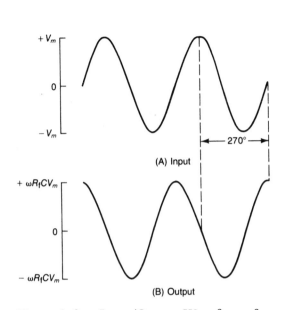

Figure 6–8. Input/Output Waveforms for Sine Wave Input Differentiator

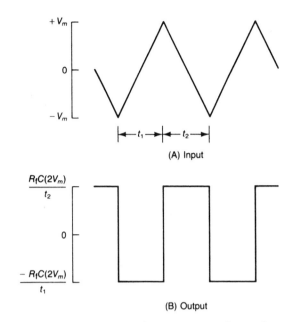

Figure 6–9. Input/Output Waveforms for Triangle Wave Input Differentiator

Triangle Wave Differentiator

The input and output waveforms for a triangle wave input differentiator are shown in Figure 6–9. The output is a square wave with the following peak value:

$$V_{out(peak)} = \frac{\pm R_f C(2V_m)}{t_1 \text{ or } t_2} \tag{6.12}$$

For a triangle wave input signal, the input frequency is

$$f = \frac{1}{t_1 + t_2} \tag{6.13}$$

The voltages for the input signals during time periods t_1 and t_2 are given by the following equations:

$$V_{in(t_1)} = -V_m + 2\frac{V_m}{t_1}T \tag{6.14}$$

$$V_{in(t_2)} = V_m - 2\frac{V_m}{t_2}T \tag{6.15}$$

The output voltages for the two time periods are as follows:

$$V_{\text{out}(t_1)} = \frac{-R_\text{f}C(2V_m)}{t_1} \tag{6.16}$$

$$V_{\text{out}(t_2)} = \frac{R_\text{f}C(2V_m)}{t_2} \tag{6.17}$$

Example 6.7

Refer to Figure 6–7. Assume a symmetrical 2 V peak-to-peak 50 Hz triangle wave input. What will be the peak output voltage?

Solution

Since the triangle wave is symmetrical, $t_1 = t_2$, and $t_1 + t_2 = T = 1/f = 1/50$ Hz $= 0.02$ seconds. Therefore, $t_1 = t_2 = 0.01$ seconds. Solve for either the $(+)$ or $(-)$ output voltage. We will solve for $(+)$, that is, $V_{\text{out}(t_2)}$:

$$\begin{aligned} V_{\text{out}(t_2)} &= \frac{R_\text{f}C(2V_m)}{t_2} = \frac{2\text{ k}\Omega \times 1\text{ }\mu\text{F} \times 2 \times 1\text{ V}}{0.01\text{ s}} \\ &= \frac{(2 \times 10^3\text{ }\Omega) \times (1 \times 10^{-6}\text{ F}) \times 2 \times 1\text{ V}}{1 \times 10^{-2}\text{ s}} = +0.4\text{ V} \end{aligned}$$

$V_{\text{out}(t_1)}$ would then be -0.4 volts, as illustrated in Figure 6–9.

6.5 Integrators

An op amp *integrator* circuit is created by interchanging the resistor and capacitor of the differentiator circuit. Figure 6–10A shows a simple *RC* integrator circuit, and Figure 6–10B shows the basic op amp circuit. In this circuit, the resistor is the input element and the capacitor is the feedback element. The integrator circuit is, effectively, the inverse of the differentiator circuit.

The integrator output voltage as a function of time is expressed by the following equation:

$$V_{\text{out}} = -\frac{1}{R_1C}tV_{\text{in}} \tag{6.18}$$

where

V_{in} = peak input voltage
$t = t_1$ or t_2

The time constant R_1C is made approximately equal to the period of the input signal to be integrated. A more practical integrator circuit can be obtained by placing a shunt resistor, R_{sh}, across feedback capacitor C, as shown in Figure 6–11A. In general, R_{sh} is made approximately ten times the value of R_1. The

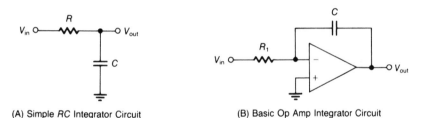

(A) Simple *RC* Integrator Circuit   (B) Basic Op Amp Integrator Circuit

Figure 6–10.   Integrators

shunt resistor limits the low-frequency gain of the circuit, which prevents integration of the dc offset voltage over the integration period. Resistor R_2 minimizes the dc offset voltage caused by the input bias current. The value for R_2 is

$$R_2 = \frac{R_1 R_{sh}}{R_1 + R_{sh}} \tag{6.19}$$

The low-frequency gain-limiting action of the shunt resistor makes Equation 6.18 valid for frequencies that are *greater* than the following cutoff frequency:

$$f_c = \frac{1}{2\pi R_{sh} C} \qquad C \downarrow f_c \uparrow \qquad C \uparrow f_c \downarrow \tag{6.20}$$

For input frequencies *lower* than the f_c of Equation 6.20, the circuit begins to take on the characteristics of the inverting amplifier with a voltage gain of

$$-\frac{V_{out}}{V_{in}} = -\frac{R_{sh}}{R_1} \tag{6.21}$$

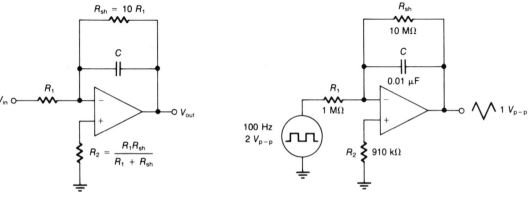

(A) Basic Integrator Circuit with Limited Low-Frequency Gain   (B) Integrator Circuit for Example 6.8

Figure 6–11.   Integrator Circuits with Limited Low-Frequency Gain

Example
6.8

Refer to Figure 6–11A. Assume a symmetrical 2 V peak-to-peak 100 Hz square wave input signal. Determine values for R_1, R_2, R_{sh}, C, and the peak output voltage.

Solution

The time constant R_1C should be approximately equal to the period of the input signal, so select a standard value for C, say 0.01 µF, and calculate R_1.

$$R_1 = \frac{1}{f_{op}C} = \frac{1}{100 \text{ Hz} \times 0.01 \text{ µF}} = \frac{1}{(1 \times 10^2 \text{ Hz}) \times (1 \times 10^{-8} \text{ F})}$$
$$= 1 \times 10^6 \text{ } \Omega = 1 \text{ M}\Omega$$

then
$$R_{sh} = 10R_1 = 10(1 \text{ M}\Omega) = 10 \text{ M}\Omega$$

Using Equation 6.19,

$$R_2 = \frac{1 \text{ M}\Omega \times 10 \text{ M}\Omega}{1 \text{ M}\Omega + 10 \text{ M}\Omega} = \frac{(1 \times 10^6 \text{ } \Omega) \times (10 \times 10^6 \text{ } \Omega)}{(1 \times 10^6 \text{ } \Omega) + (10 \times 10^6 \text{ } \Omega)} = \frac{10 \times 10^{12} \text{ } \Omega}{11 \times 10^6 \text{ } \Omega}$$
$$= 910 \text{ k}\Omega$$

The input signal is symmetrical, so $t_1 + t_2 = T = 1/f = 1/100 \text{ Hz} = 0.01 \text{ s}$, and $t_1 = t_2 = 0.005 \text{ s}$.
Using Equation 6.18,

$$V_{out} = -\frac{1}{R_1C}tV_{in} = -\frac{1}{1 \text{ M}\Omega \times 0.01 \text{ µF}}(0.005 \text{ s})(-1 \text{ V})$$
$$= -(100)(-1)(0.005) = +0.5 \text{ V}$$

The completed circuit is shown in Figure 6–12B.

The input and output waveforms for Example 6.8 are shown in Figure 6–12. Note that the negative alternation of the output triangle wave is of equal amplitude but opposite polarity to the positive alternation; that is, V_{out} for t_2 equals -0.5 V.

6.6 **Square Wave Generator**

The *square wave generator* provides an output that is constantly changing states between high and low with no input signal applied. It is, therefore, an oscillator circuit, or a multivibrator. The square wave generator is sometimes called a *relaxation oscillator* because the output frequency is set by the charging and discharging of a capacitor. Figure 6–13 shows a basic square wave generator.

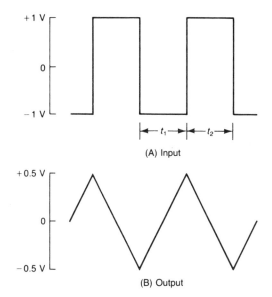

Figure 6–12. Input/Output Waveforms for Square Wave Input Integrator

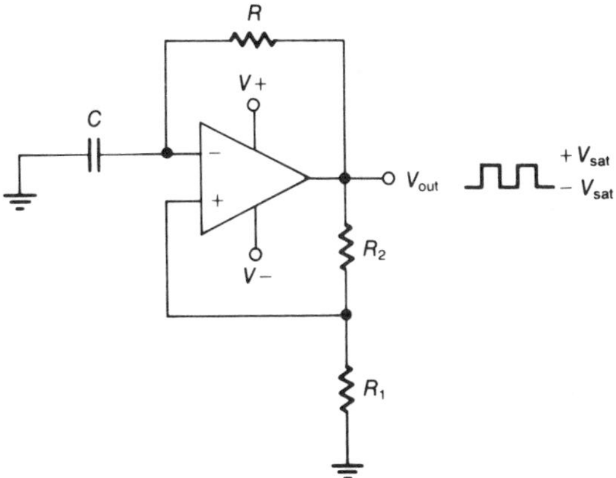

Figure 6–13. Basic Square Wave Generator

Note that there are two feedback paths for this circuit. The RC combination at the inverting input terminal determines the fundamental operating frequency of the circuit, and resistors R_1 and R_2 form a voltage divider that produces a reference voltage to the noninverting input terminal. There are two methods for determining the output frequency, depending upon the values given the resistors:

$$f_{op} = \frac{1}{2RC \ln\left(\dfrac{2R_1}{R_2} + 1\right)} \tag{6.22}$$

where

$R_1 = R/3$
$R_2 = 2$ to 10 times R_1
ln = natural logarithm

and

$$f_{op} = \frac{1}{2RC} \tag{6.23}$$

where

$R_1 = 0.86R_2$ (Figure 6–13)

The reference voltage at the noninverting input terminal causes the circuit to behave as a voltage-level detector. The output of the op amp will swing between

$+V_{\text{sat}}$ and $-V_{\text{sat}}$, based on the charge-discharge of the capacitor, which sets the threshold voltage $\pm V_{\text{th}}$ at the $(-)$ input terminal of the op amp. The saturation voltages $\pm V_{\text{sat}}$ are always about 2 V less than the supply voltages V^+ and V^-. The action of the capacitor voltage versus the op amp output voltage is illustrated in Figure 6–14.

Example
6.9

Design a square wave generator whose output frequency is 1 kHz.

Solution 1

First, select a standard value capacitor, say 0.05 μF. Then by manipulation of Equation 6.23, you can solve for R:

$$R = \frac{1}{2f_{\text{op}}C} = \frac{1}{2 \times 1 \text{ kHz} \times 0.05 \text{ μF}} = \frac{1}{2(1 \times 10^3 \text{ Hz}) \times (5 \times 10^{-8} \text{ F})}$$
$$= 10 \times 10^3 \text{ Ω} = 10 \text{ kΩ}$$

Next select $R_2 = 10R$, or 100 kΩ, and $R_1 = 0.86R_2$, or 86 kΩ. This completes the circuit design using Equation 6.23.

Solution 2

Select any standard value for C, R_1, and R_2. We will select $C = 0.01$ μF, $R_1 = 27$ kΩ, and $R_2 = 47$ kΩ. Manipulation of Equation 6.22 gives us

$$R = \frac{1}{2f_{\text{op}}C \ln\left(\dfrac{2R_1}{R_2} + 1\right)}$$

$$= \frac{1}{2 \times 1 \text{ kHz} \times 0.01 \text{ μF} \times \ln\left(\dfrac{2 \times 27 \text{ kΩ}}{47 \text{ kΩ}} + 1\right)}$$

$$= \frac{1}{2 \times (1 \times 10^3 \text{ Hz}) \times (1 \times 10^{-8} \text{ F}) \times \ln\left(\dfrac{2 \times (2.7 \times 10^3 \text{ Ω})}{4.7 \times 10^3 \text{ Ω}} + 1\right)}$$

$$\approx 6.5 \times 10^4 \text{ Ω} = 65 \text{ kΩ}$$

Either of the designed circuits will produce a 1 kHz square wave output.

6.7 **Triangle Wave Generator**

The output of the *triangle wave generator* consists of positive and negative ramp voltages and sometimes is called a *ramp generator*. The output of an integrator circuit is a triangle wave when the input is a square wave. Therefore, the basic

t_{rise} = rise time = RC $\qquad$ t_{fall} = fall time = RC

$T = t_{rise} + t_{fall} = 2\,RC$

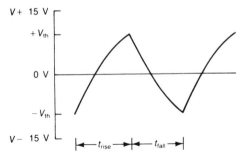

(A) Capacitor Charge-Discharge Voltage

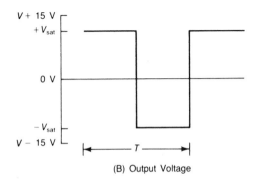

(B) Output Voltage

Figure 6–14. $\quad$ Capacitor Charge/Discharge Voltage Versus Op Amp Output Voltage

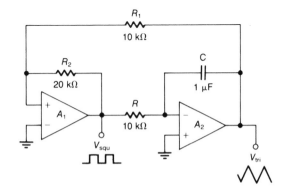

Figure 6–15. $\quad$ Basic Triangle Wave Generator

triangle wave generator circuit, shown in Figure 6–15, is a square wave generator, A_1, wired as a comparator, connected to an integrator, A_2.

The output waveforms shown in Figure 6–16 demonstrate the action of the combination circuit. When the output of the square wave generator goes positive, the output of the triangle wave generator goes negative. Conversely, when the output of the square wave generator goes negative, the output of the triangle wave generator goes positive. A generator that produces two or more different output waveforms, as this one does, is called a *function generator*. We will discuss this concept in more detail later in this chapter.

Determining Output Amplitudes

The output amplitude of the square wave generator is determined by the output swing of the generator, that is, $+V_{sat}$ to $-V_{sat}$. The output amplitude of the triangle wave generator is determined by the ratio R_1/R_2 times the output am-

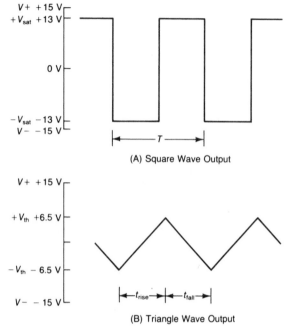

Figure 6–16. Output Waveforms of Square/
Triangle Wave Generator

plitude of the square wave generator. The frequency of oscillation for both waveforms is

$$f_{op} = \frac{1}{4RC}\left(\frac{R_2}{R_1}\right)$$ (6.24)

Example 6.10

Refer to Figure 6–15. Assume that A_1 and A_2 are (1/2)747 with a ± 15 V supply. Determine the output amplitude of each section of the generator and the oscillation frequency.

Solution

$$V_{out(squ)} = \pm V_{sat} = \pm 13 \text{ V}$$

$$V_{out(tri)} = \frac{R_1}{R_2}(V_{out(squ)}) = \frac{10 \text{ k}\Omega}{20 \text{ k}\Omega}(\pm 13 \text{ V}) = \pm 6.5 \text{ V}$$

$$f_{op} = \frac{1}{4RC}\left(\frac{R_2}{R_1}\right) = \frac{1}{4 \times 10 \text{ k}\Omega \times 1 \text{ μF}}\left(\frac{20 \text{ k}\Omega}{10 \text{ k}\Omega}\right)$$

$$= \frac{1}{4\,(1 \times 10^3 \text{ }\Omega) \times (1 \times 10^{-6} \text{ F})}\left(\frac{20 \times 10^3 \text{ }\Omega}{10 \times 10^3 \text{ }\Omega}\right) = 50 \text{ Hz}$$

6.8 Sawtooth Wave Generator

A *sawtooth wave generator* is basically a *nonsymmetrical* triangular wave generator. The time length of the negative ramp of the wave, t_2 in Figure 6–17A, is extremely short compared to the positive ramp, t_1. The shorter t_2 is compared to t_1, the more ideal the sawtooth wave signal.

One method of constructing a sawtooth wave generator is illustrated in Figure 6–17B. A single op amp connected as an integrator is used with a JFET (junction field effect transistor). An external negative trigger pulse is required to provide the input signal.

A negative voltage at the inverting input causes a positive ramp at the output. When the input pulse returns to zero volts, the JFET turns on, effectively shorting the capacitor, which causes the output voltage to drop rapidly toward zero. The period of the output signal is determined by the timing of the input pulses.

Application of the Sawtooth Waveform

The sawtooth waveform finds application in any device that uses a *cathode ray tube* (CRT). Principal among such devices are oscilloscopes and TV sets. The positive ramp (t_1) sweeps the electron beam across the face of the CRT and causes a display to appear. At the end of the display sweep, the negative ramp (t_2) returns the beam to its starting point so that the next display sweep may begin. Time interval t_2 is called the *flyback*, or *retrace*, portion of the wave. This portion of the wave is "blanked out" since it contains no information and is not observable on the screen.

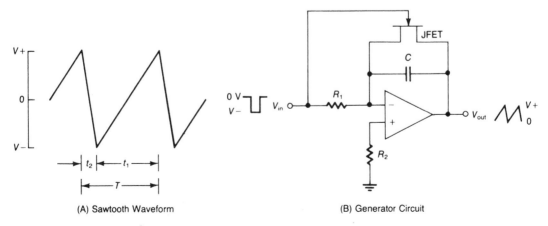

(A) Sawtooth Waveform　　　　　(B) Generator Circuit

Figure 6–17. Sawtooth Wave Generator

6.9 Staircase Generator

A simple circuit for a *staircase generator* is shown in Figure 6–18A. A square wave input signal charges capacitor C_1 to a charge (Q) equal to

$$Q = C_1(V_{in} - 0.7 \text{ V}) \tag{6.25}$$

When the reset switch is open, capacitor C_2 is charged by each input pulse, through C_1 and D_1, in equal steps as voltage changes (ΔV_{out}) so that

$$\Delta V_{out} = (V_{in} - 1.4 \text{ V}) \frac{C_1}{C_2} \tag{6.26}$$

In this manner the voltage across C_2 is built up one step at a time. At the end of a set number of pulses, or when the voltage across C_2 reaches a predetermined value, another pulse closes the switch, shorting C_2 and causing its voltage and the output voltage to go to zero. When the pulse is removed, the switch opens and the cycle begins again.

Figure 6–18B shows a similar staircase generator that uses a unijunction transistor (UJT) as a reset switch. When the charge on capacitor C_2 reaches the firing potential of the UJT emitter, the UJT conducts, discharging capacitor C_2 and driving the output voltage to zero. The UJT now acts like an open switch and the cycle begins again.

In both the generators just discussed, the output signal went negative because the input was applied to the inverting input terminal of the op amp. Figure 6–18C shows a staircase generator with a positive-going output signal. This circuit uses a separate reset pulse.

6.10 Function Generators

Recall that any signal generator that has two or more different output waveforms is considered to be a function generator. We have already looked at the square/triangle wave generator. Figure 6–19 shows a *multifunction* waveform generator that provides sine, square, and triangle waveforms. The circuit is constructed from a single 14-pin DIP IC, the LM324 quad op amp, and external components.

Amplifier A_1 is an oscillator circuit connected as a bandpass filter (Chapter 7) that feeds a sine wave to a comparator, amplifier A_2. The output of A_2 is a square wave that is fed back to the input of the bandpass filter to cause oscillation. The output of the comparator is also fed to a voltage follower, amplifier A_3. This prevents loading down the oscillator circuit and helps prevent any change in the oscillator frequency. The output of A_3 is fed to an integrator, amplifier A_4, to produce a triangle wave output.

The output frequency of this function generator can be varied from about

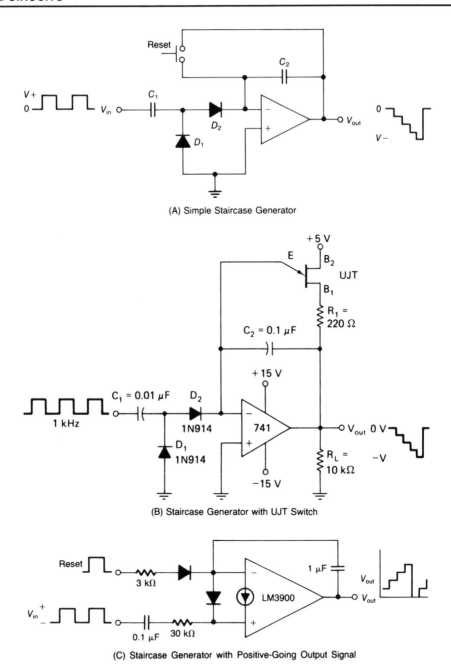

(A) Simple Staircase Generator

(B) Staircase Generator with UJT Switch

(C) Staircase Generator with Positive-Going Output Signal

Figure 6–18.   Staircase Generators (Parts B and C: Courtesy of National
Semiconductor Corporation, © copyright 1980.)

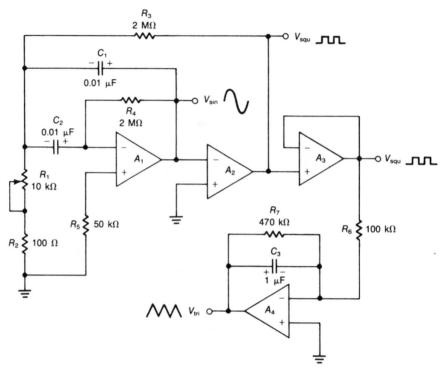

Figure 6–19. Low-Frequency Multifunction Generator

110 Hz to 1125 Hz by adjusting R_1. R_2 is simply a small-value resistor placed in the circuit to prevent shorting the feedback to ground.

The output frequency (f_{op}) for a function generator is determined by the following equation:

$$f_{op} = \frac{1}{2\pi\sqrt{R_p R_4 C_1 C_2}}$$

(6.27)

where

$$R_p = \text{the parallel resistance of } R_1 \text{ and } R_3 = \frac{R_1 R_3}{R_1 + R_3}$$

6.11 Summary

1. Two types of multivibrators used in analog systems are the astable and monostable multivibrators.

2. An astable multivibrator has a square wave output and does not require an input trigger pulse.

3. A monostable multivibrator produces a single output pulse for a pre-determined time interval and requires an input trigger pulse.

4. A differentiator circuit produces an output signal that is a derivative of its input signal. If the input signal is a sine wave, the output signal is a cosine wave that is 270° out of phase with the input signal. If the input signal is a triangle wave, the output signal is a square wave.

5. An integrator circuit is the inverse of the differentiator circuit. It is constructed simply by interchanging the resistors and capacitors in a differentiator circuit. A square wave input signal produces a triangle wave output signal.

6. A square wave generator is basically a multivibrator, sometimes called a relaxation oscillator because the output frequency is set by the charging and discharging of a capacitor.

7. A triangle wave generator, sometimes called a ramp generator, produces an output wave with positive and negative ramp voltages. The basic circuit is an integrator whose input is a square wave from a comparator.

8. A sawtooth generator output signal is a nonsymmetrical triangular wave whose negative-going ramp time interval is very short compared with its positive-going ramp time interval. The sawtooth wave is applied principally as a sweep circuit in oscilloscopes and TV sets.

9. Function generators are signal generators that have two or more different output waveforms.

6.12 Questions and Problems

6.1 List the two types of multivibrators used in analog systems.

6.2 Which type of multivibrator requires an input trigger pulse?

6.3 The astable multivibrator produces what kind of output waveform?

6.4 Describe the operation of the monostable multivibrator.

6.5 Describe the output signal of a differentiator circuit whose input signal is a sine wave.

6.6 Describe the basic circuitry differences between the differentiator and integrator circuits.

6.7 The output frequency of a relaxation oscillator is set by which components?

6.8 Describe the ramp generator operation.

6.9 Describe the sawtooth generator operation.

6.10 What is the principal application of the sawtooth generator?

6.11 Define function generators.

6.12 Refer to Figure 6–2. Assume $R_1 = 86$ kΩ, $R_f = R_2 = 100$ kΩ, and $C_1 = 1000$ pF. Determine (a) the period and (b) the frequency of oscillation.

6.13 Refer to Figure 6–3. Assume $R = 1$ MΩ and $C = 10$ μF in the feedback line to the (+) terminal. Determine the output pulse width.

6.14 Refer to Figure 6–4. Assume $R_f = 100$ kΩ, $R_1 = 20$ kΩ, $R_2 = 4$ kΩ, and $C_1 = 2$ μF. Use the approximation method to solve for the output pulse width.

6.15 Design an op amp circuit that will differentiate a 1 kHz input signal while limiting the high frequency gain to 10.

6.16 If the 1 kHz input signal of Problem 6.15

is a 0.5-volt peak sine wave, what will be the peak output voltage?

6.17 Refer to Figure 6–11B. Assume a symmetrical 3-volt peak-to-peak 400 Hz square wave input signal. Determine values for R_1, R_2, R_{sh}, C, and the peak output voltage.

6.18 Use two methods of calculation to design a square wave generator whose output frequency is 400 Hz.

6.19 Refer to Figure 6–15. Assume $R = 20$ kΩ, $R_1 = 20$ kΩ, $R_2 = 40$ kΩ, and $C = 2$ μF. A_1 and A_2 are (1/2)747 with a ± 15 volt supply. Determine the output amplitude of each section of the generator and the frequency of oscillation.

Active Filters

7.1 Introduction

Electrical filters are designed to either attenuate unwanted frequencies or to amplify desired frequencies. Attenuation is accomplished by using passive filters that use only resistors, capacitors, and inductors. Amplification is accomplished by using active filters that use resistors and capacitors around an active device, usually an op amp.

In this chapter we will review the fundamentals of filters, types of filters, filter characteristics, and filter configurations. We will distinguish between passive and active filters. We will examine first-, second-, and high-order filters. Multiple-feedback and state-variable filters will be introduced. Finally, design criteria and techniques will be thoroughly discussed.

7.2 Objectives

When you complete this chapter, you should be able to:

☐ Define and explain active filters.

☐ Explain the many uses of active filters.

☐ Define and explain a normalized response curve in terms of (a) cutoff frequency, (b) rolloff, (c) center frequency, (d) bandwidth, (e) passband, (f) stopband, (g) order, (h) decibel, (i) decade, and (j) octave.

☐ Define Butterworth response.

☐ Distinguish between Butterworth and other types of filters in terms of damping.

☐ Design and explain (a) the various orders of low-pass and high-pass filters, (b) multiple-feedback low-pass, high-pass, and bandpass filters, (c) notch filters, (d) wideband filters, and (e) state-variable filters.

7.3 Review of Filter Fundamentals

A *filter* is a device that screens out certain frequencies or passes electric current of only certain frequencies. Basically, there are two categories of filters—passive filters and active filters.

Passive filters are constructed with resistors, capacitors, and inductors. They contain no active devices such as transistors, diodes, vacuum tubes, or op amps. *Active filters* are constructed with a network of resistors and capacitors around an active device, usually an op amp. Inductors are seldom used in active filters.

Active filters offer the following advantages over passive filters:

1. *Low cost*. Inductors, which are usually expensive components, are seldom used in active filters.
2. *Ease of adjustment*. In general, active filters can be adjusted over a wide range of frequencies without appreciable change in the desired output response.
3. *No insertion loss*. The op amp in the active filter circuit overcomes any loss or attenuation of the circuit by providing amplification.
4. *Isolation*. Because of the very high input impedance and low output impedance of the op amp, there is practically no interaction between the filter and the source or load.

There are also some limitations and disadvantages in active filter circuits. They are as follows:

1. *Frequency response*. All op amps have a frequency response limit, so the type of op amp selected for the filter circuit will be the deciding factor in filter frequency response.
2. *Power supply*. Active filters require voltage power supplies to power the op amp. Many op amps require two separate supplies, or dual power supplies, although some circuits may be operated with a single power supply.

Filters are classified according to the range of frequencies they reject or allow to pass. A *low-pass* filter allows frequencies below a given value to pass and rejects frequencies above that value. A *high-pass* filter rejects lower frequencies and allows only frequencies above a given value to pass. A *bandpass* filter allows a band of frequencies between a designated high value and a designated low value to pass and rejects frequencies above and below those values. A filter that prevents a band of frequencies between two designated values from passing is variously called a *notch, band-reject,* or *band-stop* filter. Graphic representations of these various filters show the relationship between filter output in *decibels* (dB) and frequency in hertz (Hz). Figure 7–1 illustrates generalized characteristics of the filters we have mentioned, assuming a constant amplitude for all input signal frequencies. Note that unwanted frequencies are not com-

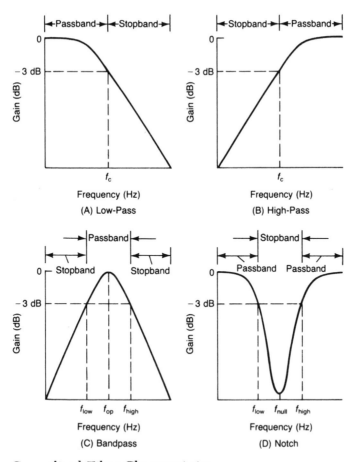

Figure 7–1. Generalized Filter Characteristics

pletely rejected, but roll off in a sharp curve. The f_c designation is the frequency cutoff and is the limiting frequency for the filter.

The range of frequencies passing through a filter with maximum gain or minimum attenuation is called the *passband*. The value of the cutoff frequency determines the width of the passband.

The *transfer function* of a filter is considered to be its *gain* or *amplitude response*. The transfer function is represented by the relationship between the filter's output voltage and its input voltage at various frequencies, expressed as V_{out}/V_{in}, a function of frequency. This ratio is more conveniently expressed in terms of decibels to eliminate the use of decimal voltage values. As a formula

$$A_{dB} = 20 \log_{10} A_v \qquad (7.1)$$

where

$$A_v = \frac{V_{out}}{V_{in}}$$

A_v may be equal to, greater than, or less than unity. A resistive network will attenuate a signal, which results in a loss, or less than unity gain. This is shown by a negative decibel value. An amplifier in the system may provide a positive decibel value, or greater than unity gain. Unity gain is represented by 0 dB.

Example 7.1	
	A low-pass filter circuit has $V_{in} = 1$ V$_{p-p}$ and $V_{out} = 3$ V$_{p-p}$. What is the decibel gain of the circuit?
	Solution
	Use Equation 7.1:
	$A_{dB} = 20 \log_{10} A_v = 20 \log_{10} 3 = 20(0.47712) = 9.5$ dB

Example 7.2	
	A high-pass filter circuit has $V_{in} = 1$ V$_{p-p}$ and $V_{out} = 0.5$ V$_{p-p}$. What is the decibel gain of the circuit?
	Solution
	$A_{dB} = 20 \log_{10} \dfrac{0.5 \text{ V}}{1 \text{ V}} = 20 \log_{10} 0.5 = 20(-0.301) \approx -6$ dB

Example 7.3

A bandpass filter circuit has unity gain. What is the decibel gain of the circuit?

Solution

$$A_{dB} = 20 \log_{10}1 = 20(0) = 0 \text{ dB}$$

Example 7.4

A notch filter circuit has $V_{in} = 1 \text{ V}_{p\text{-}p}$ and $V_{out} = 1.12 \text{ V}_{p\text{-}p}$. What is the decibel gain of the circuit?

Solution

$$A_{dB} = 20 \log_{10}1.12 = 20(0.04921) = 1 \text{ dB}$$

The frequency at which the voltage gain of the filter circuit drops to 0.707 of its maximum value is the *cutoff frequency*, f_c. The frequency or frequencies at which this occurs identifies the filter's passband limits. The cutoff frequency point is variously called the *0.707 point*, the *3 dB point*, the *break point*, and the *half-power point*. Expressed as an equation,

$$f_c \text{ point} = \text{dB gain}_{(max)} - 3 \text{ dB} \qquad (7.2)$$

or

$$f_c \text{ point} = 0.707 \frac{V_{out}}{V_{in}} \qquad (7.3)$$

where

$$\frac{V_{out}}{V_{in}} = A_{v(max)}$$

Figure 7–2 and some examples will demonstrate the origin of these designations.

Example 7.5

The frequency response curve for Example 7.1 is shown in Figure 7–2A. Determine the f_c point for the curve.

Solution

Using Equation 7.2,

$$f_c \text{ point} = \text{dB gain}_{(max)} - 3 \text{ dB} = 9.5 \text{ dB} - 3 \text{ dB} = 6.5 \text{ dB}$$

Using Equation 7.3,

$$f_c \text{ point} = 0.707 A_{v(max)} = 0.707(3) = 2.12$$

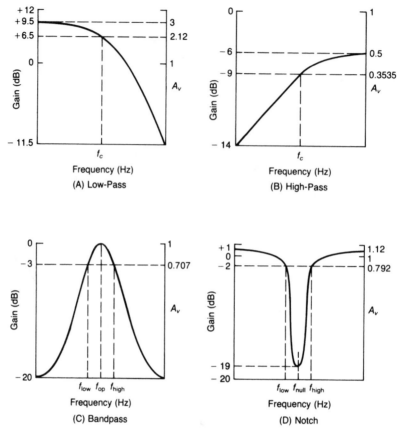

Figure 7–2. Frequency Response Curves of Filters

Example
7.6

Refer to Example 7.2 and Figure 7–2B. Determine the break point.

Solution

Using Equation 7.2,

Break point = dB gain − 3 dB = −6 dB − 3 dB = −9 dB

Using Equation 7.3,

Break point = $0.707A_v$ = $0.707(0.5)$ = 0.3535

Example 7.7	

Refer to Example 7.3 and Figure 7–2C. Find the half-power point $(P_{(1/2)})$.

Solution

Using Equation 7.2,

$\quad P_{(1/2)}$ = dB gain − 3 dB = 0 dB − 3 dB = −3 dB

Using Equation 7.3,

$\quad P_{(1/2)}$ = $0.707A_v$ = 0.707(1) = 0.707

Example 7.8	

Refer to Example 7.4 and Figure 7–2D. Find the 3 dB point and the 0.707 point.

Solution

Using Equation 7.2,

$\quad$ 3 dB point = dB gain − 3 dB = 1 dB − 3 dB = −2 dB

Using Equation 7.3,

$\quad$ 0.707 point = $0.707A_v$ = 0.707(1.12) = 0.792

Note that the bandpass and notch filter response curves have two frequency cutoff points. Also note that *the 3 dB point is always −3 dB from the maximum dB gain*, regardless of the type filter.

7.4 Order of Filters

As discussed in the previous section, the passband is that range of wanted frequencies where the amplitude response is relatively constant. The *stopband* is the range of frequencies outside the passband. In the low-pass filter, for example, the stopband is the range of frequencies above the filter's cutoff frequency.

Normalized Frequency Response Curve

A *normalized* frequency response curve is one in which the maximum amplitude response is set to unity, or 0 dB. Using a normalized frequency graph allows us to compare the relative selectivity of different orders of filters. The simplistic graph in Figure 7–3 shows the rolloff of several different orders of low-pass filters with the gain (or loss) in decibels as a function of logarithmic change in frequency. Note that the amplitude response in the stopband (above 1 kHz) decreases linearly as frequency increases logarithmically. The *order* of the filter defines the rate of this decrease, called the *rolloff* or *falloff*. Note also that the rolloff of a *first-order* filter is − 20 dB between the maximum decibel gain and 10 kHz. This can be stated as a rolloff of − 20 dB per decade, or − 6 dB per octave.

Decade and Octave Defined

A *decade* is defined as a tenfold increase or decrease in frequency. An *octave* is defined as a doubling or halving of frequency. Table 7–1 demonstrates these relationships, showing several decades and octaves above and below a cutoff frequency of 1 kHz.

Graph Evaluation

Returning to Figure 7–3, we can now more easily evaluate the graph. Note that increasing the order of the filter increases the rolloff. Also note that the designated order of the filter indicates a multiple of the first-order rolloff value.

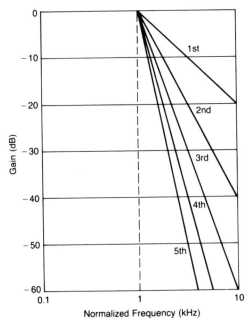

Figure 7–3. Rolloff for Various Orders of Filters

Table 7–1	Relationships between Frequency, Decade, and Octave						
				(f_c)			
Frequency (Hz)	1	10	100	1 k	10 k	100 k	1000 k
Decade	−3	−2	−1	0	+1	+2	+3
				(f_c)			
Frequency (Hz)	125	250	500	1 k	2 k	4 k	8 k
Octave	−3	−2	−1	0	+1	+2	+3

For example, a *second-order* filter has a −40 dB per decade (or −12 dB per octave) value, a *third-order* filter has a −60 dB per decade (or −18 dB per octave) value, and so on. The sharper the rolloff, the more efficient the filter.

Butterworth Response

A more realistic frequency response graph is shown in Figure 7–4A. Here we see that the cutoff frequency is 3 dB down (−3 dB) from the assumed 0 dB unity

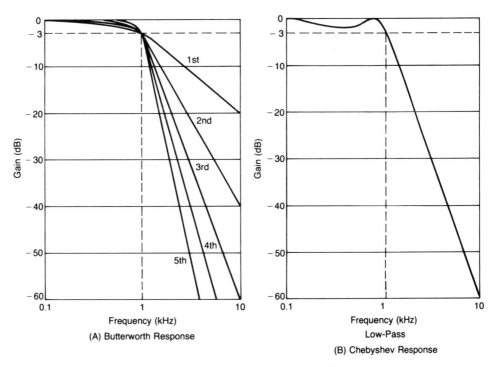

Figure 7–4. Amplitude Response Curves for Various Orders of Filters

gain. Note that the passband amplitude response, for all orders of filters, is relatively constant, or flat, with no oscillations. Such a passband response is called a *Butterworth* response.

A *Chebyshev* response is shown in Figure 7–4B. Note the oscillation, or ripple, in the passband.

Damping

The determining factor in the shaping of the filter's passband is *damping,* represented by the Greek letter alpha, α. The higher the damping, the flatter the passband. Therefore, the Butterworth response represents a highly damped filter, and a Chebyshev response represents a filter with less damping. For the low-pass and high-pass filters, we will discuss the design and evaluation of only the Butterworth response.

Filter Design Rules

To simplify the design of active filters, here are a few basic rules to follow:

1. Determine the desired cutoff frequency, f_c.
2. Select a standard capacitor value.
3. Determine the value of the resistance to be used for impedance level by the following equation:

$$R = \frac{1}{2\pi f_c C} \qquad \text{(For first-order filters)} \qquad (7.4)$$

Example 7.9

Design a low-pass filter in which $f_c = 600$ Hz.

Solution

First, select a standard capacitor, say 0.1 μF. Then determine the value for R.

$$R = \frac{1}{2\pi f_c C} = \frac{1}{(6.28)(600 \text{ Hz})(0.1 \times 10^{-6} \text{ F})}$$
$$= 2.65 \times 10^3 \ \Omega = 2.65 \text{ k}\Omega \qquad \text{(Use a standard 2.7 k}\Omega \text{ resistor.)}$$

7.5 First-Order Filters

The simplest forms of active filters are first-order low-pass, high-pass, and unity-gain filters. Such filters are built with a passive filter network, which consists of one resistor and one capacitor, connected to an op amp. First-order filters are easily constructed and inexpensive.

$f_c = 1$ kHz $\qquad R_3 = R_2(A_v - 1)$

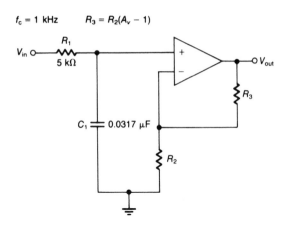

Figure 7–5. Basic Design Circuit for a First-Order Low-Pass Filter

Low-Pass Filters

A first-order low-pass filter has a rolloff of -20 dB per decade, or -6 dB per octave. Figure 7–5 shows the basic first-order low-pass filter design circuit. It is simply a low-pass passive filter (R_1C_1) connected to an active device, a noninverting amplifier, having a cutoff frequency of 1 kHz and an input impedance of 5 kΩ. The resistive feedback network of R_2 and R_3 determines the passband gain.

Starting with the basic design circuit, we can now make the necessary calculations to provide the desired filter characteristic response.

Example 7.10

Convert the standard 1 kHz filter circuit of Figure 7–5 to one with a cutoff frequency of 2 kHz.

Solution

Refer to the basic rules of filter design that we established. The first item has been predetermined; that is, the desired frequency is 2 kHz. Next, select a standard capacitor value, say 0.02 μF for C_1. Now use Equation 7.4 to determine the value for resistor R_1.

$$R_1 = \frac{1}{2\pi f_c C} = \frac{1}{(6.28)(2 \times 10^3 \text{ Hz})(0.02 \times 10^{-6} \text{ F})}$$
$$= 3.98 \times 10^3 \ \Omega = 3.98 \text{ k}\Omega \qquad \text{(Use a standard 3.9 k}\Omega \text{ resistor.)}$$

Figure 7–6 shows the completed circuit design and amplitude response curve.

f_c = 2 kHz

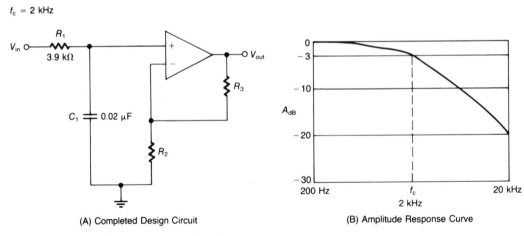

(A) Completed Design Circuit

(B) Amplitude Response Curve

Figure 7–6. Solution to Example 7.10

High-Pass Filters

A high-pass filter circuit is identical to the low-pass circuit, except that the frequency-determining resistors and capacitors (R_1C_1) are interchanged. The frequency response curve is also identical, except it is reversed. The design of a first-order high-pass filter is accomplished in the same manner as the low-pass design.

Example 7.11

Design a first-order high-pass filter with a cutoff frequency of 750 Hz.

Solution

Applying the basic rules, select a standard capacitor value and determine the resistance value. We will select a 0.1 μF capacitor for C_1, then

$$R_1 = \frac{1}{2\pi f_c C} = \frac{1}{(6.28)(750 \text{ Hz})(0.1 \times 10^{-6} \text{ F})} = 2.12 \times 10^3 \text{ }\Omega$$

$$= 2.12 \text{ k}\Omega \quad \text{(Use a standard 2 k}\Omega \text{ or 2.2 k}\Omega \text{ resistor.)}$$

Figure 7–7 shows the completed circuit design and amplitude response curve.

Unity-Gain Filters

If unity gain is desired in either the low-pass or high-pass filter, a voltage follower is used for the active device. Figure 7–8 illustrates such circuits.

f_c = 750 Hz

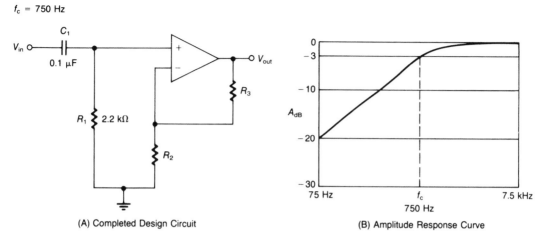

(A) Completed Design Circuit

(B) Amplitude Response Curve

Figure 7–7. Solution to Example 7.11

7.6 Second-Order Filters

The *voltage-controlled voltage source* (VCVS) circuit is the simplest second-order active filter. This circuit is sometimes called a *Sallen and Key* (S & K) filter. In this section we will examine second-order low-pass, high-pass, and equal-component VCVS filters. You will observe that the passive network consists of two resistors and two capacitors.

Low-Pass VCVS Circuit

The cutoff frequency for the VCVS second-order low-pass filter is

$$f_c = \frac{1}{2\pi\sqrt{R_1 R_2 C_1 C_2}} \qquad (7.5)$$

f_c = 1 kHz f_c = 1 kHz

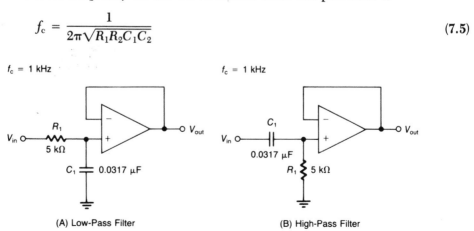

(A) Low-Pass Filter

(B) High-Pass Filter

Figure 7–8. Unity-Gain First-Order Filters

Figure 7–9 illustrates a basic VCVS second-order low-pass filter design circuit with a f_c of 1 kHz, an impedance level of 5 kΩ, and unity gain. Note that $C_2 = 2C_1$ and $R_1 = R_2$.

Scaling Techniques

A convenient method of design for the VCVS second-order filter is *scaling*. This is accomplished by first multiplying all the frequency-determining resistors of the basic circuit design by the ratio of the basic design frequency to the new desired frequency. Then select a standard value of capacitor and divide the new resistance values by the ratio of the selected capacitor value to the basic design value. Table 7–2 will help to clarify the procedure.

Example 7.12

Convert the basic design circuit of Figure 7–9 to a second-order VCVS low-pass filter with a cutoff frequency of 500 Hz and unity gain.

Solution

Follow the steps in Table 7–2. Determine the frequency ratio X:

$$X = \frac{\text{basic frequency}}{\text{desired frequency}} = \frac{1000 \text{ Hz}}{500 \text{ Hz}} = 2$$

Select a standard capacitor value and determine the capacitor ratio Y:

$$Y = \frac{\text{selected } C_1}{\text{basic } C_1} = \frac{0.05 \text{ μF}}{0.0225 \text{ μF}} = 2.22$$

$f_c = 1$ kHz

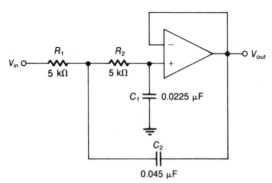

Figure 7–9.   Basic Design Circuit for a Second-Order VCVS Low-Pass Filter

Table 7–2	Scaling Procedure	

	Low-Pass	High-Pass
1.	$\dfrac{\text{Basic Frequency}}{\text{Desired Frequency}} = X$	$\dfrac{\text{Basic Frequency}}{\text{Desired Frequency}} = X$
2.	$\dfrac{\text{Selected } C_1}{\text{Basic } C_1} = Y$	$\dfrac{\text{Selected } C_1}{\text{Basic } C_1} = Y$ where: $C_1 = C_2$
3.	$XR_1 = $ New R where: $R_1 = R_2$	$X \,(\text{Basic } R_2) = $ New R_2
4.	$\dfrac{\text{New } R}{Y} = $ Required R for $R_1 = R_2$	$\dfrac{\text{New } R_2}{Y} = $ Required R_2
5.	Make $C_2 = 2$ (Selected C_1)	Make $R_1 = 2R_2$

Scale the resistances to determine the new value for R:

new $R = XR_1 = 2(5\text{ k}\Omega) = 10\text{ k}\Omega$

Determine the required value of R for R_1 and R_2:

$$\text{required } R = \frac{\text{new } R}{Y} = \frac{10\text{ k}\Omega}{2.22} = 4.5\text{ k}\Omega$$

Finally,

$$C_2 = 2(\text{selected } C_1) = 2(0.05\ \mu\text{F}) = 0.1\ \mu\text{F}$$

Standard 4.3 kΩ resistors for R_1 and R_2 will produce a cutoff frequency within 10 percent of the desired 500 Hz frequency for this circuit. Verification can be obtained by using Equation 7.5 and the new values of components, and is left as an exercise.

Figure 7–10 shows the completed circuit design and amplitude response curve.

Equal-Component Circuits

An *equal-component* second-order VCVS low-pass filter, whose basic circuit is shown in Figure 7–11, requires that the passband gain be fixed at $A_v = 1.59$. This value is derived from $A_v = 3 - \alpha$, where $\alpha = 1.41$ for a second-order Butterworth response. The circuit will not produce a Butterworth response at any other passband gain. For the configuration of Figure 7–11, using the non-inverting amplifier, the value of feedback resistor R_4 must equal 0.59 times the value of input resistor R_3.

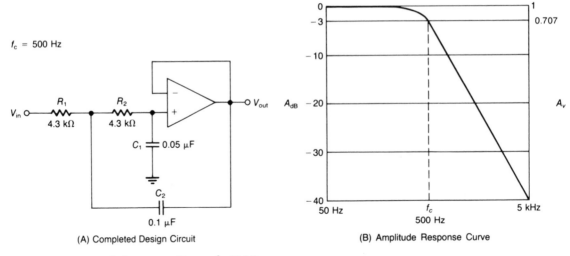

(A) Completed Design Circuit   (B) Amplitude Response Curve

Figure 7–10.   Solution to Example 7.12

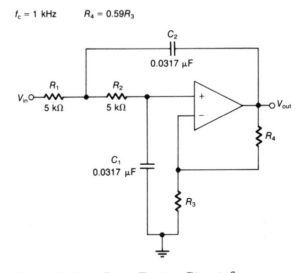

Figure 7–11.   Basic Design Circuit for an
                    Equal-Component VCVS
                    Low-Pass Filter

Equal-Component Low-Pass VCVS Circuit

Design of the equal-component second-order VCVS filter is quite simple, since $R_1 = R_2$ and $C_1 = C_2$.

Example
7.13

Design an equal-component second-order VCVS low-pass filter with a cutoff frequency of 750 Hz.

Solution

Select a standard capacitor value for C_1 and C_2. We will use 0.01 μF. Then determine the value for R_1 and R_2 by using Equation 7.4:

$$R_1 = R_2 = \frac{1}{2\pi f_c C} = \frac{1}{(6.28)(750 \text{ Hz})(0.01 \times 10^{-6} \text{ F})} = 2.12 \times 10^4 \text{ }\Omega$$

$$= 21.2 \text{ k}\Omega \quad \text{(Use standard 22 k}\Omega \text{ resistors.)}$$

Figure 7–12 shows the completed circuit design and amplitude response curve.

High-Pass VCVS Circuit

A second-order VCVS high-pass filter can be obtained by interchanging the frequency-determining resistors R_1 and R_2 with capacitors C_1 and C_2 and using the values shown in Figure 7–13. The cutoff frequency is determined in the same manner as for the low-pass:

$$f_c = \frac{1}{2\pi\sqrt{R_1 R_2 C_1 C_2}} \qquad \text{(repeat of Equation 7.5)}$$

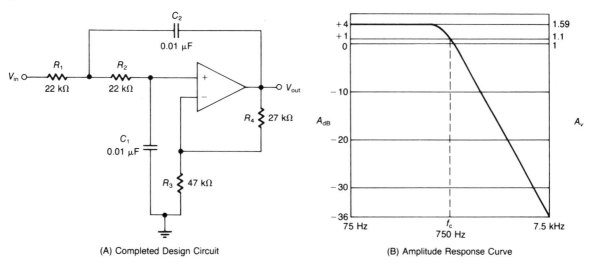

(A) Completed Design Circuit (B) Amplitude Response Curve

Figure 7–12. Solution to Example 7.13

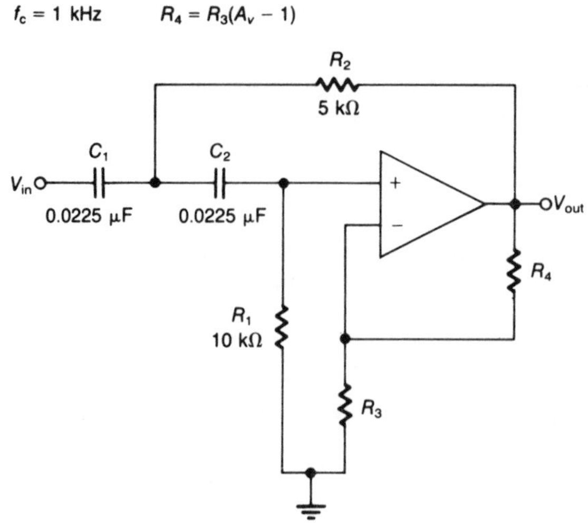

$f_c = 1$ kHz $R_4 = R_3(A_v - 1)$

Figure 7–13.   Basic Design Circuit for a
Second-Order VCVS High-
Pass Filter

Note, however, that now $C_1 = C_2$ and $R_1 = 2R_2$.

Design of the second-order high-pass VCVS filter is accomplished in the same manner as for the low-pass, by scaling.

Example 7.14

Convert the filter circuit of Figure 7–13 to have a $f_c = 3$ kHz and $A_v = 2$. Refer to Table 7–2 as necessary.

Solution

We will select $C_1 = C_2 = 0.01$ μF, so that the capacitor ratio (Y) is 0.01/0.0225 = 0.44. The frequency ratio (X) is 1 kHz/3 kHz = 0.33. Therefore,

New $R_2 = 5$ kΩ(0.33) = 1.65 kΩ

Required $R_2 = \dfrac{1.65 \text{ k}\Omega}{0.44} = 3.75$ kΩ

and

$R_1 = 2R_2 = 2(3.75 \text{ k}\Omega) = 7.5$ kΩ

Substituting these values into Equation 7.5 gives us

$$f_c = \frac{1}{2\pi\sqrt{R_1 R_2 C_1 C_2}} = \frac{1}{(6.28)\sqrt{(7.5 \text{ k}\Omega)(3.75 \text{ k}\Omega)(0.01 \text{ μF})(0.01 \text{ μF})}}$$
$$= 3 \text{ kHz}$$

This verifies the calculations since 3 kHz is the desired frequency. Figure 7–14 illustrates the completed design and amplitude response curve.

$f_c = 3$ kHz

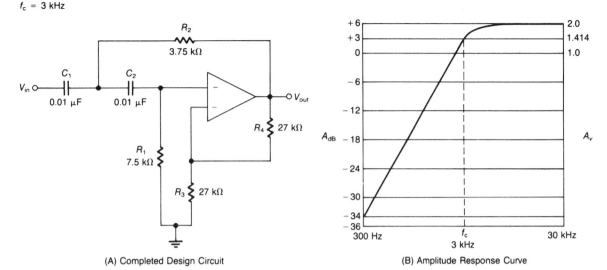

(A) Completed Design Circuit   (B) Amplitude Response Curve

Figure 7–14. Solution to Example 7.14

Equal-Component High-Pass VCVS Circuit

A basic design circuit for the equal-component second-order high-pass VCVS filter is made by interchanging the components in Figure 7–11. As with the equal-component VCVS low-pass filter, the passband gain is fixed at 1.59 in order to obtain a Butterworth response.

An obvious advantage of the equal-component filter over the unity-gain filter is the simplified component selection process. Also, $f_c = 1/(2\pi RC)$ as for the first-order filters, since $R_1 = R_2$ and $C_1 = C_2$. This simplified calculation further simplifies circuit design.

Example 7.15

Design an equal-component high-pass VCVS filter with a cutoff frequency of 2 kHz.

Solution

We will select $C_1 = C_2 = 0.01$ µF and determine the value for $R_1 = R_2$.

$$R_1 = R_2 = \frac{1}{2\pi f_c C} = \frac{1}{(6.28)(2000 \text{ Hz})(0.01 \times 10^{-6} \text{ F})} = 8 \times 10^3 \ \Omega$$
$$= 8000 \ \Omega$$

The completed circuit design and amplitude response curve are shown in Figure 7–15.

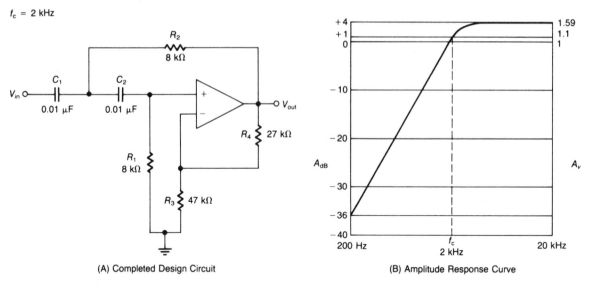

$f_c = 2$ kHz

(A) Completed Design Circuit   (B) Amplitude Response Curve

Figure 7–15.   Solution to Example 7.15

7.7 Developing High-Order Filters

It is not practical to construct efficient filters higher than second-order from a single op amp. Therefore, high-order filters must be formed by *cascading* first- and second-order filter sections. For example, a third-order filter would consist of one first-order filter and one second-order filter; a fourth-order filter would require two second-order filters; a fifth-order filter would have one first-order filter and two second-order filters, and so on.

When two or more second-order filter sections are cascaded, the sections are not identical. Each section must have a different damping factor if the filter is to produce a Butterworth response. Table 7–3 shows the damping factors

Table 7–3	High-Order Filter Damping Factors (α)			
	Order	First Section	Second Section	Third Section
	3	1.00	1.00	—
	4	1.85	0.77	—
	5	1.00	1.62	0.62
	6	1.93	1.41	0.52

required for several orders of filters. Note that all first-order sections have a damping factor of 1.00, so there is no problem with the first section of any odd-order filter. Second-order sections require more careful consideration to determine the component values necessary to produce the correct filtering action. Necessary calculations can be greatly reduced by using only equal-component VCVS second-order filter sections.

The method of deriving the damping factors is beyond the scope of this text. Advanced mathematics texts and filter design engineering manuals are recommended to those who wish to study the derivations.

Recall from earlier discussions that the relationship between the passband voltage gain A_v and the damping factor (α) of the equal-component VCVS second-order filter is

$$A_v = 3 - \alpha \tag{7.6}$$

and that the voltage gain for this type of filter is the same as for a noninverting amplifier:

$$A_v = 1 + \frac{R_f}{R_{in}} \tag{7.7}$$

Combining Equations 7.6 and 7.7 produces a simple formula for determining the values of R_f and R_{in}, as expressed by the following equation:

$$\frac{R_f}{R_{in}} = 2 - \alpha \tag{7.8}$$

where

α = the damping factor (specified in Table 7-3)

Example 7.16

Determine the values for R_f and R_{in} for the second section of a third-order low-pass filter.

Solution

Table 7-3 shows that the damping factor of the second section is 1.00. Using Equation 7.8,

$$\frac{R_f}{R_{in}} = 2 - \alpha = 2 - 1.00 = 1$$

Therefore, R_f must be equal to R_{in} for this filter section. Any standard resistor values can be used. Figure 7-16 shows a typical third-order low-pass filter, in which $R_5 = R_{in}$ and $R_6 = R_f$. Note that the second-order section is an equal-component VCVS circuit, selected for design simplification.

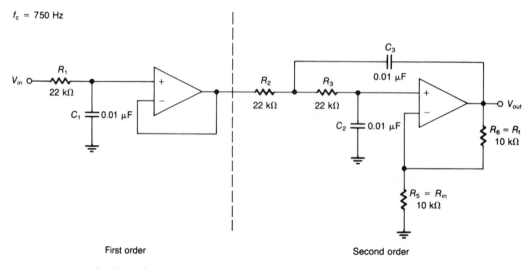

f_c = 750 Hz

First order   Second order

Figure 7–16. Third-Order Low-Pass Filter

Example 7.17

Determine the values for R_f and R_{in} for the first, second, and third sections of a sixth-order filter.

Solution

Remember that a sixth-order filter is made up of three second-order filter sections, and that we are using only equal-component VCVS sections.

First section:

$$\frac{R_f}{R_{in}} = 2 - \alpha = 2 - 1.93 = 0.07$$

If we let R_{in} = 100 kΩ, then

R_f = 100 kΩ × 0.07 ≈ 7.0 kΩ   (Use a standard 6.8 kΩ resistor.)

Second section:

$$\frac{R_f}{R_{in}} = 2 - \alpha = 2 - 1.41 = 0.59$$

If we let R_{in} = 47 kΩ, then

R_f = 47 kΩ × 0.59 ≈ 27 kΩ

Third section:

$$\frac{R_f}{R_{in}} = 2 - \alpha = 2 - 0.52 = 1.48$$

If we let $R_{in} = 15$ kΩ, then

$$R_f = 15 \text{ k}\Omega \times 1.48 \approx 22 \text{ k}\Omega$$

Figure 7–17 illustrates a typical sixth-order low-pass filter using the values calculated in Example 7.17. Interchanging the frequency-determining components would make this circuit a sixth-order high-pass filter.

Gain in High-Order Filters

The overall passband gain of high-order filters is based on the gain of the individual sections. The overall voltage gain is the PRODUCT of the voltage gain of the individual sections, and the overall decibel gain is the SUM of the individual decibel gains.

Example 7.18

Determine the overall voltage and decibel gains of the circuit in Figure 7–17.

Solutions

The voltage gain can be determined from either Equation 7.6 or 7.7 and the calculations of Example 7.17. In this case the simplest method is to use Equation 7.7. Since we have already calculated R_f/R_{in}, we can simply add *one* to the results found in Example 7.17:

$f_c = 750$ Hz

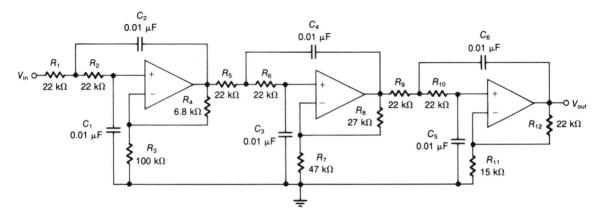

Figure 7–17.   Sixth-Order Low-Pass Filter

First section $A_v = 1 + 0.07 = 1.07$
Second section $A_v = 1 + 0.59 = 1.59$
Third section $A_v = 1 + 1.48 = 2.48$

Multiply to find the overall voltage gain:

$A_v = 1.07 \times 1.59 \times 2.48 = 4.22$

Now use Equation 7.1 to determine the overall decibel gain:

$A_{dB} = 20 \log_{10}A_v$
First section $A = 20 \log_{10}(1.07) = 0.587$ dB
Second section $A = 20 \log_{10}(1.59) = 4.028$ dB
Third section $A = 20 \log_{10}(2.48) = 7.889$ dB

Summing and rounding, we have

$A = 0.587 + 4.028 + 7.889 = 12.504 = 12.5$ dB

Using the methods demonstrated in Examples 7.17 and 7.18 will enable you to evaluate and design any order of filter.

Order of Filter Determination

Determining which order of filter to use is an important consideration when designing and constructing circuits. Practically, you should use only the order of filter necessary to meet circuit requirements. A higher order filter than necessary may be more expensive and require more space.

7.8 Bandpass Filters

A bandpass filter passes a specified range of frequencies and rejects frequencies above and below that range. Figure 7–1C shows a typical amplitude response curve for a bandpass filter. For evaluation, we are interested in the *center operating frequency* (f_{op}) and *bandwidth* (*BW*).

In general, the center operating frequency is the highest point of the amplitude response curve, or the point at which the maximum voltage gain is reached. The bandwidth is the difference between the upper frequency (f_{high}) and the lower frequency (f_{low}) where the voltage gain is down 3 dB (-3 dB), or 0.707 times its maximum voltage gain value.

$$BW = f_{high} - f_{low} \qquad (7.9)$$

In general terms, if the bandwidth is less than 10 percent of the center operating frequency, the filter is considered to be a *narrow-bandpass* filter. Conversely, if the bandwidth is greater than 10 percent of f_{op}, the filter is considered to be a *wide-bandpass* filter.

Selectivity and Circuit *Q*

A narrow bandwidth means high selectivity, and a wide bandwidth means low selectivity. The amount of selectivity is expressed by the letter Q, the quality factor, of the circuit. The relationship between the bandpass filter's bandwidth and center operating frequency is expressed as follows:

$$Q = \frac{f_{op}}{BW} \tag{7.10}$$

By manipulation of the equation, we have

$$BW = \frac{f_{op}}{Q} \tag{7.11a}$$

and

$$f_{op} = Q(BW) \tag{7.11b}$$

In general, the higher the Q of the circuit, the larger the output voltage. A comparison between high-Q and low-Q bandpass filter frequency response curves is illustrated in Figure 7–18.

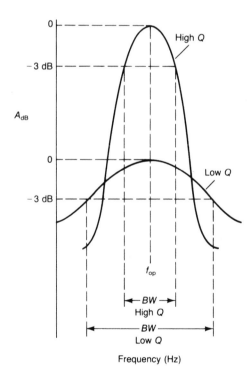

Figure 7–18.   Comparison between High-Q and Low-Q Bandpass Filters

Bandpass Filter Construction

By combining low-pass and high-pass filter networks with selected frequencies, an active bandpass filter can be constructed, as shown in Figure 7–19. Low-pass filtering is accomplished by R_1C_1, and high-pass filtering is accomplished by R_2C_2. Feedback resistor R_3 establishes the gain of the circuit, but it also affects the Q of the circuit and the center operating frequency. A large value for R_3 will result in a high f_{op} and a low Q. The center operating frequency may be found as follows:

$$f_{op} = \frac{1}{2\pi\sqrt{R_pR_3C_1C_2}} \tag{7.12}$$

where

$$R_p = \frac{R_1R_2}{R_1 + R_2}$$

If $C_1 = C_2$, then the Q of the circuit is

$$Q = 0.5\sqrt{\frac{R_3}{R_p}} \tag{7.13}$$

The center operating frequency voltage gain is

$$A_v = \frac{R_3}{R_1\left(1 + \dfrac{C_1}{C_2}\right)} \tag{7.14}$$

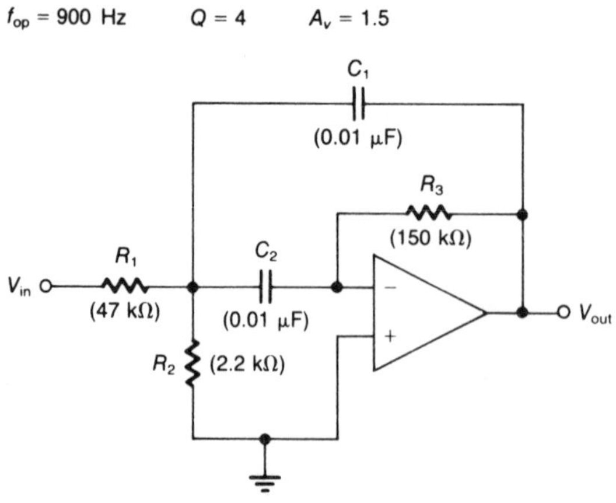

$f_{op} = 900$ Hz $Q = 4$ $A_v = 1.5$

Figure 7–19. Active Bandpass Filter (The values in parentheses are the solutions to Example 7.19.)

Designing an active bandpass filter is simplified by making $C_1 = C_2$, selecting standard capacitor values, and preselecting f_{op}, Q, and A_v. Then resistor values may be calculated as follows:

$$R_1 = \frac{Q}{2\pi f_{op} C A_v} \tag{7.15}$$

$$R_2 = \frac{Q}{2\pi f_{op} C (2Q^2 - A_v)} \tag{7.16}$$

$$R_3 = \frac{2Q}{2\pi f_{op} C} \tag{7.17}$$

Example 7.19

Design a bandpass filter with $f_{op} = 900$ Hz, $Q = 4$, and $A_v = 1.5$. Refer to Figure 7–19.

Solution

First, select a standard value for C_1 and C_2, say 0.01 μF. Then determine the resistor values using Equations 7.15, 7.16, and 7.17.

$$R_1 = \frac{Q}{2\pi f_{op} C A_v} = \frac{4}{(6.28)(900\text{ Hz})(0.01 \times 10^{-6}\text{ F})(1.5)} \approx 47\text{ k}\Omega$$

$$R_2 = \frac{Q}{2\pi f_{op} C (2Q^2 - A_v)} = \frac{4}{(6.28)(900\text{ Hz})(0.01 \times 10^{-6}\text{ F})\,2(16) - 1.5}$$
$$\approx 2.3\text{ k}\Omega$$

$$R_3 = \frac{2Q}{2\pi f_{op} C} = \frac{2(4)}{(6.28)(900\text{ Hz})(0.01 \times 10^{-6}\text{ F})} \approx 141\text{ k}\Omega$$

Figure 7–19 shows the completed design circuit with standard resistor values given in parentheses.

7.9 Wideband Filters

Sometimes, such as in voice communications systems, it is desirable to pass a wider range of frequencies than is possible with the standard bandpass filter. Typically, such systems pass the range of frequencies between 300 Hz and 3 kHz. Designing such a filter is simple; cascade a low-pass filter with a f_c of 3 kHz and a high-pass filter with a f_c of 300 Hz. The design is simplified further by using equal-component VCVS filters. It makes no difference which filter comes first in the cascading process.

**Example
7.20**

Design a wideband filter that will pass the range of frequencies between 500 Hz and 2.5 kHz.

Solution

Using equal-component second-order VCVS filters, we will first select a standard capacitor value, such as 0.022 μF. We can use the same value for both the low-pass (LP) and high-pass (HP) sections. Then, solving for R,

$$R_{LP} = \frac{1}{2\pi f_c C} = \frac{1}{(6.28)(500 \text{ Hz})(0.022 \times 10^{-6} \text{ F})} \approx 14.5 \text{ k}\Omega$$

$$R_{HP} = \frac{1}{2\pi f_c C} = \frac{1}{(6.28)(2500 \text{ Hz})(0.022 \times 10^{-6} \text{ F})} \approx 3 \text{ k}\Omega$$

Recall that the voltage gain of equal-component VCVS filters is set at 1.59. Now we can select standard values for the input and feedback resistors, such as 27 kΩ and 47 kΩ, respectively. We will use standard 15 kΩ resistors for R_{LP}. Figure 7–20 shows the completed design.

7.10 Notch Filters

The notch filter, sometimes referred to as a band-reject filter, operates in reverse of the bandpass filter; that is, it rejects a certain frequency or band of frequencies while passing all others. Typical applications are in audio and instrumentation

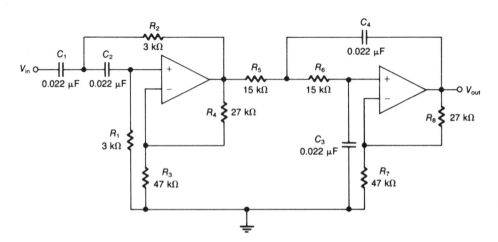

Figure 7–20. Completed Design Circuit for Example 7.20

systems where a 60 Hz line frequency or a 400 Hz motor-generated frequency can cause interference noise and on telephone lines to null out control frequencies.

Twin-T Notch Filter

A popular notch filter is the *twin-T* notch filter shown in Figure 7–21. The circuit consists of a passive twin-T filter connected to an op amp voltage follower. In general, this circuit is only good for frequencies below 1 kHz, because of stray capacitances at higher frequencies. The circuit has unity gain at frequencies above and below the *null*, or center, frequency.

Null Frequency

The null frequency is that frequency at which maximum rejection occurs, and is determined as follows:

$$f_{\text{null}} = \frac{1}{2\pi R_1 C_1} \tag{7.18}$$

where

$$R_1 = R_2 = 2R_3$$
$$C_1 = C_2$$
$$C_3 = 2C_1$$

$(f_{\text{null}} = 60 \text{ Hz})$

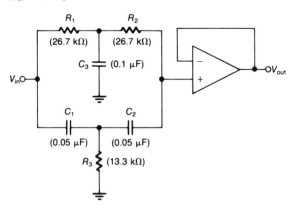

Figure 7–21. Active Twin-T Notch Filter (The values in parentheses are the solutions to Example 7.21.)

Example 7.21

Design a twin-T active notch filter with f_{null} = 120 Hz.

Solution

For ease of design, select some standard capacitor value for C_3 that will allow another standard value to be used for C_1 and C_2. Such values could be C_1 = C_2 = 0.05 μF and C_3 = $2C_1$ = 0.1 μF. Next, solve for R_1:

$$R_1 = \frac{1}{2\pi f_{null}C_1} = \frac{1}{(6.28)(120 \text{ Hz})(0.05 \times 10^{-6} \text{ F})} = 26.5 \text{ k}\Omega$$

(Use a 1 percent 26.7 kΩ resistor.) Since R_1 = R_2 = $2R_3$, $R_3 \approx 13.3$ kΩ. Use a 1 percent 13.3 kΩ resistor. Figure 7–21 shows the completed design circuit with the values calculated in this example given in parentheses.

Other Notch Filter Circuits

If gain other than unity is desired, the circuit of Figure 7–22 may be used. In this circuit, gain would be determined as for any noninverting amplifier; that is, A_v = 1 + (R_f/R_{in}).

Another notch filter design is shown in Figure 7–23 in which amplifier circuit A_1 is a basic bandpass filter, and amplifier circuit A_2 operates as a summing amplifier. At point X/A the input signal to A_2 is subtracted from the output signal

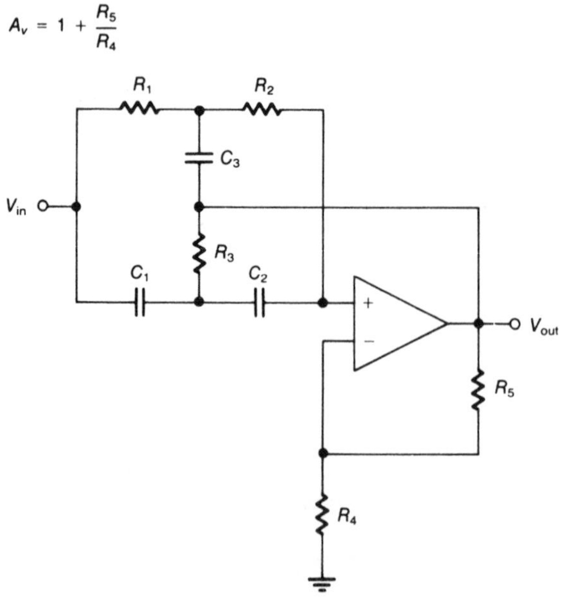

$$A_v = 1 + \frac{R_5}{R_4}$$

Figure 7–22. Active Twin-T Notch Filter with Gain

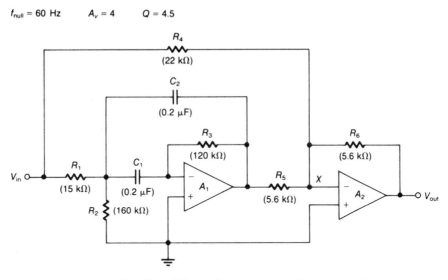

f_{null} = 60 Hz A_v = 4   Q = 4.5

Figure 7–23. Two-Op Amp Notch Filter (The values in parentheses are the solutions to Example 7.22.)

of A_1. A deep null can be obtained from this circuit by setting $C_1 = C_2$ and setting the passband gain of $A_1 = R_4/R_5$.

Example 7.22

Use Figure 7–23 to design a notch filter in which f_{null} = 60 Hz, A_v = 4, and Q = 4.5.

Solution

First we will select a standard capacitor value, such as 0.2 µF. Then using Equations 7.15, 7.16 and 7.17 where $f_{op} = f_{null}$, we solve for resistor values:

$$R_1 = \frac{Q}{2\pi f_{null}CA_v}$$

$$= \frac{4.5}{(6.28)(60 \text{ Hz})(0.2 \times 10^{-6} \text{ F})(4)} \approx 15 \text{ k}\Omega$$

$$R_2 = \frac{Q}{2\pi f_{null}C(2Q^2 - A_v)}$$

$$= \frac{4.5}{(6.28)(60 \text{ Hz})(0.2 \times 10^{-6} \text{ F})[2(20.25) - 4]} \approx 160 \text{ }\Omega$$

$$R_3 = \frac{2Q}{2\pi f_{null}C}$$

$$= \frac{2(4.5)}{(6.28)(60 \text{ Hz})(0.2 \times 10^{-6} \text{ F})} \approx 120 \text{ k}\Omega$$

Now, since the passband gain of A_1 is $A_v = 4$, the ratio R_4/R_5 must also be 4. Therefore, we will select standard resistor values such that R_4 can equal $4R_5$. We will select $R_5 = 5.6\ \text{k}\Omega$ and $R_4 = 22\ \text{k}\Omega$. Also we will want unity gain in A_2, so we will make $R_6 = R_5 = 5.6\ \text{k}\Omega$. The completed circuit design shown in Figure 7–23 gives the values for this example in parentheses.

7.11   Multiple-Feedback Filters

Multiple-feedback filters derive their name from the fact that they have an additional feedback path. Figure 7–24 shows a basic second-order low-pass multiple-feedback filter design circuit, in which R_3 is the additional feedback path. Also note that the input is connected to the inverting terminal of the op amp. The cutoff frequency for this circuit is 1 kHz and is determined by the following equation:

$$f_c = \frac{1}{2\pi\sqrt{R_2 R_3 C_1 C_2}} \qquad (7.19)$$

The passband gain is

$$A_v = \frac{R_3}{R_1} \qquad (7.20)$$

Designing the Multiple-Feedback Filter

The additional feedback component does not necessarily complicate the design of the multiple-feedback filter. For this circuit to have a Butterworth response, A_v must equal unity. Therefore, all three resistors will be equal in value. Also note that C_1 is equal to approximately $4.5 C_2$ in Figure 7–24. Scaling will produce the desired component values.

Example
7.23

Refer to Figure 7–24. Design a second-order multiple-feedback filter with f_c = 250 Hz and $A_v = 1$. (Refer to Table 7–2 as necessary.)

Solution

Since 1 kHz/250 Hz = 4, we multiply all frequency-determining resistors by 4, so that $R_1 = R_2 = 20\ \text{k}\Omega$, and with unity gain, R_3 also will equal 20 kΩ. Then we will select capacitor values such that $C_1 = 4.5 C_2$, or $C_2 = 0.022\ \mu\text{F}$ and $C_1 = 0.1\ \mu\text{F}$. Then 0.022 μF/0.015 μF = 1.47, so we divide the new resistor values by that ratio, or 20 kΩ/1.47 = 13.6 kΩ. Using 1 percent 13.7 kΩ resistors would satisfy the circuit requirements. Figure 7–25 shows the completed circuit design.

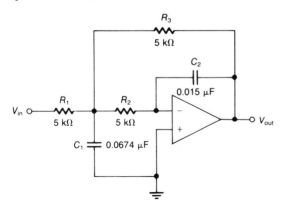

Figure 7–24. Basic Design Circuit for a
Second-Order Low-Pass
Multiple-Feedback Filter

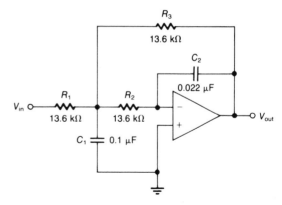

Figure 7–25. Completed Design Circuit for
Example 7.23

The multiple-feedback high-pass filter is formed by simply interchanging
the components of the low-pass filter, as shown in Figure 7–26. Here, the
additional feedback path is through C_3, and $R_2 = 4.5R_1$. The cutoff frequency
is determined by the following equation:

$$f_c = \frac{1}{2\pi\sqrt{R_1 R_2 C_2 C_3}} \qquad (7.21)$$

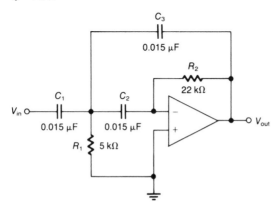

Figure 7–26. Basic Design Circuit for a
Second-Order High-Pass
Multiple-Feedback Filter

The passband gain of this circuit is determined by the ratio of two capacitors rather than by two resistors, which makes this circuit unique in op amp circuits. The gain is

$$A_v = \frac{C_3}{C_1} \tag{7.22}$$

Circuit design is accomplished in the same manner as was the design of the low-pass multiple-feedback filter, that is, by scaling the basic filter design circuit shown in Figure 7–26.

7.12 State-Variable Filters

The *state-variable* filter is a different type of multiple-feedback filter, because it uses three or four op amps, has one input, and has three outputs that provide low-pass, high-pass, and bandpass filtering simultaneously. Figure 7–27 shows the block diagram of the basic state-variable filter that consists of a summing amplifier, two integrators, and a damping network.

Designing the State-Variable Filter

For a Butterworth response from this circuit, damping is set by the resistor network, R_3–R_4. The basic design circuit for a state-variable filter is shown in Figure 7–28. Designing such a circuit may look difficult, but by making a few assumptions, we can greatly simplify the process. The best procedure is to choose standard values so that $C = C_1 = C_2$. And if we let $R_5 = R_6$ and $R_1 = R_2 = R_3 = R_7 = R_8 = R_9$, we have the following formulas,

$$R_5 = \frac{1}{2\pi f_{\text{op}}C} \tag{7.23}$$
$$R_4 = R_3(3Q - 1) \tag{7.24}$$

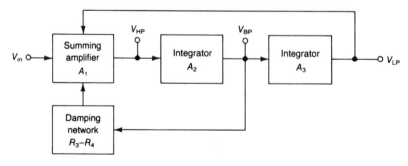

Figure 7–27. Block Diagram of a Basic State-Variable Filter

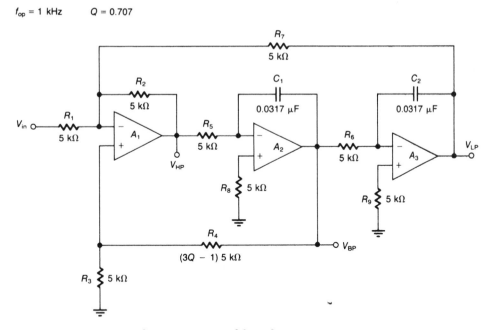

$f_{op} = 1$ kHz $Q = 0.707$

Figure 7–28.   Basic Design Circuit for a State-Variable Filter

Example 7.24

Design a state-variable filter with a Butterworth response in which $f_{op} = 500$ Hz, $Q = 0.707$, $C = 0.22$ µF, and $R_7 = 15$ kΩ.

Solution

From Equation 7.23,

$$R_5 = R_6 = \frac{1}{2\pi f_{op} C} = \frac{1}{(6.28)(500 \text{ Hz})(0.22 \times 10^{-6} \text{ F})} \approx 1.5 \text{ k}\Omega$$

From Equation 7.24 and knowing Q must equal 0.707,

$$R_4 = R_3(3Q - 1) = (15 \text{ k}\Omega)[3(0.707) - 1] = 16.8 \text{ k}\Omega$$

Figure 7–29 shows the completed circuit design.

Universal State-Variable Filter

Figure 7–30 shows a *universal state-variable* filter constructed with the LF347 wide bandwidth quad JFET-input op amp. This circuit has $f_{op} = 3$ kHz, $f_{null} = 9.5$ kHz, $Q = 3.4$, and passband gains are: $A_{v(HP)} = 0.1$, $A_{v(LP)} = 1$, $A_{v(BP)} = 1$, and $A_{v(null)} = 10$.

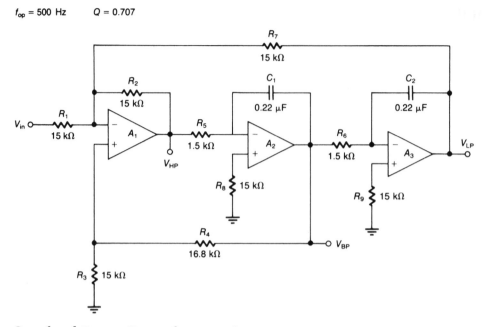

Figure 7–29. Completed Design Circuit for Example 7.24

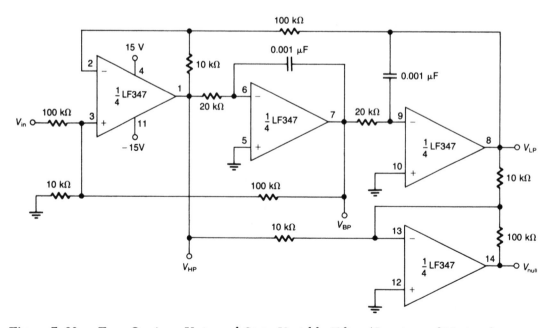

Figure 7–30. Four-Op Amp Universal State-Variable Filter (Courtesy of National Semiconductor Corporation, © copyright 1980.)

7.13 Summary

1. Active filters offer advantages over passive filters, such as lower cost, ease of adjustment, no insertion loss, and isolation between source and load.

2. Limitations and disadvantages of active filters are the frequency response limits of the op amps and the requirement for power supplies.

3. Filters are classified according to the function they perform.

4. A decade is a tenfold increase or decrease in frequency, and an octave is a doubling or halving of a frequency.

5. A first-order filter has a rolloff of 20 dB per decade, or 6 dB per octave.

6. Higher-order filters have rolloffs in direct proportion to their order; for example, a second-order has two times the rolloff of a first-order, and a fifth-order has a rolloff five times that of the first-order. These rolloffs would result in 40 dB per decade and 100 dB per decade, respectively.

7. The Butterworth response, a relatively flat response at frequencies above cutoff, occurs in highly damped filters.

8. Most filter circuit designs can be simplified by first selecting standard capacitor values and then calculating resistor values.

9. Scaling is a method of circuit design that starts with a basic circuit normalized with an input impedance of 5 kΩ and a f_c of 1 kHz. This design method is convenient in multiple-feedback circuit design.

10. A restriction on equal-component VCVS second-order filters is that the passband gain be fixed at 1.59 if a Butterworth response is to be obtained.

11. The simplest second-order filter circuit design is the equal-component second-order VCVS filter, low-pass or high-pass, since component values are equal.

12. It is not practical to use a single op amp in filters of a higher order than second-order. Such high-order filters are constructed by cascading first- and second-order filter sections.

13. The overall passband gain of high-order filters depends on the gain of the individual filter sections. The voltage gain is the product of the voltage gains of the individual sections, and the decibel gain is the sum of the decibel gains of the individual sections.

14. The order of filter to be used in any application should be determined by circuit requirements only.

15. A narrow-bandpass filter is one in which the bandwidth is less than 10 percent of f_{op}. It has a high Q, which means high selectivity, and high voltage gain.

16. A wide-bandpass filter is one in which the bandwidth is greater than 10 percent of f_{op}. It has a low Q, which means low selectivity, and low voltage gain.

17. Wideband filters can be constructed by cascading a low-pass filter and a high-pass filter. It makes no difference which filter is first in the cascade.

18. Notch filters can be constructed by connecting a passive twin-T filter to an op amp voltage follower. Such a circuit is good for frequencies below 1 kHz.

19. Multiple-feedback filters have more than one feedback path, and the input is connected to the inverting terminal of the op amp.

20. State-variable filters use three op amps and provide three outputs that consist of low-pass, high-pass, and bandpass filtering.

21. The universal state-variable filter uses four op amps and provides four outputs that consist of low-pass, high-pass, bandpass, and notch filtering.

7.14 Questions and Problems

7.1 Define *active filter*.

7.2 State the distinguishing features of (a) the low-pass filter and (b) the high-pass filter.

7.3 Design a first-order low-pass filter with $f_c = 100$ Hz and $A_v = 3$. Determine the decibel gain of the circuit and the break point.

7.4 Design a first-order high-pass filter with $f_c = 400$ Hz and $A_v = 2$. Determine the decibel gain of the circuit and the break point.

7.5 Design a second-order VCVS low-pass filter with $f_c = 100$ Hz. What is the decibel gain and the break point?

7.6 Design a second-order high-pass VCVS filter with $f_c = 400$ Hz. What is the decibel gain and the break point?

7.7 A second-order VCVS high-pass filter section with $f_c = 300$ Hz is cascaded with a second-order low-pass filter section with $f_c = 3$ kHz. (a) What kind of filter circuit is this? (b) What is the overall voltage gain? (c) What is the overall decibel gain?

(d) What is the bandwidth? (e) What is the overall order of the circuit?

7.8 What would you use to construct a sixth-order high-pass filter?

7.9 What would be the overall voltage gain and decibel gain of the circuit of Problem 7.8? (HINT: Refer to Table 7–3.)

7.10 A bandpass filter has $Q = 5$ and $BW = 200$ Hz. What is f_{op}?

7.11 A bandpass filter has $f_{op} = 2$ kHz and $f_{low} = 1.8$ kHz. What is (a) f_{high} and (b) Q?

7.12 Describe the state-variable filter and identify its sections by function.

7.13 A 60 Hz notch filter has unity gain and a bandwidth of 14 Hz. What is the lower cutoff frequency and the upper cutoff frequency?

7.14 What is the dB break point for the notch filter of Problem 7.13?

7.15 Design a state-variable filter with a Butterworth response where $f_{op} = 750$ Hz. Select your own values for components.

Chapter 8

Signal Processing Circuits

8.1 Introduction

Signal processing and conditioning are important functions of electronic circuits. The op amp, with its associated external components, is an excellent device to use as a basic signal processing and conditioning circuit. We have studied many such circuits in previous chapters, but in this chapter we will look at the types of circuits used in process control systems. Such systems require that certain signals be conditioned, or processed, in some form so that the resulting signal can be used as a control signal. Chapter 9 will provide a more in-depth study of process control systems.

8.2 Objectives

When you complete this chapter, you should be able to:

- [ ] Define the comparator and explain its operation.
- [ ] Explain the operation of the Schmitt trigger and distinguish the difference between it and a comparator.
- [ ] Design a peak detector and explain its operation.
- [ ] Design a window detector and explain its operation.

☐ Select a precision rectifier for a specific application.

☐ Discuss instrumentation amplifiers and their uses.

☐ Discuss transducers, their uses, and list several types of transducers.

8.3 Comparators

The comparator as a sensing device was discussed briefly in Chapter 3. Recall that the comparator is not used to amplify signals, but is a circuit that compares an input voltage at one terminal to a reference voltage, V_{ref}, at the other terminal. The reference voltage can be positive, negative, or zero. The output voltage of the op amp comparator swings to $\pm V_{sat}$, depending upon the comparison of the two input voltages. The output signal polarity is dependent upon the input terminal to which V_{ref} is applied. A simple op amp comparator is shown in Figure 8–1A, in which the zero volt reference voltage is applied to the $(-)$ input terminal. If V_{in} is greater than V_{ref}, then V_{out} swings to $+V_{sat}$. If V_{in} is less than V_{ref}, then V_{out} swings to $-V_{sat}$. Figure 8–1B shows the idealized results of this action.

Saturation Voltage

Saturation voltage, $\pm V_{sat}$, is normally about one volt less than the applied power supply voltages of the op amp. For example, the op amp in Figure 8–1A has supply voltages of $+15$ V and -15 V applied to it. The saturation voltage would then be approximately ± 14 V, as shown in Figure 8–1B.

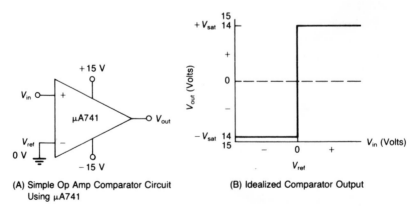

(A) Simple Op Amp Comparator Circuit
Using μA741

(B) Idealized Comparator Output

Figure 8–1.   Simplified Comparator Circuit

Constructing Comparator Circuits

A comparator can be constructed with any op amp operated in the *open-loop* mode, that is, with no feedback. However, many applications of comparators require interfacing with digital devices, so their output swing must be compatible with the logic levels of devices they drive. Those logic levels are normally between 0 V and +5 V, so the simple op amp comparator of Figure 8–1A could not be used in such an application, since the output swing of ±14 V would destroy the driven digital device. Therefore, manufacturers provide linear ICs with special characteristics, designed especially to operate as comparators.

Comparator Definitions

Comparators have their own set of terms and definitions. Several of those definitions are as follows:

Input bias current—the average of the two input currents with no signal applied.

Input offset current—the absolute value of the difference between the two input currents for which the output will be driven to change states.

Input offset voltage—the absolute value of the voltage between the input terminals required to make the output voltage greater than or less than some specified value.

Input voltage range—the range of voltage on the input terminals (common-mode) over which the offset specifications apply.

Logic threshold voltage—the voltage at the output of the comparator at which the driven logic circuitry changes its digital state. This definition relates to the operation of digital devices.

Response time—the interval between the application of an input step function and the time when the output crosses the logic threshold voltage. (This parameter is similar to the propagation delay of standard op amps.)

Strobe current—the current out of the strobe terminal when it is at the zero logic level.

Strobed output level—the dc output voltage, independent of input conditions, with the voltage on the strobe terminal equal to or less than the specified low state.

Strobe "ON" voltage—the maximum voltage on either strobe terminal required to force the output to the specified high state independent of the input voltage.

Strobe "OFF" voltage—the minimum voltage on the strobe terminal that will guarantee that it does not interfere with the operation of the comparator.

Strobe release time—the time required for the output to rise to the logic threshold voltage after the strobe terminal has been driven from zero to the high logic level.

Strobing—a method used to enable or to disable the comparator. When strobed OFF, the comparator is disabled and the output will not respond to an input signal. Conversely, when strobed ON, the comparator is enabled and the output responds to an input signal. Figure 8–2 shows a strobed LM311 comparator. Additional definitions may be obtained from manufacturer's data books.

Effects of Slew Rate

The open-loop operation of an op amp increases the bandwidth and unity gain of the device. No compensation circuit is needed; therefore, the device has an increased slew rate that results in a faster output change. The faster change in output means the comparator has the ability to more closely follow rapid changes of signals at the input. The time it takes for a comparator output to change from one supply voltage to the other is determined by dividing the change in output voltage, ΔV, by the slew rate, SR, of the device. As a formula, this relationship is expressed by the following equation:

$$t = \frac{\Delta V}{SR} \tag{8.1}$$

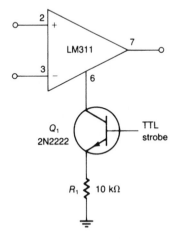

(*Note:* Do not ground strobe pin.)

Figure 8–2. Strobed LM311 Comparator (Courtesy of National Semiconductor Corporation, © copyright 1980.)

Example
8.1

Refer to Figure 8–1A. The slew rate of the μA741C is 0.5 V/μs. With the supply voltages in the figure, how long will it take for the output to swing from $+V_{\text{sat}}$ to $-V_{\text{sat}}$?

Solution

Use Equation 8.1:

$$t = \frac{\Delta V}{SR} = \frac{+14\text{ V} - (-14\text{ V})}{0.5\text{ V/μs}} = \frac{28\text{ V}}{0.5\text{ V}}\text{ μs} = 56\text{ μs}$$

The rate of change in Example 8.1 is very slow, so the circuit using the μA741C would be limited. Now let us look at an op amp with a higher slew rate.

Example
8.2

An LM318 op amp has a typical slew rate of 70 V/μs. If supply voltages of ±15 V are applied, how long will it take for the output to swing from $+V_{\text{sat}}$ to $-V_{\text{sat}}$?

Solution

Use Equation 8.1:

$$t = \frac{\Delta V}{SR} = \frac{+14\text{ V} - (-14\text{ V})}{70\text{ V/μs}} = \frac{28\text{ V}}{70\text{ V}}\text{ μs} = 0.40\text{ μs}$$

Obviously, the LM318 is preferable over the μA741C for use as a comparator, because the time required for the output swing is about 140 times faster in the LM318. It should be just as obvious then that devices designed specifically to operate as comparators, with their faster response times, would be even more desirable.

Evaluating a Comparator Circuit

As previously discussed, a comparator is used to detect a changing voltage on one input with a reference voltage on the other input. Refer to Figure 8–3A. Assume that a sine wave is applied to the noninverting input terminal and the inverting input terminal is grounded, resulting in a 0 V reference voltage. When the input signal is positive, the output is at $+V_{\text{sat}}$. When the input signal swings through the zero point going negative, the output swings to $-V_{\text{sat}}$. The output is in phase with the input. Figure 8–3B shows the input/output relationship. If

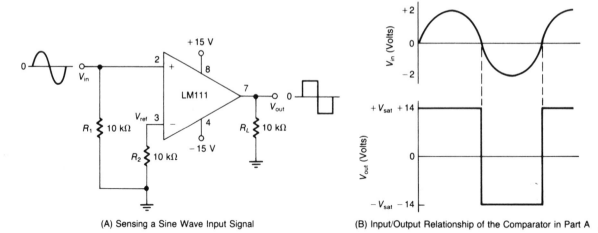

(A) Sensing a Sine Wave Input Signal   **(B)** Input/Output Relationship of the Comparator in Part A

Figure 8–3. Noninverting Comparator Circuit (Part A: Courtesy of National Semiconductor Corporation, © copyright 1980.)

the input signals were reversed, the output would be 180° out of phase with the input.

Zero Detector

The circuit in Figure 8–3A is sometimes referred to as a *zero detector*, because each time the input signal crosses the zero point, the output swings to the opposite polarity. The circuit also is sometimes called a *squaring circuit*, because the sine wave at the input is converted to a square wave output.

Phase Difference Detection

The common-mode rejection characteristic of the comparator provides the ability to detect phase differences. If two input signals are of the same frequency, as shown in Figure 8–4A, the output results detect the phase differences caused by a differential voltage at the inputs whenever the two signals are out of phase, as shown in Figure 8–4B. When V_2 is more positive than V_1, output will be $+V_{sat}$, and when V_1 is more positive than V_2, the output will be $-V_{sat}$. When the two input signals are in phase, the CMRR (common-mode rejection ratio) characteristic causes the output to be zero.

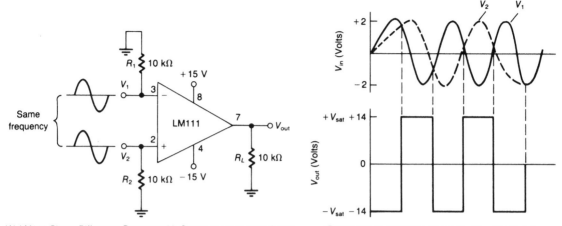

(A) LM111 Phase-Difference Detector with Common-mode Input Signal

(B) Input/Output Relationship of the Circuit in Part A

Figure 8–4.   Phase-Difference Detector Circuit (Part A: Courtesy of National Semiconductor Corporation, © copyright 1980.)

8.4 Schmitt Trigger

If noise is present at either of the input terminals of a comparator, false output signals are generated. Figure 8–5 shows the output results of a noninverting zero-crossing comparator with noise riding on the sine wave input. Positive feedback, shown in Figure 8–6, overcomes the noise problem. The feedback is

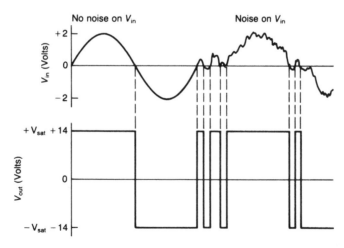

Figure 8–5.   Effect of Noise on a Zero-Crossing Detector

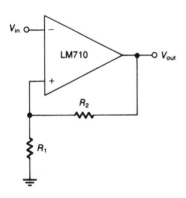

Figure 8–6. LM710 Comparator with Positive Feedback (Courtesy of National Semiconductor Corporation, © copyright 1980.)

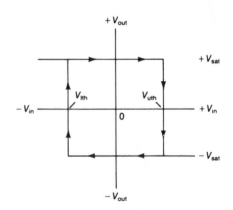

Figure 8–7. Input/Output Characteristics of Hysteresis Voltage

accomplished by taking a small fraction of the output voltage and feeding it back to the (+) terminal, creating a reference voltage dependent upon V_{out}. Such a configuration is called a *Schmitt trigger*.

When $V_{\text{out}} = +V_{\text{sat}}$, the positive feedback voltage is called the *upper threshold voltage*, $+V_{\text{uth}}$, and is expressed as follows:

$$V_{\text{uth}} = V_{\text{out}}\left(\frac{R_1}{R_1 + R_2}\right) + V_{\text{ref}} \tag{8.2}$$

Conversely, when $V_{\text{out}} = -V_{\text{sat}}$, the positive feedback voltage is called the *lower threshold voltage*, V_{lth}, and is expressed as follows:

$$V_{\text{lth}} = V_{\text{ref}} - V_{\text{out}}\left(\frac{R_1}{R_1 + R_2}\right) \tag{8.3}$$

Example 8.3

Refer to Figure 8–6. Assume $R_1 = 1\ \text{k}\Omega$, $R_2 = 100\ \text{k}\Omega$, $V_{\text{ref}} = 0\ \text{V}$, and $\pm V_{\text{sat}} = \pm 5\ \text{V}$. Find V_{uth} and V_{lth}.

Solution

Use Equation 8.2:

$$V_{\text{uth}} = V_{\text{out}}\left(\frac{R_1}{R_1 + R_2}\right) + V_{\text{ref}} = 5\ \text{V}\left(\frac{1\ \text{k}\Omega}{1\ \text{k}\Omega + 100\ \text{k}\Omega}\right) + 0\ \text{V}$$

$$= 49.5\ \text{mV}$$

Use Equation 8.3:

$$V_{lth} = V_{ref} - V_{out}\left(\frac{R_1}{R_1 + R_2}\right) = 0 \text{ V} - 5 \text{ V}\left(\frac{1 \text{ k}\Omega}{1 \text{ k}\Omega + 100 \text{ k}\Omega}\right)$$
$$= -49.5 \text{ mV}$$

Hysteresis

The difference in voltage between the upper threshold voltage and the lower threshold voltage is called the *hysteresis voltage*, V_H. It is expressed in equation form as follows:

$$V_H = V_{uth} - V_{lth} \tag{8.4}$$

Hysteresis produces two switching points instead of one. The resultant switching action is sometimes called *snap action*, because of the rapid change of output states when the threshold voltages have been reached; in effect, the output voltage snaps to the opposite state.

Hysteresis can be illustrated on a single graph, as in Figure 8–7. In this graph we show V_{in} on the horizontal axis and V_{out} on the vertical axis. When V_{in} is less than V_{uth}, $V_{out} = +V_{sat}$. When V_{in} increases to V_{uth}, the output snaps to $-V_{sat}$. When V_{in} drops back to V_{lth}, V_{out} snaps back to $+V_{sat}$. With a 0 V reference and symmetrical $\pm V_{sat}$, the output voltages will be symmetrical.

By designing the circuit so that hysteresis voltage is greater than the peak-to-peak noise voltage of the input signal, false output signal swings are eliminated. Therefore, we can say that V_H provides the *limits* of the noise levels that the circuit can withstand.

Example 8.4

Refer to Example 8.3. Calculate V_H.

Solution

Use Equation 8.4:

$$V_H = V_{uth} - V_{lth} = 49.5 \text{ mV} - (-49.5 \text{ mV}) = 99 \text{ mV}$$

Therefore, the circuit of Figure 8–6 can withstand a maximum peak-to-peak noise voltage of 99 mV superimposed on the input signals without false output signal switching.

8.5 Peak Detector

The *peak detector* shown in Figure 8–8A is a circuit that *detects* and *remembers* the peak value of an input signal. Momentarily shorting capacitor C_1 to ground with RESET switch S_1 sets the output to zero. When a positive voltage is applied to the noninverting input terminal, the output voltage of the op amp forward biases diode D_1, and C_1 is charged until the inverting input voltage equals the noninverting input voltage.

The time that the peak value is held is determined by the time constant of capacitor C_1 and load impedance. Placing a buffer amplifier with its high input impedance between the peak detector and the load, as shown in Figure 8–8B, will extend the holding time. (Peak detector action is similar to that of the sample-and-hold circuit to be discussed in the next chapter.)

**Example
8.5**

Refer to Figure 8–8A. Calculate the peak value holding time.

Solution

Using the standard time constant formula $t = RC$,

$$t = 10 \text{ M}\Omega \times 100 \text{ }\mu\text{F} = (10 \times 10^6 \text{ }\Omega)(100 \times 10^{-6} \text{ F}) = 1000 \text{ s}$$

Peak detectors can provide either positive or negative peak detection, as shown in Figure 8–9. Here we use LM311 voltage comparators with LM310 voltage followers to construct the circuits. Note that the diode used in the basic circuits previously discussed is not required for these circuits.

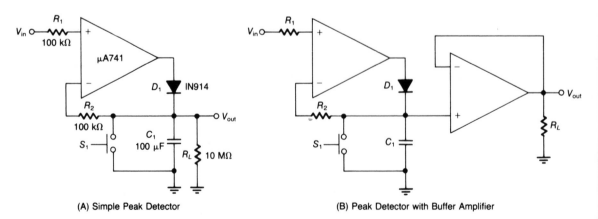

(A) Simple Peak Detector   (B) Peak Detector with Buffer Amplifier

Figure 8–8. Peak Detectors

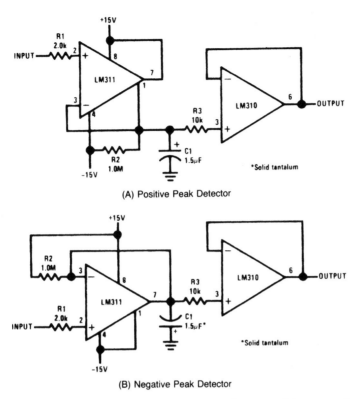

(A) Positive Peak Detector

(B) Negative Peak Detector

Figure 8–9. Standard Peak Detectors (Courtesy of National Semiconductor Corporation, © copyright 1980.)

8.6 Window Detector

A circuit called a *window detector*, designed to monitor an input voltage and indicate predetermined upper and lower limits, is shown in Figure 8–10A. Such a circuit is sometimes referred to as a *double-ended limit detector*.

A typical application for the window detector in Figure 8–10A is for limiting IC logic power supplies to the standard +5.0 V and 0 V. For this circuit, V_{out} = +5 V when V_{in} is greater than V_{lth} and less than V_{uth}, and V_{out} = 0 V when V_{in} is less than V_{lth} and greater than V_{uth}. The output of this single monolithic chip is compatible with RTL (resistor-transistor logic), DTL (diode-transistor logic), and TTL (transistor-transistor logic). The device is also capable of driving lamps and relays at currents up to 25 mA.

Another example of the window detector is shown in Figure 8–10B, where standard op amps are used. This circuit monitors an input voltage of +5 V, ±10 percent, and lights indicator light-emitting diode (LED) L_1 when the input voltage exceeds these limits. V_{uth} is +5.5 V and V_{lth} is +4.5 V. If V_{in} exceeds V_{uth}, the output of A_1 swings negative, forward biasing the LED, and L_1 lights. If V_{in}

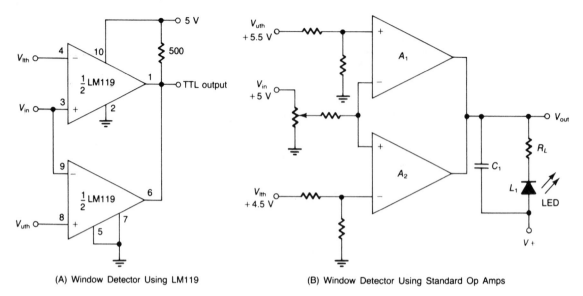

(A) Window Detector Using LM119   (B) Window Detector Using Standard Op Amps

Figure 8–10.   Window Detectors (Part A: Courtesy of National Semiconductor Corporation, © copyright 1980.)

falls below V_{lth}, the output of A_2 swings negative, forward biasing the LED, and L_1 lights. The indicator lamp could just as well be a correction voltage or a switch to turn OFF a system when limits are exceeded.

8.7 Precision References

In many signal processing and conditioning applications, standard discrete reference zener diodes are not adequate to provide the precision reference needed. Therefore, monolithic ICs have been developed to overcome this problem. Figure 8–11 shows the schematic and the functional block diagram of the LM3999 *precision reference* IC. Constructed on a single monolithic chip, the circuit consists of a temperature stabilizer circuit and an active reference zener circuit. The active circuitry reduces the dynamic impedance of the zener to about 0.5 Ω and allows the zener to operate over the current range of 0.5 mA to 10 mA with essentially no change in voltage or temperature coefficient.

The IC configuration offers many advantages over the discrete zener. It is easy to use, it is free of voltage shifts due to lead stress, and warm-up time is short because the unit is temperature stabilized. Further, in many cases the LM3999 can replace references in existing equipment with a minimum of wiring changes.

The LM3999 precision reference can be used in almost any application in place of ordinary zeners and improve performance. Some ideal applications are analog-to-digital converters, precision voltage or current sources, and precision power supplies. Figure 8–12 shows one such typical application, that of a pre-

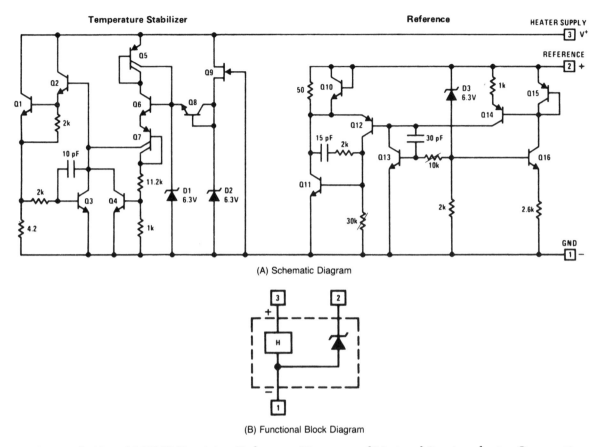

(A) Schematic Diagram

(B) Functional Block Diagram

Figure 8–11. LM3999 Precision Reference (Courtesy of National Semiconductor Corporation, © copyright 1980.)

cision voltage reference. In this case the precision output voltage is 1.01 V with an input voltage that ranges between 15 V and 20 V. A precision 5 V logic drive voltage can be obtained by adjusting the output adjust potentiometer.

8.8 Instrumentation Amplifiers

A most useful circuit for precision measurement and control is the *instrumentation amplifier* (IA). A buffered-input IA consisting of two input voltage followers, precision resistors, and a differential amplifier is shown in Figure 8–13A, and the standard schematic symbol is shown in Figure 8–13B. The voltage followers offer extremely high input impedance with low error to the differential amplifier, which provides gain and high CMRR. If R_5 is made variable, any common-mode voltage can be balanced out. A single resistor, R_1, is used to set the voltage gain of the circuit. The input impedance does not change as gain is

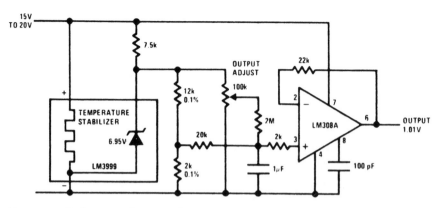

Figure 8–12. Precision Voltage Reference (Courtesy of National Semiconductor Corporation, © copyright 1980.)

varied, and V_{out} depends only on the difference between the input voltages of the differential amplifier.

High-Impedance Low-Drift IA

A practical high-impedance low-drift IA is shown in Figure 8–13C. In this circuit, resistor R_3 at pin 3 of amplifier A_3 can be trimmed to boost CMRR to 120 dB. Circuit offset voltage V_{os} is adjusted by the 25 kΩ V_{os} adjust resistor between pins 1 and 5 of amplifier A_2. Circuit voltage gain A_v is adjusted by resistor R_1, and is determined as follows:

$$A_v = \frac{R_3}{R}\left(\frac{2R_2}{R_1} + 1\right) \tag{8.5}$$

Output voltage of the circuit is determined as follows:

$$V_{out} = A_v(V_1 - V_2) \tag{8.6}$$

Example
8.6

In Figure 8–13C, let $R = 10$ kΩ, $R_2 = 47$ kΩ, and $R_3 = 100$ kΩ. Assume that resistor R_1 is adjusted to 7 kΩ. $V_1 = 5.010$ V and $V_2 = 5.005$ V. Calculate (a) A_v and (b) V_{out}.

Solution

a.   Use Equation 8.5:
$$A_v = \frac{R_3}{R}\left(\frac{2R_2}{R_1} + 1\right) = \frac{100 \text{ k}\Omega}{10 \text{ k}\Omega}\left(\frac{2(47 \text{ k}\Omega)}{7 \text{ k}\Omega} + 1\right) = 10(14.43) = 144.3$$

b.   Use Equation 8.6:
$$V_{out} = A_v(V_1 - V_2) = 144.3(5.010 \text{ V} - 5.005 \text{ V})$$
$$= 144.3(5 \text{ mV}) \approx 722 \text{ mV}$$

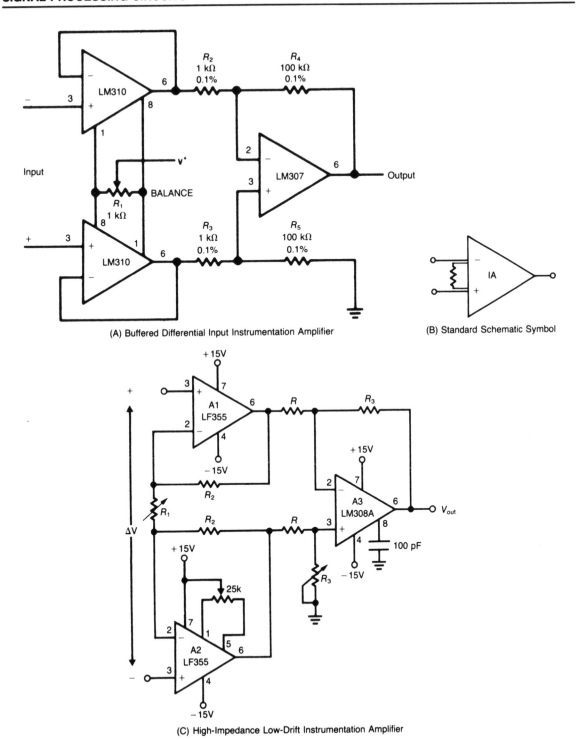

(A) Buffered Differential Input Instrumentation Amplifier

(B) Standard Schematic Symbol

(C) High-Impedance Low-Drift Instrumentation Amplifier

Figure 8–13. Instrumentation Amplifiers (Parts A and C: Courtesy of National Semiconductor Corporation, © copyright 1980.)

Transducers

A *transducer* is a device that produces an electrical output signal proportional to an applied physical stimulus, that is, a device that converts an environmental change to a resistance change. The basis of many transducers is the *resistance strain gauge*. Load cells for use in electric weighing, fluid pressure transducers, and pressure-difference transducers are typical examples of resistance strain gauge transducers. Examples of other types of transducers are thermistors, photoconductive cells, thermocouples, phonograph pickups, and microphones.

Applications for IAs

In some applications, such as providing the input to an oscillograph (chart recorder) trace pen, it may be desirable to control the output voltage of the IA to some reference level other than zero. This can be easily accomplished by adding a reference voltage in series with the (+) input terminal resistor of amplifier A_3 in Figure 8–13C, assuming that the outputs of A_1 and A_2 are equal to 0 V. We can then assume that the inputs to A_3 are equal to 0 V, as shown in Figure 8–14. We must also set $R = R_3$ in this application.

In this circuit, V_{ref} is inserted in series with R_3, which is equal to R, so that V_{ref} is divided by two and is applied to the (+) input terminal of A_3. The noninverting amplifier provides a gain of 2, and $V_{out} = V_{ref}$. Adjusting V_{ref} will then provide any desired V_{out} reference level, allowing placement of the trace pen of the oscillograph to any desired starting point. In practice, for stability and accuracy in such a circuit, the output of a voltage follower circuit is used as V_{ref}.

Another typical IA application is shown in Figure 8–15. This circuit is used to monitor a medical patient and provides protection from electrical shock. The

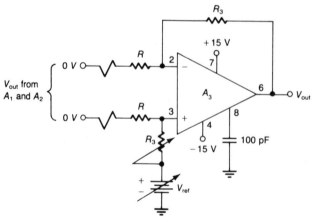

Figure 8–14. Referenced Output Voltage

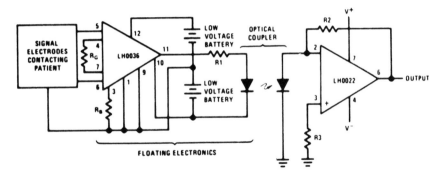

Figure 8–15. Isolation Amplifier for Medical Telemetry (Courtesy of National Semiconductor Corporation, © copyright 1980.)

input circuit, that is, the electronics contacting the patient, is floating (not grounded). The output of the floating circuit is coupled to the IA by an optoisolator. This assures that the power supply voltage is safely isolated from the patient.

Other typical applications of IAs are the thermocouple amplifier illustrated in Figure 8–16 and the bridge amplifiers shown in Figures 8–17 and 8–18. The cold junction adjustment in Figure 8–16 allows setting the thermocouple amplifier system to some predetermined temperature level prior to making mea-

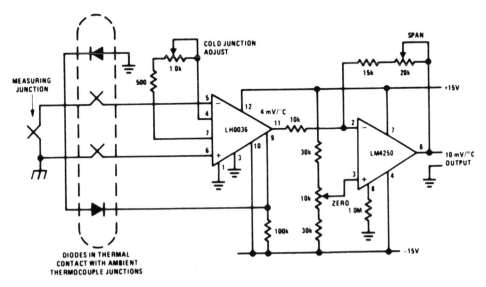

Figure 8–16. Thermocouple Amplifier with Cold-Junction Compensation (Courtesy of National Semiconductor Corporation, © copyright 1980.)

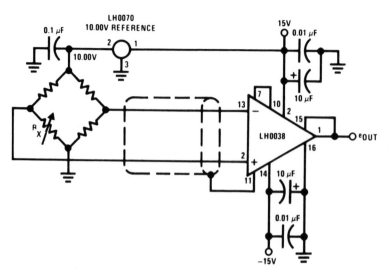

Figure 8–17. X-1000 Bridge Amplifier (Courtesy of National Semiconductor Corporation, © copyright 1980.)

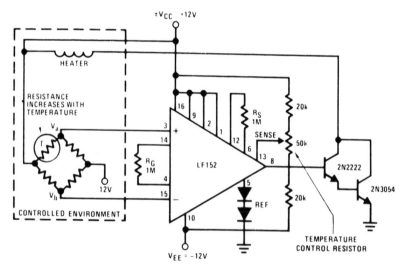

Figure 8–18. Temperature Control Circuit (Courtesy of National Semiconductor Corporation, © copyright 1980.)

surements of the physical changes under evaluation. Measuring the nozzle temperatures of a rocket engine is an example of this application.

The bridge circuit in Figure 8–17 provides a method of measuring electrical signals due to changes in a physical property, such as temperature, pressure, stress, or strain, that cause changes in a transducer. Variation of the physical property causes a change in R_X, resulting in an unbalanced bridge. The output of the IA feeds a monitoring or recording device, such as an oscillograph, data recorder, or an analog or digital meter.

Figure 8–18 shows a temperature control circuit. In this circuit, increases in temperature cause an increase in the temperature-dependent resistor of the bridge circuit. Under balanced conditions, $V_{SENSE} - V_{ref}$ appears across R_S (series resistor), $V_a - V_b$ appears across R_G (gain resistor) and $I_{R_G} = I_{R_S}$.

V_{SENSE}, voltage applied to pin 13, is fixed by the temperature control resistor and R_G/R_S is constant. The LF152 is used as a comparator with a feedback loop closed through the heater and the temperature dependent resistor. If $V_a - V_b$ is greater than V_{SENSE} (R_G/R_S), the output goes high, turning "ON" the heater. If $V_a - V_b$ is less than V_{SENSE} (R_G/R_S), the output goes low, turning "OFF" the heater.

8.9 Summary

1. A comparator is used for comparing two voltages at the input, not for amplifying signals.

2. A comparator can be constructed with an op amp used in the open-loop mode.

3. Devices designed specifically to operate as comparators have different characteristics than standard op amps.

4. Comparators are used in applications such as interfacing with digital devices. In such applications, the comparator must provide a stable drive voltage, normally $+5$ V.

5. Comparators are sometimes called squaring circuits because of the rapid output voltage swings between $+V_{sat}$ and $-V_{sat}$ that square off the peaks.

6. The Schmitt trigger is a variation of a standard comparator. It has positive feedback that prevents swings in output voltages caused by noise at the input.

7. Hysteresis voltage in a Schmitt trigger is the difference between the upper threshold voltage and the lower threshold voltage. It is this hysteresis voltage that causes the snap action of the output voltage.

8. A peak detector detects and remembers the peak value of an input signal. Peak detectors can be constructed to detect either positive or negative peak values.

9. A window detector is a circuit designed to indicate predetermined upper and lower limits of an input signal. Another name for such a circuit is a double-ended limit detector.

10. Precision reference ICs are used in place of ordinary zener diodes to provide improved performance and accurate references that are free from voltage shifts due to lead stress and temperature variations.

11. Instrumentation amplifiers are useful circuits for precision measurement and control.

12. A typical instrumentation amplifier consists of two input voltage followers and a differential amplifier output. The voltage gain of a typical IA is set by a single resistor, and output voltage depends only upon the difference between the input voltages.

13. Control of the output voltage of an IA can be accomplished by inserting a reference voltage in series with the (+) input of the differential amplifier. Practically, this reference voltage will be the output of a voltage follower.

14. For medical applications, the IA is optically isolated from the ungrounded patient for safety of the patient.

15. Typical applications of IAs are measuring changes in physical properties such as temperature, stress, strain, and pressure.

8.10 Questions and Problems

8.1 List five uses of the comparator.

8.2 What is the basic function of the comparator?

8.3 Why are standard op amps not always suitable for use as comparators?

8.4 What is meant by strobing?

8.5 The slew rate of an op amp is 50 volts per microsecond. Calculate the time it will take for the output to swing from $+V_{sat}$ to $-V_{sat}$ if the power supply is ± 15 V.

8.6 A comparator with a sine wave input signal produces what output waveform?

8.7 What possible problem is created in the output of a comparator by noise on either of the input terminals?

8.8 What method is used to eliminate the problem referred to in Problem 7?

8.9 Define upper and lower threshold voltage.

8.10 Define hysteresis voltage and explain its purpose.

8.11 Refer to Figure 8–6. Assume that $R_1 = 10$ kΩ, $R_2 = 100$ kΩ, $V_{ref} = 2$ V, and $\pm V_{sat} = \pm 5$ V. Calculate V_{uth} and V_{lth}.

8.12 Are the upper and lower threshold voltages calculated in Problem 11 symmetrical? Explain.

8.13 Calculate V_H for Problem 11.

8.14 A peak detector has a 10 μF capacitor and a 20 MΩ load impedance. What is the peak value holding time?

8.15 Assume a need to have a peak detector hold a peak value for five minutes. Assume a load impedance of 12 MΩ. Calculate the capacitor value needed.

8.16 A circuit constructed with standard op amps must light a green indicator lamp when the input voltage exceeds $+5$ V and light a red indicator lamp when the input voltage drops to $+4.5$ V. What is such a circuit called? (HINT: It has two names.)

8.17 List three advantages of precision reference ICs over standard reference zener diodes.

8.18 Refer to Figure 8–13C. Assume $R_1 = 22$ kΩ, $R_2 = 68$ kΩ, $R = 27$ kΩ, $R_3 = 180$ kΩ, $V_1 = 5.005$ V, and $V_2 = 5.0025$ V. Calculate (a) A_v and (b) V_{out}.

8.19 Define transducer.

8.20 List four typical applications of instrumentation amplifiers.

8.21 Why is it necessary to use an optoisolator between the electronics connected to a patient and the IA in medical applications?

8.22 What is the purpose of inserting a reference voltage in series with the noninverting input of the differential amplifier stage of the instrumentation amplifier?

Conversion between Technologies

9.1 Introduction

In today's electronic world few systems are purely analog or purely digital. The proliferation of microcomputer process control systems demands a marriage of the two technologies, in which each may be used to its fullest efficiency. As you will recall from earlier discussions, the two technologies are unique in their characteristics. Therefore, there must be a method for interfacing the two. This interfacing is accomplished with *analog-to-digital converters* (ADCs) and *digital-to-analog converters* (DACs).

In this chapter we will examine the DAC first, because many ADCs use DACs as part of the conversion process. We will also examine a circuit that is important to the conversion process, the sample-and-hold circuit. Finally, we will look at several applications of the converters.

9.2 Objectives

When you complete this chapter, you should be able to:

☐ Discuss the need to convert analog signals to digital form and digital signals to analog form.

☐ Describe the principles of operation of the DAC and the ADC.

☐ State the various types of converters available.

☐ Describe the principles of operation of sample-and-hold circuits.

☐ List the advantages and disadvantages of the various types of converters.

☐ Define the major terminology used in the description of converters.

9.3 Digital-to-Analog Converters

To accommodate today's industrial process control systems, it is necessary to combine analog and digital techniques. *Process control* involves the measurement of such physical quantities as temperature, liquid flow rate, pressure, light intensity, speed, strain, and vibration. The state of those quantities is then converted to electrical signals by transducers. The transduced signals are in analog form and must be converted to binary form so that a digital computer can process the information. The output signals of the computer are in binary form and may have to be converted back into analog form to be useful in making adjustments in a system that requires corrective action.

Data Acquisition

A general data acquisition system block diagram is shown in Figure 9–1. The transduced analog signal is amplified and filtered. If more than one transduced

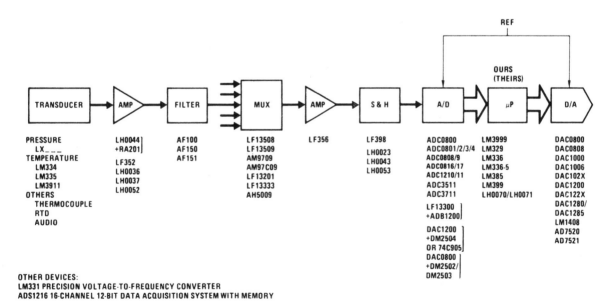

Figure 9–1. General Data Acquisition System Block Diagram (Courtesy of National Semiconductor Corporation, © copyright 1980.)

signal is being converted, a multiplexer is used and the output is again amplified. To provide a steady input to the ADC, a sample-and-hold circuit is used. The ADC output is then digitally processed and the processor output goes to the final stage, the DAC.

Numbering System

Figure 9–2 shows a typical parts numbering system for DACs and ADCs. This is the National Semiconductor system, but most manufacturers use a similar system.

ADC and DAC Definitions

Before proceeding to the various methods and circuits involved in the conversion process, let us define ADC and DAC terms.

Least significant bit (LSB)—the digital input bit carrying the lowest nu-

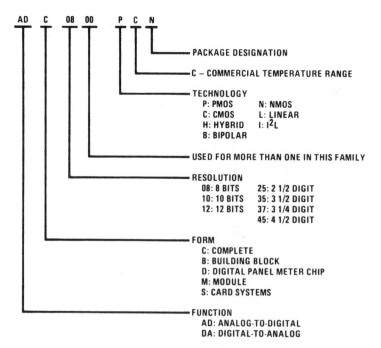

AD C 08 00 P C N

- PACKAGE DESIGNATION
- C – COMMERCIAL TEMPERATURE RANGE
- TECHNOLOGY
  P: PMOS   N: NMOS
  C: CMOS   L: LINEAR
  H: HYBRID I: I^2L
  B: BIPOLAR
- USED FOR MORE THAN ONE IN THIS FAMILY
- RESOLUTION
  08: 8 BITS   25: 2 1/2 DIGIT
  10: 10 BITS   35: 3 1/2 DIGIT
  12: 12 BITS   37: 3 1/4 DIGIT
                45: 4 1/2 DIGIT
- FORM
  C: COMPLETE
  B: BUILDING BLOCK
  D: DIGITAL PANEL METER CHIP
  M: MODULE
  S: CARD SYSTEMS
- FUNCTION
  AD: ANALOG-TO-DIGITAL
  DA: DIGITAL-TO-ANALOG

Figure 9–2. Converter Products Part Numbering System (Courtesy of National Semiconductor Corporation, © copyright 1980.)

merical weight or the analog output level shift associated with this bit, which is the smallest possible analog output step.

Most significant bit (MSB)—the digital input bit carrying the highest numerical weight or the analog output level shift associated with this bit.

Resolution—an indication of the number of possible analog output levels a DAC will produce. Usually, it is expressed as the number of input bits. For example, a 12-bit binary DAC will have $2^{12} = 4096$ possible output levels (including zero) and it has a resolution of 12 bits.

Quantization uncertainty—a direct consequence of the resolution of the converter. All analog voltages within a given range are represented by a single digital output code. There is, therefore, an inherent conversion error even in a perfect ADC. Quantization uncertainty can only be reduced by increasing resolution.

Absolute accuracy—a measure of the deviation of the analog output level from the ideal value under any input combination. Accuracy can be expressed as a percentage of full-scale range (FSR), a number of bits, or a fraction of the LSB. Accuracy may be of the same, higher, or lower order of magnitude as the resolution. Possible error in individual bit weight may be cumulative with combinations of bits and may change due to temperature variations. Usually, the accuracy of the DAC is expressed in terms of nonlinearity, differential nonlinearity, and zero and gain drift due to temperature variations.

Nonlinearity (linearity error)—a measure of the deviation of the analog output level from an ideal straight line transfer curve drawn between zero and full scale (commonly referred to as endpoint linearity).

Differential nonlinearity—a measure of the deviation between the actual output level change from the ideal (1 LSB) output level change for a one-bit change in input code. A differential nonlinearity of ± 1 LSB or less guarantees monotonicity; that is, the output always increases for an increasing input.

Monotonicity—a characteristic of the DAC that requires a nonnegative output step for an increasing input digital code. It demands no back steps or sign changes of the DAC transfer characteristic slope.

Gain drift—a measure of the change in full-scale analog output, with all bits ones, over the specified temperature range expressed in parts per million of full-scale range per °C (PPM of FSR/°C).

Offset drift (unipolar or bipolar)—a measure of the change in analog output, with all bits zeros, over the specified temperature range expressed in parts per million of full-scale range per °C (PPM of FSR/°C).

Settling time—the total time measured from a digital input change to the time the analog output reaches its new value within a specified error band.

Compliance—compliance voltage is the maximum output voltage range that can be tolerated and still maintain the specified accuracy.

Power supply sensitivity—a measure of the effect of power supply changes on the DAC full-scale output.

Four-Bit DAC

A simplified block diagram of a four-bit DAC is shown in Figure 9–3. The analog output can be either voltage or current. Assuming $V_{out} = 1$ V for the LSB, the results of the four-bit DAC are shown in Table 9–1. The table shows that each digital input contributes a different amount to the analog output. Each input is weighted according to its position in the binary number, with A being the LSB and D being the MSB. It should be obvious, then, that the analog output is the weighted sum of the digital inputs.

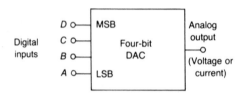

Figure 9–3.   Simplified Four-Bit DAC

Table 9–1	Voltage Output Table for a Four-Bit DAC			
D	**C**	**B**	**A**	V_{out} **(Volts)**
0	0	0	0	0
0	0	0	1	1
0	0	1	0	2
0	0	1	1	3
0	1	0	0	4
0	1	0	1	5
0	1	1	0	6
0	1	1	1	7
1	0	0	0	8
1	0	0	1	9
1	0	1	0	10
1	0	1	1	11
1	1	0	0	12
1	1	0	1	13
1	1	1	0	14
1	1	1	1	15

Example 9.1

Assume a four-bit DAC produces V_{out} = 0.5 V for a digital input of 0001. Find the value of V_{out} for the digital inputs of (a) 0011, (b) 1010, and (c) 1111.

Solutions

Since 0.5 V is the weight of the LSB, bit A, then the weights of the other bits are bit B = 1 V, bit C = 2 V, and bit D = 4 V. Therefore,

a. 0011 = 0.5 V + 1 V = 1.5 V
b. 1010 = 1 V + 4 V = 5 V
c. 1111 = 0.5 V + 1 V + 2 V + 4 V = 7.5 V

Resolution

Now let us take a closer look at resolution by referring to Example 9.1. We know that the resolution for this example is, by definition, 4 bits. But there is more to resolution than the definition allows. Resolution is always equal to the weight of the LSB, which is also referred to as *step size*, since it is the amount V_{out} will change as the digital input goes from one step to the next. Therefore, resolution can also be expressed as the amount of voltage or current per step. Many manufacturers specify DAC resolution as the number of bits, but a more useful method of expression is as a percentage of full-scale range (FSR), that is,

$$\% \text{ resolution} = \frac{\text{Step size}}{\text{FSR}} \times 100\% \qquad (9.1)$$

Example 9.2

Determine the percentage of resolution for the DAC of Example 9.1.

Solution

Use Equation 9.1:

$$\% \text{ resolution} = \frac{\text{Step size}}{\text{FSR}} \times 100\% = \frac{0.5 \text{ V}}{7.5 \text{ V}} \times 100\% = 6.67\%$$

Another equation that can be used to determine percent resolution is

$$\% \text{ resolution} = \frac{1}{\text{Total number of steps}} \times 100\% \qquad (9.2)$$

Using this equation, the result of Example 9.2 would be as follows:

$$\% \text{ resolution} = \frac{1}{15} \times 100\% = 6.67\%$$

The following points should be obvious now:

1. The larger the number of bits, the smaller the percentage of resolution.
2. Increasing the number of bits increases the number of steps to reach full-scale.
3. Each step is a smaller part of the full-scale range.

Summing Amplifier DAC

One of the simplest DACs is a circuit with which you are already familiar, an op amp summing amplifier. Figure 9–4A shows the circuitry needed to implement the DAC of Figure 9–3. Recall that a summing amplifier multiplies each input voltage by the ratio of the feedback resistor R_f to the corresponding input resistor R_{in}. Therefore, assuming a 5 V input voltage for Figure 9–4A we have the following general equation:

$$V_{out} = -V_{in}\frac{R_f}{R_{in}} \qquad (9.3)$$

And

$$\text{Input } A = 5 \text{ V}(2 \text{ k}\Omega/10 \text{ k}\Omega) = -1 \text{ V}$$
$$\text{Input } B = 5 \text{ V}(2 \text{ k}\Omega/5 \text{ k}\Omega) = -2 \text{ V}$$
$$\text{Input } C = 5 \text{ V}(2 \text{ k}\Omega/2.5 \text{ k}\Omega) = -4 \text{ V}$$
$$\text{Input } D = 5 \text{ V}(2 \text{ k}\Omega/1.25 \text{ k}\Omega) = -8 \text{ V}$$

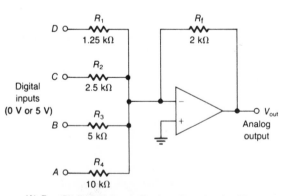

(A) Four-Bit DAC Using an Op Amp Summing Amplifier

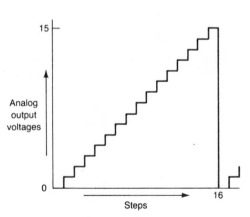

(B) Ideal Analog Output Voltage Steps for Part A

Figure 9–4. Four-Bit DAC

Recall that the minus sign only indicates that the input to the op amp is applied to the inverting terminal. Table 9–2 and Figure 9–4B represent the ideal output values for the circuit shown in Figure 9–4A. The minus signs have been eliminated in Table 9–2, and the voltage steps in Figure 9–4B are shown as positive-going to prevent confusion.

Example 9.3

Design a four-bit DAC using an op amp summing amplifier. Resolution = LSB = 0.250 V and V_{in} = 5 V.

Solution

Select a standard value for the feedback resistor R_f, say 10 kΩ. Determine the ratio of the LSB and the input voltage:

$$\frac{0.025 \text{ V}}{5 \text{ V}} = 0.05 \qquad \frac{.25 v}{5 v} = .05$$

From this ratio, the value for input A resistor $R_{in(A)}$ can be determined.

Table 9–2

Ideal Input/Output Conditions for Figure 9–4A

Digital Inputs				Individual V_{out} (Volts)				
D	C	B	A	D	C	B	A	Sum of V_{out}
0	0	0	0	0	0	0	0	0
0	0	0	1	0	0	0	1	1
0	0	1	0	0	0	2	0	2
0	0	1	1	0	0	2	1	3
0	1	0	0	0	4	0	0	4
0	1	0	1	0	4	0	1	5
0	1	1	0	0	4	2	0	6
0	1	1	1	0	4	2	1	7
1	0	0	0	8	0	0	0	8
1	0	0	1	8	0	0	1	9
1	0	1	0	8	0	2	0	10
1	0	1	1	8	0	2	1	11
1	1	0	0	8	4	0	0	12
1	1	0	1	8	4	0	1	13
1	1	1	0	8	4	2	0	14
1	1	1	1	8	4	2	1	15

$$\frac{R_f}{0.05} = \frac{10 \text{ k}\Omega}{0.05} = 200 \text{ k}\Omega$$

Since we know that each succeeding input must double the voltage of the LSB (because of resolution), then each succeeding resistor must be one-half the value of the preceding resistor. Therefore, to determine the values for the remaining input resistors, divide successive resistors by two:

$$R_{in(B)} = \frac{R_{in(A)}}{2} = \frac{200 \text{ k}\Omega}{2} = 100 \text{ k}\Omega$$

$$R_{in(C)} = \frac{R_{in(B)}}{2} = \frac{100 \text{ k}\Omega}{2} = 50 \text{ k}\Omega$$

$$R_{in(D)} = \frac{R_{in(C)}}{2} = \frac{50 \text{ k}\Omega}{2} = 25 \text{ k}\Omega$$

The completed circuit is shown in Figure 9–5, and Table 9–3 shows the input/output results. The minus sign has been eliminated for simplicity.

Precision-Level Amplifier

We have been discussing DACs with ideal input/output results, assuming that the input and feedback resistors are precision values and that the input voltages are precisely 0 V or 5 V. While precision resistors can be obtained, input voltages taken directly from the outputs of flip-flops (FF) or logic gates are not precise values but vary over a given range. Therefore, it is necessary to place a *precision-level amplifier* (PLA) between each digital input and its corresponding input resistor to the summing amplifier. A precision 5 V reference supply feeds the level amplifiers, and the result is very precise output levels of 0 V or 5 V, depending on whether the digital inputs are LOW or HIGH. Figure 9–6 shows the complete four-bit DAC with precision-level amplifiers.

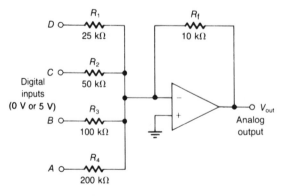

Figure 9–5. Completed Circuit for
Example 9.3

Table 9–3	Input/Output Results for Example 9.3				
	D	C	B	A	V_{out} (Volts)
	0	0	0	0	0
	0	0	0	1	0.250
	0	0	1	0	0.500
	0	0	1	1	0.750
	0	1	0	0	1.000
	0	1	0	1	1.250
	0	1	1	0	1.500
	0	1	1	1	1.750
	1	0	0	0	2.000
	1	0	0	1	2.250
	1	0	1	0	2.500
	1	0	1	1	2.750
	1	1	0	0	3.000
	1	1	0	1	3.250
	1	1	1	0	3.500
	1	1	1	1	3.750

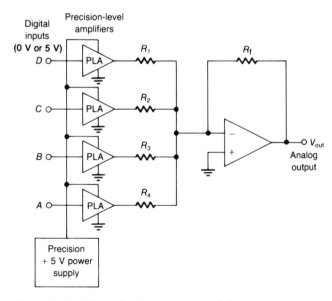

Figure 9–6.   Four-Bit DAC with Precision Amplifiers

R-2R Ladder

Another popular approach to the summing amplifier DAC is illustrated in Figure 9–7. The resistive network has been transformed into a ladder form, often referred to as the *R-2R ladder*. This type of circuit is particularly useful where a large number of digital outputs are involved. The circuit finds popular application in constant-input/constant-output audio and RF level controls.

Note that there are only two values for the resistors in the ladder, R and $2R$. This is true regardless of the number of stages. Feedback resistor R_f can be any value selected to provide the desired output scale level.

Output voltage is determined as follows:

$$V_{out} = K(V_{ref})(F) \tag{9.4a}$$

where

K = decimal value of binary input
F = scale factor

$$F = \frac{R_f}{R} \tag{9.4b}$$

A short explanation will acquaint you with the method for determining the decimal value of a binary number. We will look only at the fractional portion of the number, that is, the portion of the number to the right of the binary point. Each bit has a specified weight as follows:

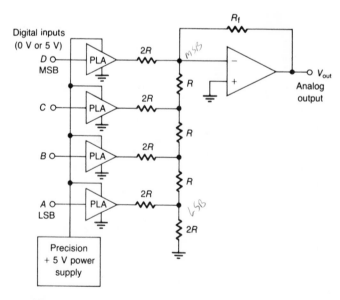

Figure 9–7. Ladder-Type Summing Amplifier

$$
\begin{array}{lll}
\text{LSB} & 1 = 1 \\
\text{binary point} & . = . \\
\quad\text{bit 1} & 1 = 0.5 \\
\quad\text{bit 2} & 1 = 0.25 \\
\quad\text{bit 3} & 1 = 0.125 \\
\quad\text{bit 4} & 1 = 0.0625
\end{array}
$$

To determine the decimal value of a binary number, add together all weights for bits that contain a binary one. For example, the decimal value of the binary number 0.0101 is calculated as follows:

$$0.0 + 0.25 + 0 + 0.625 = 0.875$$

Example 9.4

Refer to Figure 9–7. Assume $R_f = R = 10 \text{ k}\Omega$, then $2R = 20 \text{ k}\Omega$. Further assume a digital input of 0.0110. What is the output voltage?

Solution

Converting the binary input to decimal value, we have 0.375. Then using Equation 9.4a,

$$V_{out} = K(V_{ref})(F) = 0.375(5 \text{ V})(1) = 1.875 \text{ V}$$

The answer in Example 9.4 holds true, since $F = 1$ in this case. Now assume R_f is changed to 5 kΩ and all other factors remain the same. Now,

$$F = \frac{R_f}{R} = \frac{5 \text{ k}\Omega}{10 \text{ k}\Omega} = 0.5$$

Therefore,

$$V_{out} = (0.375)(5 \text{ V})(0.5) = (1.875 \text{ V})(0.5) \approx 0.938 \text{ V}$$

9.4 Sample-and-Hold Circuits A/D

Converting a voltage from analog to digital form requires that the voltage be sampled periodically and then held constant as an input to an ADC. A circuit that performs this function is called a *sample-and-hold* (S & H) circuit.

Figure 9–8A shows a basic sample-and-hold circuit, with the op amp connected as a voltage follower. When the switch is closed, the capacitor charges to $V_{in(max)}$. After the switch is opened, the capacitor remains charged and V_{out} will be at the same potential as the capacitor. The sampled voltage will be held temporarily, the time being determined by leakage in the circuit.

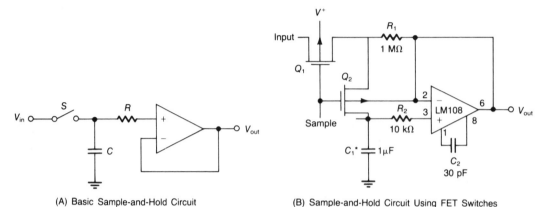

*Teflon, polyethylene, or polycarbonate dielectric capacitor. Worst case drift is less than 2.5 mV/s.

(A) Basic Sample-and-Hold Circuit   (B) Sample-and-Hold Circuit Using FET Switches

Figure 9–8.   Sample-and-Hold Circuits (Part B: Courtesy of National Semiconductor Corporation, © copyright 1980.)

The basic circuit cannot sample rapidly changing voltages, so circuits using FET switches, such as that shown in Figure 9–8B, are used. When a pulse is present at the sample input, the FETs are turned on and act as low resistances to the input signal. When the sample pulse is absent, the FETs are turned off and act as high impedances. The desired voltage is then held by capacitor C_1, which is isolated from the output by the high input impedance op amp. The capacitor cannot discharge significantly between pulses.

A typical monolithic sample-and-hold circuit is the LF198 shown in the functional diagram in Figure 9–9. This device utilizes BIFET (bipolar field effect transistor) technology to obtain ultra-high dc accuracy with fast acquisition of signal and low droop rate. The overall design guarantees no feed-through from input to output in the hold mode even for input signals equal to the supply voltages. The device will operate from ±5 V to ±18 V supplies.

Sample-and-Hold Definitions

A brief description of terminology will help you to more fully understand the S & H circuits.

Mode control signal—the signal pulse that allows the FETs to turn on, thus allowing an input signal to be sampled. This signal is usually a clock pulse set at some predetermined frequency.

Acquisition time—the time required to acquire a new analog voltage with an output step of 10 V. The acquisition time is not just the time required for the output to settle, but also includes the time required for all internal nodes to settle so that the output assumes the proper value when switched to the hold mode.

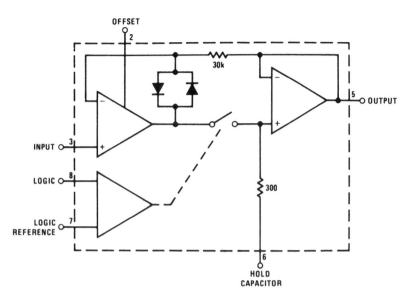

Figure 9–9. Functional Diagram of the LF198 Monolithic Sample-and-Hold Circuit (Courtesy of National Semiconductor Corporation, © copyright 1980.)

Aperture time—the delay required between "hold" command and an input analog transition, so that the transition does not affect the held output.

Hold settling time—the time required for the output to settle within 1 mV of final value after the "hold" logic command.

Dynamic sampling error—the error introduced into the held output due to a changing analog input at the time the "hold" command is given. Error is expressed in millivolts with a given hold capacitor value and input slew rate. Note that this error term occurs even for long sample times.

Gain error—the ratio of output voltage swing to input voltage swing in the sample mode expressed as a percent difference.

Hold step—the voltage step at the output of the sample-and-hold when switching from sample mode to hold mode with a steady (dc) analog input voltage. Logic swing is 5 V.

Droop (hold voltage drift)—the variation or drift due to the charge leakage out of the holding capacitor through the amplifier input terminals and the switch. Droop rate is usually expressed in millivolts per second (mV/s).

9.5 Analog-to-Digital Converters

The analog-to-digital converter is used in the process of converting an analog signal to an equivalent digital signal. The conversion process is much more complex than the converse DAC operation discussed in Section 9.3. A number of different methods for the ADC process have been developed. In this section we will examine four of those methods—comparator, digital ramp, successive approximation, and dual slope.

Comparator ADC

There are several types of analog-to-digital converters, some of which use circuitry with which you are already familiar. One such device is the *comparator ADC*. As the name implies, this device uses op amp comparators to convert the analog signals to digital form. A comparator ADC is shown in Figure 9–10. This is the fastest converter we will consider, but it requires more hardware, that is, more comparators, than other ADCs.

The resistors are interconnected to form a voltage divider that establishes the reference voltage, V_{ref}, input to each comparator. Recalling comparator operation, remember that if the input voltage, V_{in}, is greater than V_{ref} by an amount equal to the threshold voltage, V_{th}, for any comparator, the output of that com-

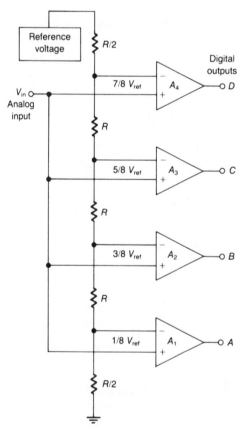

Figure 9–10. Comparator ADC

parator will saturate to $+V_{sat}$, resulting in a binary one. If V_{in} is smaller than V_{ref} by an amount equal to V_{th}, that comparator's output will be $-V_{sat}$ or a binary zero.

The resolution for the comparator ADC is determined by the number of comparators used. For example, the ADC in Figure 9–10 has a resolution of $V_{ref}/4$. If we used eight comparators, then the resolution would be $V_{ref}/8$. In other words, increasing the number of comparators improves resolution.

Digital-Ramp ADC

The *digital-ramp ADC*, also called a *counting ADC*, is one of the simplest methods of analog-to-digital conversion. It is the slowest of all the ADCs, but it requires the least hardware. A block diagram of the digital-ramp ADC is shown in Figure 9–11.

In this circuit, a positive START pulse resets the counter to zero and inhibits the AND gate. In this way no clock pulses (clp) can pass through to the counter while the START pulse is HIGH. When the START pulse goes LOW, the AND gate is enabled and pulses are allowed to pass to the counter. Now assume some positive value for V_{in} that causes the output of the comparator to go HIGH and place a binary one at one of the inputs to the AND gate. This means the clock pulse determines the output of the AND gate, and, therefore, the input to the counter. Each clock pulse increases the counter output by an increment of one, which causes the DAC output to increase by the amount of voltage associated with the LSB, in equal steps. When the DAC output voltage reaches a value equal to or greater than V_{in}, the comparator output goes LOW, placing a binary zero on the AND gate input and turning it OFF. The binary counter output

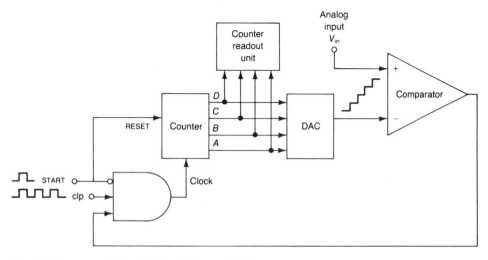

Figure 9–11.   Block Diagram of the Digital-Ramp ADC

terminals then reflect the value of the signal at V_{in}, and the conversion is completed.

The time necessary for the conversion to take place is determined by the frequency of the clock pulse (f_{clp}). The frequency is limited by the speed with which the counter and the DAC can respond to a pulse.

Example
9.5

Assume the following values for the ADC of Figure 9–11: f_{clp} = 1 MHz, V_{th} = 1 mV, and DAC full-scale output = 2.55 V with an eight-bit input. Determine (a) the binary count of the counter output when V_{in} = 1.28 V, (b) the conversion time, and (c) the resolution of the ADC.

Solutions

a.   Since the DAC has an eight-bit input and a full-scale output of 2.55 V, the total possible number of steps is $2^8 - 1$ = 255, and the step size is as follows:

$$\text{Step size} = \frac{\text{FSR}}{\text{Number of steps}} = \frac{2.55 \text{ V}}{255} = 10 \text{ mV} \qquad (9.5)$$

Therefore, the DAC output increases by 10 mV per step as the counter counts up from zero. With V_{in} = 1.28 V, the DAC output must reach 1.281 V or more before the comparator will switch LOW, therefore,

$$\frac{1.281 \text{ V}}{10 \text{ mV}} = 128.1 = 129 \text{ steps}$$

You must go to the next full step when there is a decimal fraction in the calculation. The binary count will be 10000001, which is the binary equivalent of 129_{10}, representing the desired digital equivalent of V_{in} = 1.28 V.

b.   With f_{clp} = 1 MHz, each clock pulse requires 1 μs. Since 129 steps were required at 1 μs per step, the total conversion time is 129 μs.

c.   Resolution = step size of the DAC = 10 mV. As a percent, it is expressed as follows:

$$\frac{10 \text{ mV}}{2.55 \text{ V}} \times 100\% = 0.39\%$$

Successive-Approximation ADC

The *successive-approximation* ADC is the most widely used ADC. It goes through a sequence of approximations to obtain the digital representation of the analog input voltage. The conversion time is much shorter than for the counter types and is a fixed value independent of the value of V_{in}. However, this type of ADC does require more complex control circuitry. A block diagram of a four-bit successive-approximation ADC is shown in Figure 9–12.

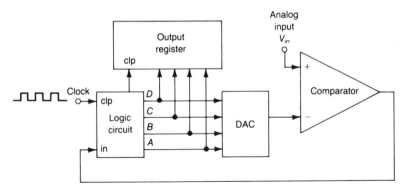

Figure 9–12.   Block Diagram of a Successive-Approximation ADC

This type of ADC uses a register instead of a counter to provide the input for the DAC. The binary values of the register are operated on in the following manner:

1. A binary number is entered into the register with the MSB set to one and all other bits set to zero. This produces a value at the DAC output equal to the weight of the MSB. If this value is greater than V_{in}, the comparator goes LOW, resetting the MSB to zero. If the value is less than V_{in}, the MSB stays at one.
2. The next significant bit is set to one and the comparison is again made as in the first step. If the new value is greater than V_{in}, this bit is reset to zero; otherwise, it remains a one.
3. This sequence continues for all bits in the binary number of the register. The process requires one clock pulse period per bit.
4. After all bits have been tested, the output of the logic circuit provides a binary one to the clock pulse terminal of the register that records the digital representation of the analog input voltage.

A comparison of the maximum conversion times of an eight-bit digital-ramp ADC to an eight-bit successive-approximation ADC, both using a clock frequency of 1 MHz, will demonstrate the difference in speeds. The digital-ramp ADC has a maximum conversion time of

$$2^N \times 1 \ \mu s = 2^8 \times 1 \ \mu s = 256 \ \mu s$$

where

$$N = \text{number of bits}$$

The successive-approximation ADC has a maximum conversion speed of the number of bits times the clock period, or

$$8 \times 1 \ \mu s = 8 \ \mu s$$

It is obvious that the successive-approximation ADC is the choice for high-speed applications. Also, the higher the number of bits to be operated on, the greater the advantage in speed.

Dual-Slope ADC

The *dual-slope ADC* is shown in the block diagram in Figure 9–13. This type of ADC requires no DAC, but it does require a close tolerance reference voltage. A constant V_{in} is obtained by using sample-and-hold circuits, as discussed earlier in this chapter. This circuit makes use of both an integrator and a comparator.

The logic circuit controls electronic switches S_1 and S_2, and resets the counter. A READ signal is applied to the logic circuit, causing the counter to reset, S_1 to close at the V_{in} position, and S_2 to open. The integrator output is

$$V_{out(int)} = -\frac{V_{in}t}{RC} \tag{9.6}$$

Since V_{in} is held constant by a sample-and-hold circuit, $V_{out(int)}$ will have the waveform shown in Figure 9–14A. This voltage is then applied to the comparator input. Since the reference voltage is zero, the comparator output becomes $+V_{sat}$, which is adjusted to represent a binary one. This voltage is then one input of

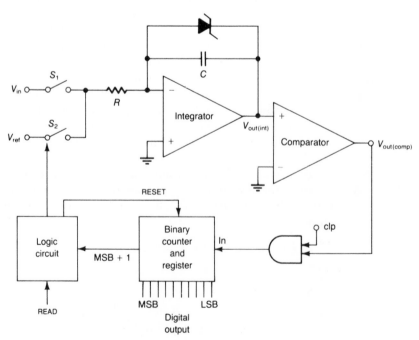

Figure 9–13. Block Diagram of a Dual-Slope ADC

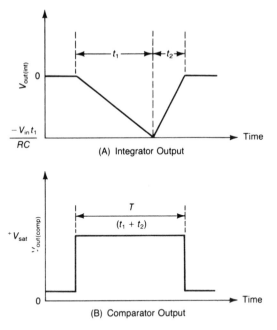

Figure 9–14. Integrator and Comparator
Output Waveforms

the AND gate, and the output of the AND gate will be the same as the clock
pulse. The counter counts the clock pulses and the process continues until all
bits, with the exception of the MSB + 1, are set to zero. The binary one at the
MSB + 1 causes the logic circuit to throw S_1 to the V_{ref} position, S_2 to close,
and at this instant,

$$V_{out(int)} = -\frac{V_{in}t_1}{RC} \tag{9.7}$$

as shown in Figure 9–14A. The integrator now integrates the V_{ref-} voltage and
causes the voltage to rise linearly toward zero (Figure 9–14A). When the comparator input becomes slightly greater than zero, the comparator output switches
to $-V_{sat}$, or binary zero, as shown in Figure 9–14B, closing the AND gate and
stopping the clock pulse count. We can now read the output bit count that is
stored in the register.

The bit count is proportional to $t_2 - t_1$. Time t_1 is a fixed interval, the time
it takes to go from 000 . . . 0 to 100 . . . 0 (with the 1 being $N + 1$). If the
number of output bits (2^N) and the clock frequency (f_{cl}) are known, then

$$t_1 = \frac{(2^{N-1} - 1)}{f_{cl}} \tag{9.8}$$

and

$$t_2 = \frac{V_{in}}{V_{ref-}}(t_1) \tag{9.9}$$

Also, the number of clock pulses (#clp), or the binary output, will be

$$\#clp = t_2 f_{cl} \tag{9.10}$$

Example 9.6

Refer to Figure 9–13. The counter of this ADC has a 10-bit output and a clock frequency of 1 MHz. $V_{ref-} = 6$ V and $V_{in} = 4$ V. Find (a) t_1, (b) t_2, and (c) the binary output.

Solutions

a. Use Equation 9.8:
$$t_1 = \frac{(2^{N-1} - 1)}{f_{cl}} = \frac{(2^{10-1} - 1)}{1 \text{ MHz}} = \frac{511}{1 \times 10^6} = 511 \text{ μs}$$

b. Use Equation 9.9:
$$t_2 = \frac{V_{in}}{V_{ref-}}(t_1) = \frac{4 \text{ V}}{6 \text{ V}}(511 \text{ μs}) = 341 \text{ μs}$$

c. Use Equation 9.10:
$$\#clp = t_2 f_{cl} = 341 \text{ μs} \times 1 \text{ MHz} = 341 = 0101010101$$

9.6 Applications

There are a great number of DAC and ADC applications. In this section we will examine just five common ones—digital voltmeter, process control, amplifier/attenuator, blending system, and transistor curve tracer.

Digital Voltmeter

The dual-slope ADC is widely used in digital voltmeters (DVMs) where high conversion speeds are not important to the operation. DVMs can be constructed with any of the ADCs we have discussed, but the dual-slope is most frequently used. This ADC is especially useful for making low-level voltage measurements because the input signal is averaged. This means that any high-frequency noise pulses will be averaged out, resulting in high accuracy. Also, this circuit does not require precision resistors to attain its high accuracy. A typical DVM block diagram is shown in Figure 9–15.

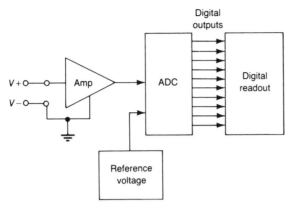

Figure 9–15. Block Diagram of a Digital Voltmeter (DVM)

Process Control

A typical *closed-loop process control system* is shown in Figure 9–16. Notice the transduced analog signal. This signal is converted by the ADC into binary form, which is processed by the computer. The computer provides two output signals: one output produces a visual display, and the other output feeds binary signals to the DAC. The DAC output is an analog control voltage that can cause the

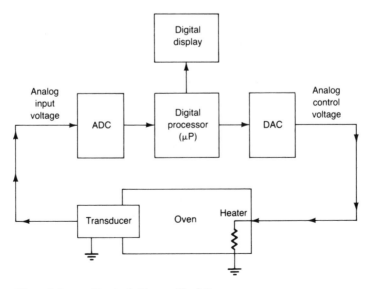

Figure 9–16. Closed-Loop Digital Controlled System

heater in the oven to increase or decrease in temperature as required to provide the exact desired oven temperature.

Amplifier/Attenuator

An unusual application of the DAC is the *digitally controlled amplifier/attenuator* illustrated in Figure 9–17. The input voltage is applied through an on-chip feedback resistor. Op amp A_1 automatically adjusts the $V(\text{ref})_{\text{in}}$ voltage such that $I_{\text{out }1}$ is equal to the input current (V_{in}/R_f). The magnitude of this $V(\text{ref})_{\text{in}}$ voltage depends on the digital word that is in the DAC register. $I_{\text{out }2}$ then depends upon both the magnitude of V_{in} and the digital word. Op amp A_2 converts $I_{\text{out }2}$ to a voltage, V_{out}, that is given by the following equation:

$$V_{\text{out}} = V_{\text{in}}\left(\frac{1023 - N}{N}\right) \tag{9.11}$$

where

$$0 < N < 1023$$

Note that $N = 0$ (or a digital code of all zeros) is not allowed because this will cause the output amplifier to saturate at either $\pm V_{\text{sat}}$ depending on the sign of V_{in}.

To provide a digitally controlled divider, op amp A_2 can be eliminated.

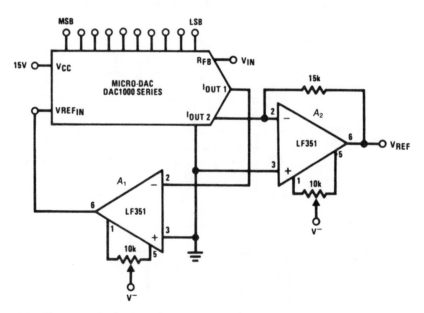

Figure 9–17. Digital Controlled Amplifier/Attenuator (Courtesy of National Semiconductor Corporation, © copyright 1980.)

Ground the I_{out2} pin of the DAC and V_{out} will be taken from the op amp A_1 (which also drives the V_{ref} input of the DAC). The expression of V_{out} will now be

$$V_{out} = -\frac{V_{in}}{M} \tag{9.12}$$

where

M = digital input (expressed as a fractional binary number)
$0 < M < 1$

Blending System

Another process control system is the blending system shown in Figure 9–18. The control system on the left includes the computer, the DAC, and the ADC. On the right is the controlled object, the blender. The two analog outputs of the DAC actuate the HOT and COLD pumps. The blended fluid flows through a discharge tube that contains a flowmeter and a temperature probe. The controlled variables, flow and temperature, provide the analog inputs to the ADC. An external clock regulates the sampling times.

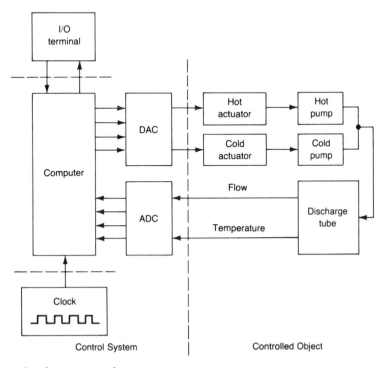

Figure 9–18.   Blending Control System

Transistor Curve Tracer

A block diagram of a transistor curve tracer using a DAC is illustrated in Figure 9–19. The output of the Schmitt trigger cycles the four-bit counter continuously from zero through 15, providing the capability of 16 curves of increasing levels of base current I_B.

The DAC can be adjusted for a desired level per bit by varying the voltage level, V^+. For example, adjusting V^+ for 5 μA per bit will produce I_B characteristic curves ranging from $I_B = 0$ to $I_B = 75$ μA in 16 equal 5 μA steps. After the count of 15, I_B returns to zero, and the curves retrace themselves, providing a continuous visual display on the oscilloscope.

There are many more uses for DAC and ADC circuits than we have the space to list. Manufacturers' data books will provide ample information on applications for the selection of devices for specific purposes.

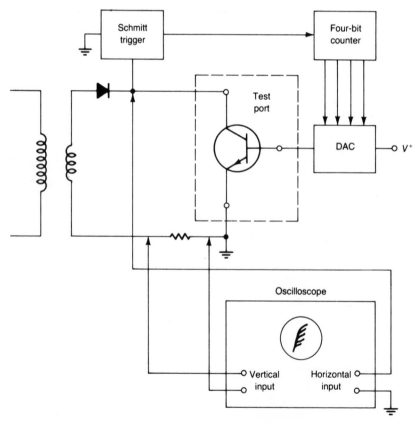

Figure 9–19.   Block Diagram of a Typical Transistor Curve Tracer

9.7 Summary

1. Combining analog and digital signals is necessary to accommodate industrial process control systems.

2. Transducers are used to convert to electrical signals the state of such physical quantities as temperature, liquid flow, pressure, light intensity, speed, strain, and vibration.

3. The DAC is a relatively simple circuit that provides an analog output of either voltage or current.

4. The ADC is a more complex circuit and generally requires a DAC as part of the conversion process.

5. The analog output of a DAC is the weighted sum of the digital inputs.

6. One of the simplest DACs is an op amp summing amplifier.

7. A variation of the summing amplifier is the R-$2R$ ladder circuit.

8. The R-$2R$ circuit is popular in audio and RF level control circuits.

9. Sample-and-hold circuits are used to provide a steady input voltage to an ADC.

10. The comparator ADC provides the fastest conversion time of the various types of ADCs, but it requires more hardware than other ADCs.

11. The successive-approximation ADC is the most widely used type, but it requires more complex circuitry than other types.

12. The counting, or digital-ramp, ADC is one of the simplest types of ADCs. It is the slowest of all types, but it requires much less hardware than the others.

13. The dual-slope ADC requires no DAC, but it does require a very stable reference voltage, usually attained by using a sample-and-hold circuit.

14. The dual-slope ADC uses both an integrator and a comparator.

15. One of the most widely used applications of ADCs is in digital voltmeters.

9.8 Questions and Problems

9.1 List five physical quantities that can be measured in process control systems.

9.2 What is the purpose of a transducer in a process control system?

9.3 Why must analog signals be converted to binary form in a process control system?

9.4 Draw a simplified block diagram of a process control system using both DACs and ADCs.

9.5 In a binary DAC, what output level shift is created by the MSB?

9.6 What is the resolution of a 10-bit binary DAC?

9.7 What is the possible number of output levels with a 10-bit binary DAC?

9.8 How can quantization uncertainty be reduced?

9.9 An eight-bit DAC produces 0.25 V for a digital input of 00000001. Find the value

of V_{out} for the digital inputs of
(a) 00001010, (b) 00000111, (c) 00001111,
(d) 00010010, and (e) 11111111.

9.10 What is the percentage of resolution for
the DAC in Problem 9?

9.11 What is the step size of the DAC in Problem 9?

9.12 Design a four-bit DAC using an op amp
summing amplifier, where $V_{in} = 5$ V and
the resolution is 0.5 V.

9.13 Explain the operation of a sample-and-
hold circuit.

9.14 Which of the several ADCs described in
this chapter offers the fastest conversion
time? the slowest?

9.15 Refer to Figure 9–11. Assume $f_{clp} = 100$
kHz, comparator $V_{th} = 1$ mV, and DAC
FSR = 10.23 V with a 10-bit input. Determine (a) the binary count of the

counter output when $V_{in} = 3.41$ V,
(b) the conversion time, and (c) the resolution of the ADC.

9.16 What is the maximum conversion time
for a 10-bit digital-ramp ADC with a clock
frequency of 100 kHz?

9.17 What is the maximum conversion time
for a 10-bit successive-approximation
ADC with a clock frequency of 100 kHz?

9.18 What conclusion can you make when
comparing the conversion times of the
ADCs of Problems 16 and 17?

9.19 What type of ADC requires no DAC?

9.20 An ADC of the type referred to in Problem 19 has a 10-bit output and a clock
frequency of 100 kHz. $V_{ref} = 4$ V and
$V_{in} = 2$ V. Find (a) t_1, (b) t_2, and (c) the
binary output.

Modulation and Demodulation Circuits

10.1 Introduction

The concepts of modulation and demodulation are important for the electronics technician. Circuits that involve these concepts are used in many of the systems and devices with which you will be required to work.

This chapter will introduce several methods of modulation and demodulation, with the emphasis on demodulation. Modulation will be covered in greater detail in communications courses. We will examine various devices used in the processes of modulation and demodulation. The phase-locked loop (PLL) will be explored, and compandors will be introduced. Finally, various applications of modulation and demodulation circuits will be discussed.

10.2 Objectives

When you complete this chapter, you should be able to:

☐ Define the differences between amplitude modulation (AM), frequency modulation (FM), and phase modulation (PM).

☐ Draw a block diagram of a typical AM transmitter and explain the operation of its modulator circuit.

☐ Draw a block diagram of a typical AM receiver and explain the operation of its detector circuit.

☐ Draw a block diagram of a typical FM transmitter and explain the operation of the modulator.

☐ Draw a block diagram of a typical FM receiver and explain the operation of the discriminator.

☐ List several applications for compandors.

☐ Draw a block diagram of a PLL and describe the operation of its four functional blocks.

☐ List the three modes of a PLL and describe the operating ranges of the PLL.

☐ Describe the operation of the VCO.

☐ Discuss the differences between digital and analog PLLs.

☐ Design a frequency synthesizer from a standard PLL circuit.

☐ List several applications for PLL circuits.

10.3 Modulation

The term *modulation* implies variation or shaping, and the term *demodulation* implies removal of variation or shaping. For example, audio signals can be transmitted by changing some characteristics of a higher-frequency carrier wave. If this change is in the amplitude, with frequency held constant, the process is called *amplitude modulation* (AM). If the change is in frequency or phase angle, with amplitude held constant, it is called *frequency modulation* (FM), or *phase modulation* (PM), respectively.

A communications system transmits intelligence (information) from one point to another. Circumstances often dictate the frequency of transmission. For example, the basic telephone system uses a direct current in a circuit as a carrier current, and the carrier is varied in amplitude at a rate of 500 Hz per second, thus it is said to be amplitude-modulated.

If the transmitted signal is a radio frequency (RF) ac carrier wave of some frequency made to vary in amplitude at a certain rate, we again have amplitude modulation. If the RF ac carrier wave is made to vary in frequency, then we have frequency modulation. These are the basic principles of AM and FM broadcast transmissions.

Amplitude Modulation

Let us first explore amplitude modulation. The block diagram of Figure 10–1 shows an AM transmitter with audio modulation used for radiotelephone communication. Radiotelephone indicates voice communications, but it also can in-

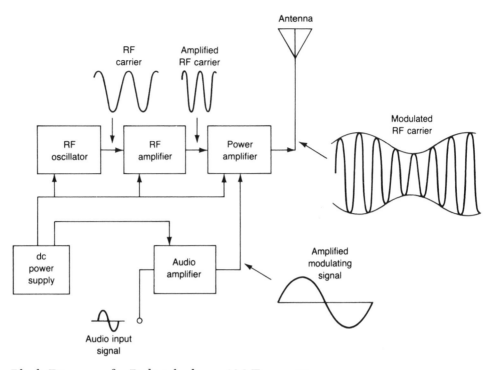

Figure 10–1. Block Diagram of a Radiotelephone AM Transmitter

clude music. The oscillator stage generates the RF carrier wave. Crystal-controlled oscillators are typically used for this stage.

The output of the oscillator is amplified by the RF amplifier stage. This stage performs two functions: first, it provides a large enough signal to drive the power amplifier (PA); and second, it acts as a buffer that separates the oscillator from the PA to reduce loading on the oscillator circuit and to increase frequency stability.

The PA drives the antenna with the amount of current needed for the desired power output. The more current that is provided, the stronger the radiated signal, and, therefore, the greater the distance the signal can be transmitted. Also, the PA is modulated by the audio signal from the audio amplifier. Therefore, the amplitude variations of the modulated RF signal are present in the output of the PA and, consequently, in the antenna circuit.

A radio frequency must be used for the transmitted carrier frequency because audio frequencies cannot be transmitted over any great distance. But by means of modulation, the carrier provides the transmission of the RF wave, while the audio modulating signal has the desired intelligence. Figure 10–2 shows the relationships between the unmodulated RF carrier and the modulated signal being transmitted. Simply stated, the basic sine waves of the RF carrier have been converted into a complex waveform. The waveform is still an RF signal,

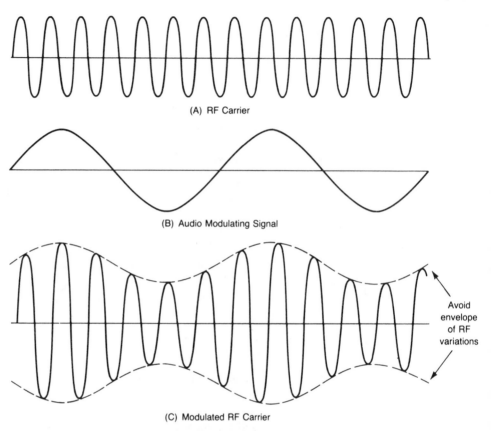

(A) RF Carrier

(B) Audio Modulating Signal

(C) Modulated RF Carrier

Avoid
envelope
of RF
variations

Figure 10–2. Relationships between Unmodulated and Modulated Signals

but it contains modulating signal variations. The RF signal allows long-distance transmission, and the modulation has the desired intelligence.

Modulation Factor

The variation in the AM signal compared with the unmodulated carrier is called the *modulation factor,* or *index of modulation,* designated by the letter *m*. Multiplying *m* by 100 results in the percent modulation, expressed as follows:

$$\%m = \frac{I_{\max} - I_{\min}}{I_{\max} + I_{\min}} \times 100 \tag{10.1}$$

where

I = current of the modulated carrier on the antenna

If voltages are used, substitute V for I in the formula.

Two examples of amplitude modulation are shown in Figure 10–3, where Part A illustrates the unmodulated carrier. The varying amplitudes of the RF carrier wave results in an outline that corresponds to the audio modulating signal. The outline, as shown by the dashed lines at the top and bottom of the waveforms in Parts B and C, is called the modulation envelope. Note that both envelopes have the same variations, so they are symmetrical around the center axis of the carrier.

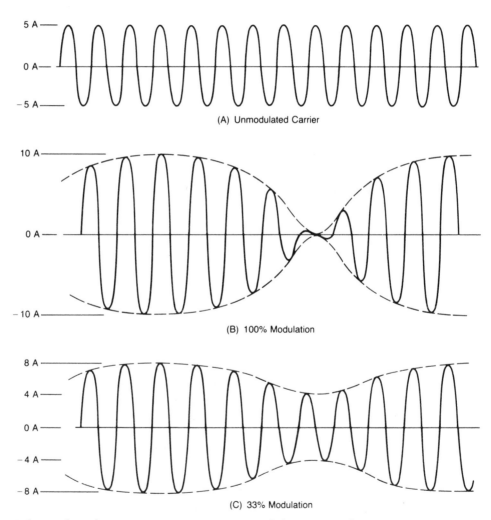

(A) Unmodulated Carrier

(B) 100% Modulation

(C) 33% Modulation

Figure 10–3.   Relationships between Various Percent Modulation Signals

Example 10.1

Assume a carrier current of 5 A. If the maximum current is 10 A and the minimum current is 0 A, what is the percent modulation?

Solution

Use Equation 10.1:

$$\%m = \frac{I_{max} - I_{min}}{I_{max} + I_{min}} \times 100 = \frac{10\text{ A} - 0\text{ A}}{10\text{ A} + 0\text{ A}} \times 100 = \frac{10\text{ A}}{10\text{ A}} \times 100$$
$$= 1 \times 100 = 100\%$$

Note that the RF signal for 100% modulation shown in Figure 10–3B varies from zero level to twice the level of the unmodulated carrier.

Example 10.2

Now assume $I_{max} = 8$ A and $I_{min} = 4$ A. What is the $\%m$?

Solution

Use Equation 10.1:

$$\%m = \frac{I_{max} - I_{min}}{I_{max} + I_{min}} \times 100 = \frac{8\text{ A} - 4\text{ A}}{8\text{ A} + 4\text{ A}} \times 100 = \frac{4\text{ A}}{12\text{ A}} \times 100$$
$$= 0.33 \times 100 = 33\%$$

Figure 10–3C shows the results of Example 10.2.

Overmodulation

In order for the most audio signal to be recovered by a receiver circuit, called the demodulator, or detector, it is important to have a high percentage modulation at the transmitter. For that reason modulation is generally maintained close to but not at 100%. However, the complexity of audio signals can cause the modulation to exceed 100%, a situation called *overmodulation*. This situation is demonstrated by Figure 10–4. Here the carrier signal is intermittently turned on and off, the effect of which is the generation of new frequencies that interfere with nearby channels and cause distortion of the transmitted intelligence.

Frequency Modulation

Frequency modulation differs from amplitude modulation in the following way. The FM signal has a constant amplitude but varying frequencies above and below

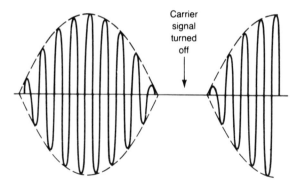

Figure 10–4. Overmodulated Signal

the center frequency of the carrier, and the AM signal has a varying amplitude but a constant carrier frequency. Figure 10–5 shows an FM signal. The changes in frequency of the RF carrier are produced by an audio-modulating voltage. The amount of change, called frequency deviation, increases with an increase in the audio-modulating voltage. The modulating voltage amplitude, not its frequency, determines the frequency deviation.

Frequency Deviation

Frequency deviation is the amount of change from the center frequency. The center frequency, also called the rest frequency, is the frequency of the transmitted RF carrier without modulation, that is, when the modulating voltage is at its zero value. The peak audio-modulating voltage produces the peak frequency deviation. The total frequency deviation above and below center frequency is called the *frequency swing*.

A major advantage of FM transmission is that the signal is practically noise free, since most types of noise produce AM variations in a signal. In FM the desired signal is in the variations of frequency, therefore the noise does not affect the signal appreciably. However, to recover the desired signal without noise at

Center frequency = 100 kHz
Frequency deviation = 30 kHz

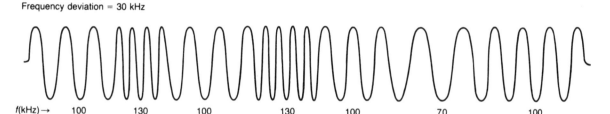

f(kHz) →   100   130   100   130   100   70   100

Figure 10–5. FM Signal

the receiver, you must have an FM detector circuit that responds to the changes in carrier frequency to recover the original audio signal, and a limiting circuit to remove amplitude variations in the FM signal. These circuits will be discussed later in this chapter.

Phase Modulation

Phase modulation is, in effect, a modified version of FM, because any change in phase is equivalent to a change in frequency. The phase modulation method generally uses crystal-controlled oscillators to shift the phase angle of the RF output in step with the audio-modulating voltage. The RF carrier from the crystal oscillator is coupled into a reactive network where the audio input voltage shifts the bias above and below its average dc value, causing the phase angle of the RF carrier to be shifted from its average value. This action results in phase modulation.

10.4 Demodulation

The modulated signal is transmitted, then is captured by a receiver in which it is demodulated, or detected. This process involves stripping the carrier wave from the intelligence information (audio signal) and amplifying the original signal to some usable level.

At the receiver, the information contained in the modulated signal must be separated from the carrier and passed on to a speaker or display device such as a television screen. This separation process is called demodulation, or detection. The circuits used for this process depend on the form of modulation used in the transmission of the signal. For AM signals, the circuit is called an *envelope detector*, and for FM signals, the circuit is referred to as a *discriminator*. Television signals are composites in which the audio information is FM, and the video information is AM. At the receiver, the audio and video signals are separated and channeled to different sections of the receiver to their proper audio and video demodulators.

A block diagram of an AM superheterodyne receiver is illustrated in Figure 10–6. The AM signal that operates in the 540–1600 kHz range is received by the antenna, is coupled into the RF amplifier, then coupled to the mixer where it is combined with a signal that is generated by a local oscillator. The oscillator and mixer stages combine to form the frequency-converter stage that produces an *intermediate frequency* (IF) of 455 kHz. Converting the received RF signal to an IF simplifies the design of the amplifier portion of the receiver. The IF signal is amplified and fed to the detector circuit, where the desired information is removed from the IF carrier.

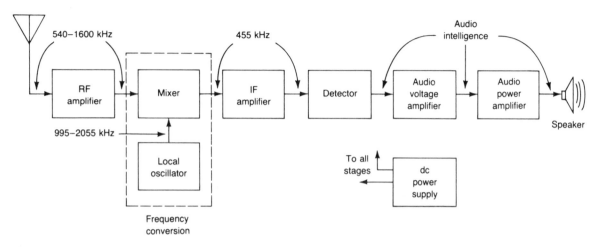

Figure 10–6. Block Diagram of an AM Superheterodyne Radio Receiver

Diode Detector

A commonly used AM detector circuit is the *diode detector*, shown in Figure 10–7. The diode is used as a half-wave rectifier of the AM IF signal. The low-pass filter, composed of the RF parallel network, filters out the 455 kHz signal, but allows the audio signal to pass to the audio amplifier. As a result, the detector output has the desired audio intelligence.

Automatic Gain Control

In addition to the intelligence signal, there is an average value dc level at the detector output that is used as a control signal for an *automatic gain control* (AGC) system. A feedback signal is sent to the RF amplifier to increase the amplifier gain when the dc level is low, or to decrease the gain when the dc level is high. In this manner, the output signal at the speaker is held relatively constant.

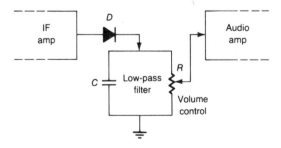

Figure 10–7. Diode Detector Circuit

Single-Chip Radio Device

Many of the functions that previously required many discrete components and much board space are available now in a single linear IC. One such single package device is the LM3820, illustrated in Figure 10–8. This is a three-stage AM radio IC that consists of an RF amplifier, an oscillator, a mixer, an IF amplifier, an AGC detector, and a zener regulator. The device was originally designed for use in slug-tuned automobile radio applications, but is also suitable for capacitor-tuned portable radios.

The circuit shown in Figure 10–9 that uses the LM3820 can be used as a starting point for portable radio designs. Loopstick antenna L1 is used in place of L0, and the RF amplifier is used with a resistor load to drive the mixer. A double-tuned circuit at the output of the mixer provides selectivity, while the remainder of the gain is provided by the IF section, which is matched to the diode through a unity-turn ratio transformer. R_{AGC} may be used in place of C_{AGC} to bypass the internal AGC detector and to provide more recovered audio. Coil specifications for Figure 10–9 are given in Figure 10–10.

An AM automobile radio design using the LM3820 is shown in Figure 10–11. Tuning of both the input and the output of the RF amplifier and the mixer is accomplished with variable inductors (dashed lines). Selectivity is improved through the use of double-tuned interstage transformers. Input circuits are inductively tuned to prevent microphonics and to provide a linear tuning motion to facilitate push-button operation.

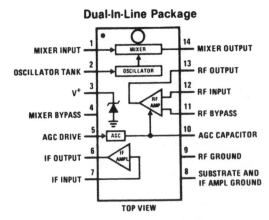

Figure 10–8. Connection Diagram of an LM3820 AM Radio Receiver System (Courtesy of National Semiconductor Corporation, © copyright 1980.)

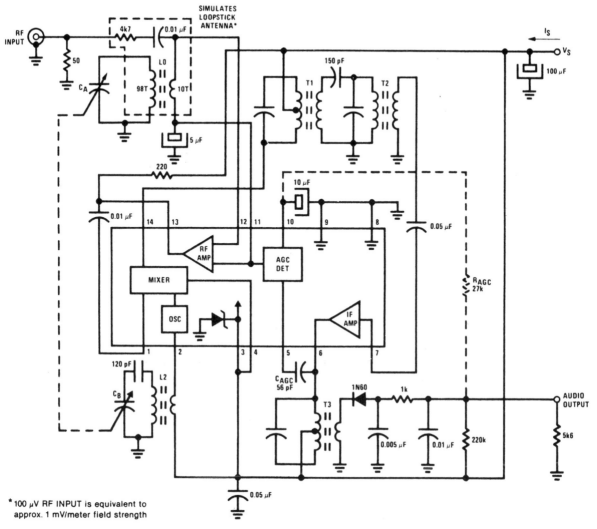

Figure 10–9. Capacitor-Tuned AM Portable Radio Design Using the LM3820 (Courtesy of National Semiconductor Corporation, © copyright 1980.)

FM Radio Receiver

Superheterodyne circuits are used in FM receivers as well as in AM receivers. The IF in FM receivers is much higher at 10.7 MHz, because the RF carrier frequencies for FM radio broadcasting are in the 88–108 MHz band. Figure 10–12 shows a block diagram of an FM radio receiver.

The received signal is fed from the antenna into an RF tuner that includes the RF amplifier, local oscillator, and mixer stages. The RF amplifier improves

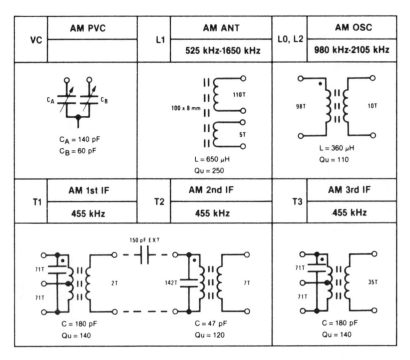

VC	AM PVC	L1	AM ANT	L0, L2	AM OSC
			525 kHz-1650 kHz		980 kHz-2105 kHz

T1	AM 1st IF	T2	AM 2nd IF	T3	AM 3rd IF
	455 kHz		455 kHz		455 kHz

Figure 10–10. Coil Specifications for Figure 10–9 (Courtesy of National Semiconductor Corporation, © copyright 1980.)

the signal-to-noise ratio (S/N), and the frequency-converter section converts the RF signal to an IF signal at 10.7 MHz. The mixer output is the input to the IF section.

Limiter Stage

The *limiter stage* is an IF amplifier tuned to the 10.7 MHz IF. The limiter provides a relatively constant output level for different input levels. The stage operates between saturation and cutoff, and it usually has signal bias that automatically adjusts itself to the amount of signal.

Detector Stage

The *detector stage* receives the amplified IF signal from the limiter. This circuit recovers the audio-modulating signal by allowing the frequency variations in the signal to provide equivalent variations in amplitude that can be rectified by a diode. This stage generally uses two diodes in a balanced detector circuit, such as the *Foster-Seeley discriminator* or the *ratio detector*. The Foster-Seeley discriminator and ratio detector circuits are not used in constructing linear ICs, therefore they will not be discussed further.

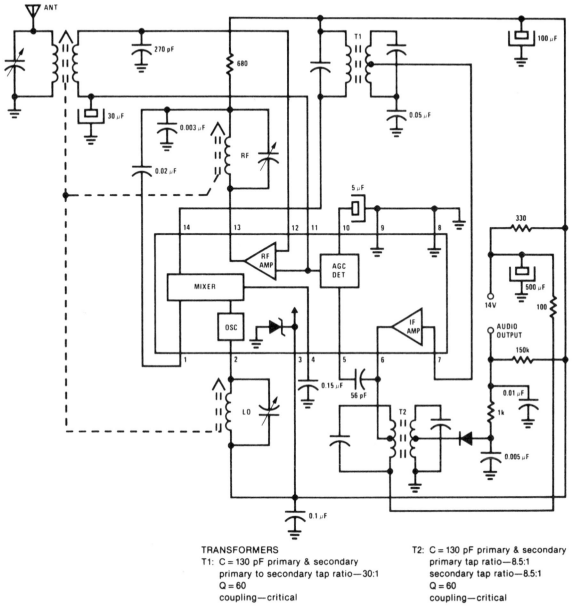

Figure 10–11. Slug-Tuned AM Car Radio Design Using the LM3820 (Courtesy of National Semiconductor Corporation, © copyright 1980.)

Quadrature Detector

For today's FM systems, a discriminator circuit called the *quadrature detector* has been designed. The quadrature detector circuit is used in most modern ICs that contain FM discriminators.

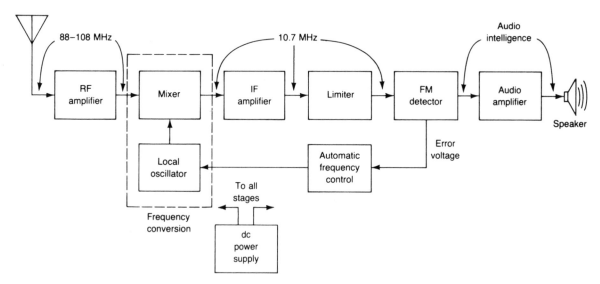

Figure 10–12. Block Diagram of an FM Radio Receiver

As shown in Figure 10–13, the quadrature detector circuit is basically a difference amplifier formed by Q_1 and Q_2, with the output taken at the collector of Q_2. With the current in R_5 being constant, any change in the current of Q_1 results in an equal but opposite change in Q_2. The quadrature circuit provides the audio output when an FM carrier is applied to the input.

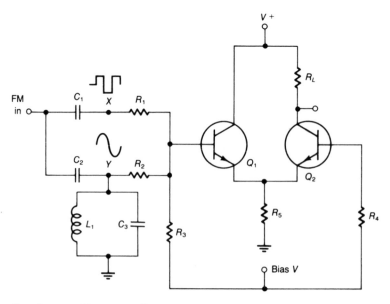

Figure 10–13. Quadrature Detector Circuit

Capacitor C_1 is large enough in value that no phase shift of the input signal is produced at point X. The value of C_2 is small enough so that its reactance to the carrier frequency is large when compared with the tuned circuit impedance, resulting in a phase shift at point Y. The waveform at Y is the FM carrier sine wave, and the waveform at X is the squared-off constant amplitude wave from the limiter. The two waveforms shown in Figure 10–14A are combined at the base of Q_1. When both waveforms are negative, the detector output is high, as shown in Figure 10–14B.

Frequency Shifts

With modulation, the carrier frequency shifts and causes a phase shift of the signal at Y. This shift occurs because of the tuned circuit going off resonance, which causes a change in the phase relationship between the tuned circuit and capacitor C_2. This change in phase at Y causes a variance in interval of the output when both waveforms are negative. This variance in interval has the effect of varying the output pulse width, and, therefore, the average value of the output voltage. The change in the pulse average voltage is proportional to the amplitude of the original modulation, since it is the amplitude of the modulating signal voltage that causes the frequency to change. The audio can now be separated from the pulse output through a low-pass filter.

Using linear IC devices that contain quadrature detectors requires a minimum of external components. One such device is the CA2111A shown in Figure 10–15. This circuit contains a limiter-amplifier, a quadrature detector, and an emitter-follower stage that provides a low-impedance output for driving an external audio amplifier. The detector circuit can be tuned easily because of the external coil.

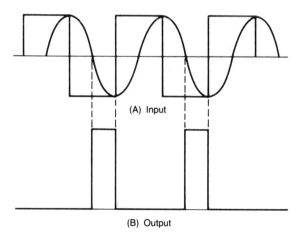

(A) Input

(B) Output

Figure 10–14. Input/Output Relationships of a Quadrature Detector

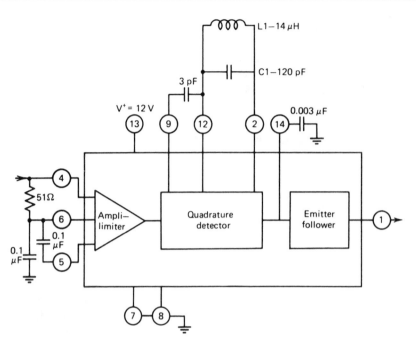

Figure 10–15.   CA2111A Quadrature Detector IC (Courtesy of RCA Corporation, © copyright 1981.)

A Modern FM-IF System

Taking modern sophistication one step further results in such ICs as the LM3189N illustrated in Figure 10–16. This IC provides all the functions of a comprehensive FM-IF system. It includes a three-stage FM-IF amplifier/limiter configuration with level detectors for each stage, a doubly-balanced quadrature FM detector, and an audio amplifier that features the optional use of a *muting* (squelch) circuit.

The advanced design of this IC provides such desirable features as programmable delayed AGC for the RF tuner, an AFC drive circuit, and an output signal to drive a tuning meter or provide stereo switching logic. In addition, internal power supply regulators maintain a nearly constant current drain over the voltage supply range of $+8.5 - 16$ V. This circuit is ideal for high-fidelity operation, providing single-coil tuning capability with the external detector coil.

10.5 Compandors

A *compandor* is a gain control device that is used for dynamic gain expansion or compression. The name is derived from its two functions—compression and expansion. Applications are in telephone subscriber and trunk carrier systems,

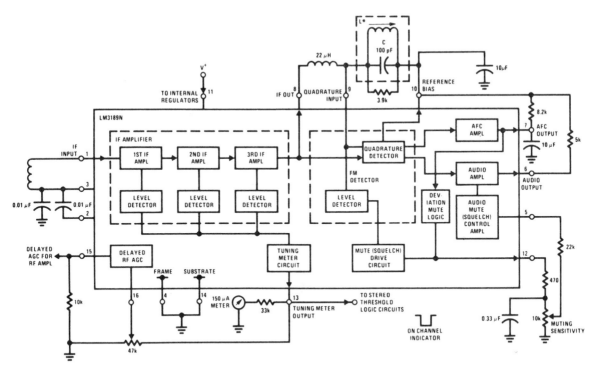

All resistance values are in ohms
* L tunes with 100 pF (C) at 10.7 MHz, $Q_0 \cong 75$
(Toko No. KACS K586HM or equivalent)

Figure 10–16. LM3189N FM-IF System (Courtesy of National Semiconductor Corporation, © copyright 1980.)

communications systems, high-quality audio systems, and video recording systems.

Compandor Building Blocks

The NE570/571 compandor is shown in Figure 10–17. The block diagram shows the internal building blocks for one-half of the dual-channel device. Each channel has a full-wave rectifier, a variable gain cell, an op amp, and a bias system. The arrangement of these blocks in the IC results in a circuit that can perform well with few external components, yet can be adapted to many diverse applications.

Rectifier Stage

The full-wave rectifier rectifies the input current that flows from the rectifier input (pin 2) to an internal summing node that is biased at V_{ref}. The rectified current is averaged on an external filter capacitor tied to the C_{rect} terminal (pin

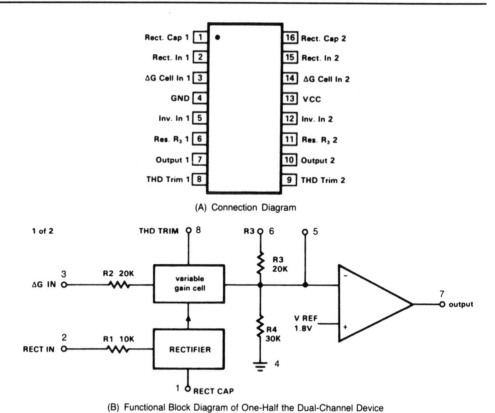

(A) Connection Diagram

(B) Functional Block Diagram of One-Half the Dual-Channel Device

Figure 10–17. NE570/571 Dual-Channel Compandor (Courtesy of Signetics, a subsidiary of U.S. Philips Corporation, © copyright 1981.)

1), and the average value of the input current controls the gain of the variable gain cell (ΔG cell). The gain will thus be proportional to the average value of the input signal for capacitively coupled inputs. Since capacitively coupled inputs have no error-producing offset voltage, any error will come from the internally supplied bias current of the rectifier, which is less than 0.1 µA. Expressed in equation form,

$$G \propto \frac{|V_{in} - V_{ref}|\ \text{av}}{R_1} \tag{10.2a}$$

or

$$G \propto \frac{|V_{in}|\ \text{av}}{R_1} \tag{10.2b}$$

The speed with which gain changes to follow changes in input signal levels is determined by the rectifier filter capacitor at pin 1. A small value capacitor will yield rapid response but will not fully filter low-frequency signals. Any ripple on the gain control signal will modulate the signal passing through the variable gain cell. In an expandor or compressor application, this would lead to third harmonic distortion, so there is a trade-off to be made between fast attack and decay times, and distortion. For step changes in amplitude, the change in gain with time is as follows:

$$G(t) = (G_{\text{initial}} - G_{\text{final}})\, e^{-t/\tau} + G_{\text{final}} \tag{10.3}$$

where

$$\tau = 10\ \text{k}\Omega \times C_{\text{rect}}$$

Gain Cell Stage

The variable gain cell is a current in-current out device with the ratio $I_{\text{out}}/I_{\text{in}}$ controlled by the rectifier. I_{in} is the current flowing from the G input (pin 3) to an internal summing node biased to V_{ref}. For capacitively coupled inputs,

$$I_{\text{in}} = \frac{V_{\text{in}} - V_{\text{ref}}}{R_2} \tag{10.4a}$$

or

$$I_{\text{in}} = \frac{V_{\text{in}}}{R_2} \tag{10.4b}$$

The output current, I_{out}, is fed to the summing node of the op amp.

A compensation scheme built into the ΔG cell compensates for temperature and cancels out odd harmonic distortion. The only distortion that remains is even harmonics, and they exist only because of internal offset voltages. The THD (trim-high distortion) trim terminal (pin 8) provides a means for nulling the internal offsets for low distortion operation.

Op Amp Stage

The op amp (which is internally compensated) has the noninverting input tied to V_{ref} and the inverting input connected to the ΔG cell output as well as brought out externally (pin 5). Resistor R_3 is brought out from the summing node (pin 6) and allows compressor or expandor gain to be determined only by internal components.

Output Stage

The output stage is capable of ± 20 mA output current. This allows a $+13$ dBm ($3.5\ V_{\text{rms}}$) output (pin 7) into a $300\ \Omega$ load that, with a series resistor and proper transformer, can result in $+13$ dBm with a $600\ \Omega$ output impedance.

Reference Voltage

A band gap reference provides the reference voltage for all summing nodes, a regulated supply voltage for the rectifier and ΔG cell, and a bias current for the ΔG cell. The low temperature coefficient (tempco) of this type of reference provides very stable biasing over a wide temperature range.

Programmable Analog Compandor

A programmable analog compandor is the NE572 shown in Figure 10–18. This is a dual-channel high-performance device in which either channel may be used

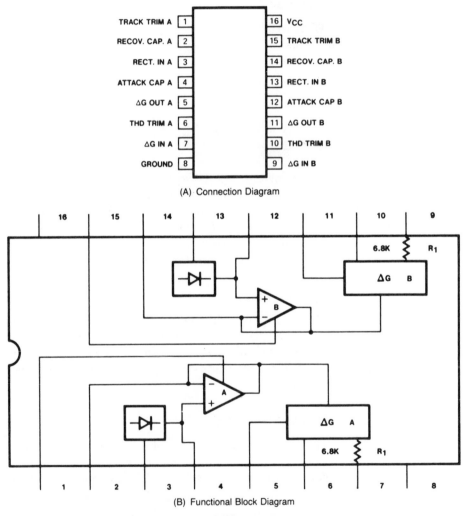

(A) Connection Diagram

(B) Functional Block Diagram

Figure 10–18. NE572 Programmable Analog Compandor (Courtesy of Signetics, a subsidiary of U.S. Philips Corporation, © copyright 1981.)

for dynamic gain compression or expansion. Each channel has a full-wave rectifier to detect the average value of input signal, a linearized temperature-compensated variable gain cell, and a dynamic time-constant buffer. The buffer permits independent control of dynamic attack and recovery time with minimum external components and improved low-frequency gain control ripple distortion.

The NE572 is designed to reduce noise in high-performance audio systems, but it can be used in a wide variety of such applications as a voltage control amplifier, a stereo expandor, an automatic level control, a high-level limiter, a low-noise gate, and a state-variable filter.

10.6 Phase-Locked Loops

The *phase-locked loop* (PLL) is basically an electronic feedback loop that consists of a phase detector (or comparator), a low-pass filter, a dc amplifier, and a voltage-controlled oscillator (VCO). The basic block diagram of a PLL is shown in Figure 10–19. The components between the input and output are considered to be in the forward path of the loop, and the single connection between the VCO and the phase detector is the feedback path.

The purpose of the PLL circuit is to make a variable-frequency oscillator lock in at the frequency and phase angle of a standard frequency (f_s) used as a reference. The oscillator will then have the same frequency accuracy as the referenced standard.

PLL Functional Blocks

To better understand the operation of the PLL, we will examine each of the four functional blocks of the PLL—the phase detector, the low-pass filter, the dc amplifier, and the voltage-controlled oscillator (VCO).

Phase Detector

All PLL systems, whether analog or digital, use a phase detector, or comparator, circuit to generate the dc control voltage. The basic difference between the analog and digital PLLs is the type of phase detector used. In general, digital

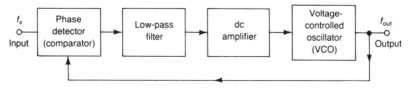

Figure 10–19. Block Diagram of a PLL

systems use either an exclusive-OR gate, or some type of edge-triggered phase detector, and the analog system uses a double-balanced mixer. Basically, the analog phase detector uses two diodes in a balanced rectifier circuit. The phase detector output voltage, V_{out}, is proportional to the phase difference between the two inputs, and is determined as follows:

$$V_{out} = K_c \Delta \phi \qquad (10.5)$$

where

K_c = phase detector conversion gain in volts/radian (V/rad)
$\Delta \phi$ = the input phase difference in radians (rad)
1 rad = 57.3°

Example 10.3

A phase detector has a conversion gain of 20 mV and an input phase difference of 0.15 rad. Calculate the phase detector output voltage.

Solution

Use Equation 10.5:

$$V_{out} = K_c \Delta \phi = 20 \text{ mV/rad} \times 0.15 \text{ rad} = 3 \text{ mV}$$

Example 10.4

Calculate the phase detector output for a conversion gain of 15 mV and an input phase difference of 20°.

Solution

First, convert the phase difference from degrees to radians:

$$\frac{20°}{57.3°/\text{rad}} = 0.349 \text{ rad}$$

Next, use Equation 10.5:

$$V_{out} = K_c \Delta \phi = 15 \text{ mV/rad} \times 0.349 \text{ rad} = 5.2 \text{ mV}$$

Low-Pass Filter

The low-pass filter network can be either passive or active. It serves two major functions: first, it removes the ac signal variations of the two oscillators, or traces of higher frequency noise, from the rectified dc output voltage of the phase detector; and second, it controls the lock, capture, bandwidth, and transient response of the loop. That is, the filter determines the dynamic performance of the loop.

dc Amplifier

The dc amplifier circuit amplifies the filtered dc control voltage to the desired level for better control and in the polarity needed for the varactor in the VCO.

VCO

The VCO circuit uses a varactor to set the oscillator frequency. The output frequency of the VCO is directly proportional to its input control voltage, which keeps the VCO locked into the frequency and phase of the referenced standard oscillator. The VCO is also termed a *voltage-to-frequency converter*.

The VCO circuit is used for electronic tuning of the oscillator frequency. The *varactor*, also called a *varicap*, is a semiconductor capacitive diode that operates on the principle of a varying capacitance that is inversely proportional to the amount of reverse dc voltage applied. The reverse voltage applied can be either negative at the anode or positive at the cathode of the diode. The more reverse voltage that is applied, the wider the depletion area of the PN junction. The effect of the widened depletion area is equivalent to an increased distance between capacitor plates, which produces less capacitance.

PLL Operating Modes

There are three operating modes for a PLL. They are *free-running, capture,* and *phase-lock* (sometimes called lock-in or tracking). If the VCO output frequency (f_{out}) is too far from the standard frequency (f_s), the PLL cannot lock in the oscillator. Without such lock-in, the VCO is in the free-running mode. Once the control voltage from the dc amplifier starts to change the VCO frequency, the oscillator is in the capture mode. When f_{out} is exactly the same as f_s, the VCO is in lock-in and the PLL is in the phase-locked mode. The PLL will remain in the phase-locked mode as long as the dc control voltage is applied.

Lock Range and Capture Range

The frequency range over which the PLL can follow the incoming signal is called the *lock range*. The bandwidth over which capture is possible is called the *capture range*. The capture range can never be wider than the lock range. To keep the capture range narrower than the lock range, the phase detector is used to compare the two frequencies. Any difference in phase causes an error signal at the output of the phase detector. This error signal is a dc voltage that is proportional to the difference in frequency and phase of the standard oscillator and the VCO. The dc error voltage is used to correct the VCO frequency by forcing it to change in a direction that reduces the frequency difference between the input oscillator and the VCO.

The low-pass filter circuit removes the ac variations of the two oscillators, or traces of higher frequency noise from the rectified dc output of the phase detector. The output of the filter is a filtered dc control voltage that is fed to the dc amplifier, where it is increased in amount to provide better control. The amplifier output provides the desired dc level and polarity for the control voltage needed for the varactor in the VCO.

The VCO uses the varactor to set the oscillator frequency. Input from the dc amplifier keeps the VCO locked into the frequency of the reference oscillator. Although locked in, there is always a finite phase difference between the input and output.

A General-Purpose VCO

A popular monolithic IC VCO is the LM566C shown in Figure 10–20. This is a general-purpose VCO that may be used to generate square and triangle waves. The frequency of the square and triangle waves is a very linear function of a control voltage and of an external resistor and capacitor. The LM566C operates over a supply voltage range of 10–24 V, and provides very linear modulation characteristics, high-temperature stability, and excellent power supply voltage rejection (PSRR). It has a 10-to-1 frequency range with fixed capacitor, and frequency is programmable by means of current, voltage, resistor, or capacitor. It can be operated from either a single supply or a split ($\pm$) power supply. Its applications include FM modulation, signal generation, function generation, frequency shift keying (FSK), and tone generation.

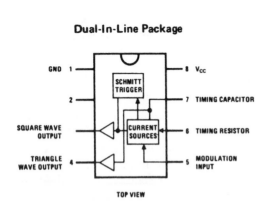

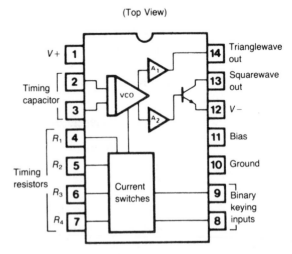

Figure 10–20. Block Diagram of the LM566C VCO (Courtesy of National Semiconductor Corporation, © copyright 1980.)

Figure 10–21. Connection and Block Diagram for the XR-2207 VCO (Courtesy of Raytheon Company, Semiconductor Division, © copyright 1978.)

A Four-Block VCO

Another monolithic IC VCO is the XR-2207 shown in Figure 10–21. The circuit comprises four functional blocks: a variable-frequency oscillator that generates the basic periodic waveforms; four current switches that are actuated by binary keying inputs; and two buffer amplifiers for simultaneous triangle and square wave outputs, available over a range of 0.01 Hz to 1 MHz. The internal switches transfer the oscillator current to any of four external timing resistors to produce four discrete frequencies that are selected according to the binary logic levels at the keying terminals. This device is ideally suited for FM, FSK, and sweep or tone generation, as well as for PLL applications.

10.7 PLL Applications

There are many applications for the PLL circuit. PLLs are ideally suited for routine applications as AM and FM detectors, FSK decoders, signal conditioners, prescalars for frequency counters, tone detectors, and touch-tone decoders. In general, when a PLL is used to control the frequency of an oscillator, such as in radio and TV receiver sound systems, it is referred to as *automatic frequency control* (AFC). In the RF tuner section of TV receivers the PLL is referred to as *automatic fine tuning* (AFT), and in the horizontal synchronizing circuit, it is called the horizontal AFC.

Frequency Synthesis

An important application of the PLL is in communication systems, where a crystal-controlled oscillator, such as the MC4024 shown in Figure 10–22, is the

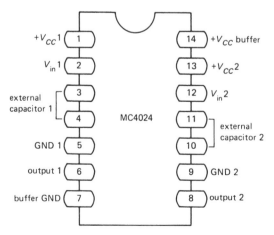

Figure 10–22. MC4024 VCO (Courtesy of Breton Publishers, © copyright 1982.)

standard source for a reference. The PLL circuit then provides an oscillator without a crystal with the same frequency stability as the crystal reference oscillator. This procedure is called *frequency synthesis,* that is, putting together, or mixing, two frequencies to provide the desired output.

A basic frequency synthesizer is illustrated in Figure 10–23, where a divide-by-N ($\div N$) counter is inserted in the feedback path of a PLL. The output frequency of the synthesizer is

$$f_{\text{out}} = Nf_{\text{in}} \tag{10.6}$$

where

N = the divider value of the $\div N$ counter

f_{in} = input frequency (generally from a crystal-controlled oscillator)

The synthesizer phase detector produces a dc control voltage that is proportional to the phase difference between the input frequency, f_{in}, and the $\div N$ counter output, f_{out}/N. The counter generates a single output pulse for every N input pulse. The dc control voltage from the phase detector, after filtering and amplification, then controls the output frequency of the VCO. The output signal from the $\div N$ counter is equivalent to the reference input frequency, except for the following small phase difference:

$$f_{\text{in}} = \frac{f_{\text{out}}}{N} \tag{10.7}$$

Example 10.5

A frequency synthesizer has an input frequency of 750 kHz. The $\div N$ counter is set at a count of 50. Calculate the synthesizer frequency output.

Solution

Use Equation 10.6:

$$f_{\text{out}} = Nf_{\text{in}} = 50 \times 750 \text{ kHz} = 37.5 \text{ MHz}$$

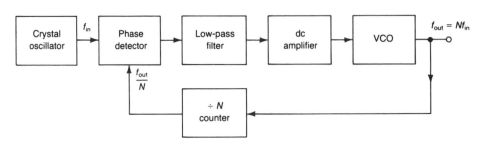

Figure 10–23.   A Frequency Synthesizer

Example 10.6

What is the input frequency of the phase detector if the output frequency of a synthesizer is 200 MHz and the $\div N$ counter is set at a count of 400?

Solution

Use Equation 10.7:

$$f_{\text{in}} = \frac{f_{\text{out}}}{N} = \frac{200 \text{ MHz}}{400} = 500 \text{ kHz}$$

Data Communications

The XR-2211 shown in Figure 10–24 is a monolithic PLL system designed especially for use in data communications. It operates over a wide supply voltage range of 4.5–20 V and a wide frequency range of 0.01 Hz to 300 kHz. It can accommodate analog signals between 2 mV and 3 V, and can interface with conventional logic families. The circuit consists of a basic PLL for tracking an input signal frequency within the passband, a quadrature phase detector that provides carrier detection, and an FSK voltage comparator that provides FSK demodulation. External components are used to independently set carrier frequency, bandwidth, and output delay. Design equations and typical applications of the XR-2211 are available in the data sheets in Appendix E.

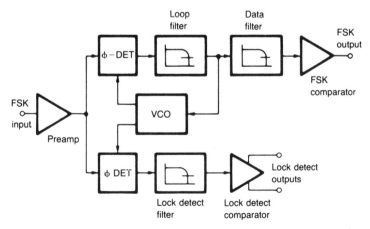

Figure 10–24. Functional Block Diagram of the XR-2211 PLL (Courtesy of Raytheon Company, Semiconductor Division, © copyright 1978.)

560 Series PLL

A group of monolithic analog PLL devices are the 560 series, which use the double-balanced mixer. You will recall that the output of this circuit is an average dc voltage that is proportional to the phase difference between its two inputs. We will discuss the most popular of the series, the 565 and the 567, illustrated in Figure 10–25 and Figure 10–26, respectively.

The 565 is a general-purpose PLL that contains a stable, highly linear VCO for low distortion FM demodulation, and a double-balanced phase detector with good carrier suppression. The VCO frequency is set with an external resistor and capacitor, and a tuning range of 10:1 can be obtained with the same capacitor. The bandwidth, response speed, capture, and pull-in range can be adjusted over a wide range with an external resistor and capacitor. Inserting a digital frequency divider between the VCO and the phase detector will provide frequency synthesis.

Other applications of the 565 are data and tape synchronization, FSK demodulation, FM demodulation, tone decoding, frequency multiplication and division, telemetry reception, and signal generation. For design considerations, see the data sheets in Appendix F.

The 567 is a general-purpose tone decoder designed to provide a saturated transistor switch to ground when an input signal is present within the passband. The bandwidth, center frequency, and output delay are independently determined by external components. Typical applications of the 567 are in touch-tone decoding, precision oscillators, frequency monitoring and control, wideband FSK demodulation, ultrasonic controls, carrier current remote controls, communications paging decoders, and 0°–180° phase shifting. Diagrams for the applications

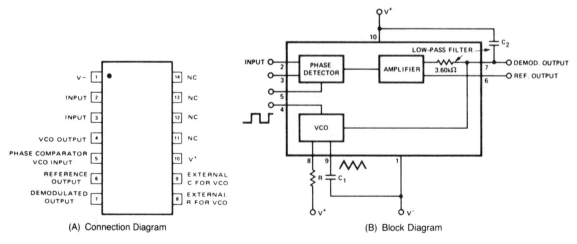

(A) Connection Diagram   (B) Block Diagram

Figure 10–25.   SE/NE565 PLL (Courtesy of Signetics, a subsidiary of U.S. Philips Corporation, © copyright 1982.)

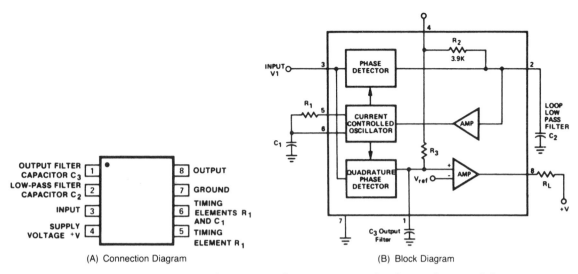

(A) Connection Diagram   (B) Block Diagram

Figure 10–26. SE/NE567 PLL (Courtesy of Signetics, a subsidiary of U.S. Philips Corporation, © copyright 1982.)

are shown in the data sheets in Appendix G. Many additional applications for modulators, demodulators, and PLLs will be explored in Chapter 12.

10.8 Summary

1. Amplitude modulation (AM) is the result of varying a carrier wave in amplitude at a certain rate.

2. Frequency modulation (FM) is the result of varying frequencies above and below the center frequency of a carrier wave, while maintaining a constant amplitude.

3. Phase modulation (PM) is a modified form of FM.

4. Demodulation, or detection, is the process of extracting the audio intelligence from the carrier wave and passing it on to a speaker or display device, such as a television screen.

5. AM demodulators are called envelope detectors, and FM demodulators are called discriminators.

6. Television signals are composite signals that consist of AM video information and FM audio information.

7. The diode detector is a commonly used AM demodulator, and the quadrature detector circuit is commonly used in FM demodulation.

8. The name *compandor* is a contraction of the two functions that the device performs—compression and expansion.

9. Phase-locked loops (PLL) are basically closed loops that provide an electronic feedback to lock in a desired frequency. A PLL consists of a phase detector, a low-pass filter, a dc amplifier, and a voltage-controlled oscillator (VCO).

10. The phase detector in a PLL compares an input reference frequency and the output frequency of the VCO, and generates a dc control voltage proportional to the phase difference of the two input signals.

11. The low-pass filter in a PLL can be either passive or active. It removes noise from the two oscillator signals, and it controls the lock, capture, bandwidth, and transient response of the loop.

12. The dc amplifier in a PLL provides the desired level of control voltage in the polarity needed for the VCO.

13. The VCO in a PLL uses a varactor to set oscillator frequency. The varactor is a semiconductor capacitive diode that operates on the principle of a varying capacitance that is inversely proportional to the amount of applied dc reverse voltage.

14. There are three operating modes for the PLL: free-running, capture, and phase-lock (sometimes referred to as lock-in or tracking).

15. Lock range is the frequency range over which the PLL can follow incoming signals.

16. Capture range is the bandwidth over which capture is possible. Capture range is never wider than lock range.

17. Changing the $\div N$ count in a frequency synthesizer can provide many different frequencies while using a single reference crystal-controlled oscillator.

18. There are many monolithic IC PLL circuits available that require only a few external components to provide a multitude of applications.

19. Frequency shift keying (FSK) means that the carrier frequency is changed from one value to another instead of keying the transmitter on and off.

10.9 Questions and Problems

10.1 What is the purpose of modulation in a radio transmitter?

10.2 What function does the detector in a radio receiver perform?

10.3 State the major advantage of FM over AM.

10.4 Name two FM demodulator circuits.

10.5 Draw a block diagram of (a) an AM radio receiver and (b) an FM radio receiver.

10.6 Briefly describe the operation of (a) an AM receiver and (b) an FM receiver.

10.7 Draw a block diagram of a PLL and describe the operation of the four functional blocks.

10.8 What is the purpose of the varactor in the VCO circuit? Explain its operation.

10.9 What is the output frequency of a synthesizer if the input frequency is 1.5 MHz and the $\div N$ counter is set at a count of 20?

10.10 What is the input frequency of the phase detector if the output frequency of a synthesizer is 350 MHz and the counter is set to a count of 240?

10.11 Calculate the phase detector output voltage for a conversion gain of 0.02 V/rad and an input phase difference of 0.3 rad.

10.12 Calculate the phase detector output voltage for a conversion gain of 30 mV/rad and an input phase difference of 23°. (HINT: 1 radian = 57.3°)

10.13 What is the percent modulation of a signal with a carrier frequency of 3 A, a maximum current of 6 A, and a minimum current of 0 A?

10.14 What is the percent modulation of a signal with a carrier frequency of 2 A, a maximum current of 6 A, and a minimum current of 4 A?

10.15 What will be the result at the receiver if a transmitted signal exceeds 100% modulation? What is this situation called?

10.16 List eight applications for the PLL circuit.

10.17 List five applications for the programmable compandor.

Timers

11.1 Introduction

Among existing linear integrated circuits, perhaps the most popular is the 555 IC timer. Introduced in 1972 by Signetics Corporation, the 555 is a versatile, reliable, low-cost device that is easy to use in a variety of applications and generally requires simple connections to a few external low-cost components. The device can operate from supply voltages ranging from 4.5 V to 18 V, making it compatible with both TTL circuits and op amp circuits.

More sophisticated and wide-ranging versions of the simple IC timer have been developed. The 556 is a dual 555 timer package, and the 2240 is a programmable timer/counter combination that contains a 555 timer plus a programmable binary counter in a single 16-pin package. The LM322 is a precision timer that can operate from unregulated power supplies over a range of 4.5–40 V. The timing range for the single 555 timer is microseconds to hours, while the timing range for the timer/counter is microseconds to days. Cascading the devices can extend the timing range of both to months or even years.

Our study of IC timers will begin with the functional 555 and its modes of operation, followed by other timers and applications.

11.2 **Objectives**

When you complete this chapter, you should be able to:

☐ State the basic characteristics of the 555, 322, and 2240 IC timers.

☐ Explain the operation of a monostable timer circuit.

☐ Explain the operation of an astable timer circuit.

☐ Determine the output pulse width of a monostable timer circuit and the factors that affect the pulse width.

☐ Determine the output pulse width and frequency of an astable timer circuit and the factors that affect them.

☐ Determine duty cycle and how to change it.

☐ Design and construct both monostable and astable timer circuits using the 555, 322, and 2240 IC timers.

☐ Design and construct a variety of application circuits using the 555, 322, and 2240 IC timers.

☐ Program the outputs of the 2240 programmable timer/counter IC.

☐ List many applications for the different types of IC timers.

11.3 **555 IC Timer**

The 555 monolithic timing circuit shown in the block diagram of Figure 11–1 is a very stable controller for producing accurate time delays or oscillations. Terminals are provided for *triggering* (pin 2) and *reset* (pin 4) if desired. In the time delay mode, the delay time is precisely controlled by one external resistor and capacitor. For stable operation in the oscillator mode, the free-running frequency and the duty cycle are both accurately controlled with two external resistors and one capacitor. The output is capable of sinking or sourcing 200 mA, is compatible with TTL logic circuits, and can drive relays and indicator lamps.

Modes of Operation

The 555 IC timer has two modes of operation. It can operate either as a monostable (one-shot) multivibrator or as an astable (free-running) multivibrator. It is available in two package styles, the eight-pin MINI DIP and the eight-pin TO-100, as shown in Figure 11–2. To understand how the 555 timer operates, we will briefly review the purpose of each of its terminals.

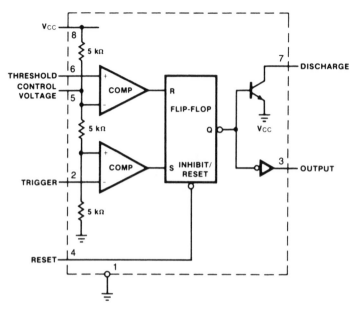

Figure 11-1. 555 IC Timer (Courtesy of Fairchild Camera & Instrument Corporation, Linear Division, © copyright 1982.)

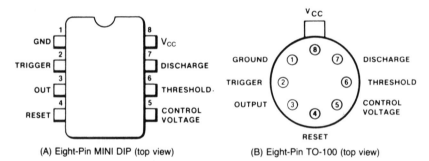

(A) Eight-Pin MINI DIP (top view)   (B) Eight-Pin TO-100 (top view)

Figure 11-2. Connection Diagrams for the 555 IC Timer (Part A: Courtesy of Fairchild Camera & Instrument Corporation, Linear Division, © copyright 1982; Part B: Courtesy of National Semiconductor Corporation, © copyright 1980.)

Terminals and Their Purposes

Refer to Figure 11-1 as we identify the terminals and their functions. Pin 1 is the common, or ground, terminal, and pin 8 is the positive supply voltage terminal, V_{CC}, which can be any voltage between $+4.5$ V and $+18$ V. This means that the 555 timer can be powered by +5 V digital logic supplies, by +15 V linear IC supplies, and by car and dry cell batteries.

Pin 3 is the output terminal, and can either sink or source current, as shown in Figure 11–3. Either a grounded or a floating (ungrounded) supply load can be connected. A floating supply load is ON when the output is LOW, and OFF when the output is HIGH. A grounded load is ON when the output is HIGH, and OFF when the output is LOW. For most applications, both types of loads are not required at the same time.

Technically, the maximum sink or source current is 200 mA, but in normal operation, the current is realistically about 40 mA. For load currents below 25 mA, the LOW output voltage is typically 0.1 V above ground, and the HIGH output voltage is typically 0.5 V below V_{CC}.

Pin 4 (Figure 11–1) is the RESET terminal, and has an overriding function. This terminal allows the 555 to be disabled and overrides command signals on the TRIGGER input. When not used, pin 4 should be tied to V_{CC} to avoid any possibility of false resetting. If the potential on the RESET terminal is reduced below 0.4 V, or grounded, both output pin 3 and discharge pin 7 are forced LOW, holding the output LOW.

Pin 5 is the CONTROL VOLTAGE terminal. An external voltage applied to pin 5 will change both threshold and trigger voltages, since this terminal allows direct access to the upper comparator and indirect access to the lower comparator through a 2:1 voltage divider (Figure 11–1). Use of this terminal is optional, but it does offer great flexibility through modification of the timing period, and it can be used to modulate the output waveform. When not used, it is practical to connect a 0.01 μF bypass capacitor between pin 5 and ground. This bypasses

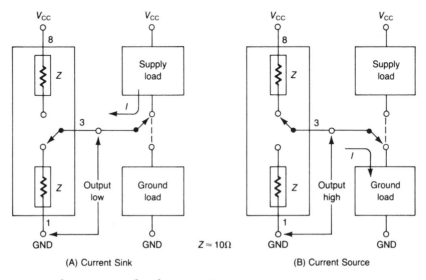

Figure 11–3.   Terminal 3 Outputs for the 555 IC Timer

noise or ripple voltages from the power supply, thus minimizing their effect on threshold voltage.

Pin 7 is the DISCHARGE terminal, used to discharge an external timing capacitor during the LOW output period. When the output is HIGH, pin 7 acts as an open circuit and allows the timing capacitor to charge at a rate determined by the RC time constant of external resistors and capacitors. In certain applications, this terminal can be used as an auxiliary output terminal, with current sink capabilities similar to those of output pin 3.

Pins 2 and 6, TRIGGER and THRESHOLD terminals, respectively, will be considered together because the two possible operating states and two possible memory states of the 555 are determined by both pins. Pin 2 is the input to the lower comparator, where it is compared with a lower threshold voltage, V_{lth}, that is equal to 1/3 V_{CC}, determined by the voltage divider (Figure 11–1). This pin is used to set the latch, which in turn causes a HIGH output. Triggering is accomplished by bringing the input level on pin 2 from above to below V_{lth}. Pin 6 is one input (pin 5 is the other) to the upper comparator, where it is compared with a higher threshold voltage, V_{uth}, that is equal to 2/3 V_{CC}, determined by the voltage divider. This pin is used to reset the latch, which in turn causes a LOW output. Resetting is accomplished by bringing the input level on pin 6 from below to above V_{uth}. The voltage range that can safely be applied to pins 2 and 6 is between V_{CC} and ground.

11.4 556 Timer

The 556 dual timing system shown in Figure 11–4 is simply two 555 timers in a single 14-pin DIP. The legends for the pins are the same as for the 555, but there are two separate circuits with different pin designations. Care should be exercised when using the 556 dual circuit, so that correct pin connections are made. For safety and accuracy, always refer to the data sheet when making connections.

Design Considerations

One of the most important considerations for a design using timer circuits is the selection of appropriate components for timing resistor R_A and timing capacitor C. If the inherent performance capability of the timer is to be achieved, the component selection process must be done very carefully.

Resistor Selection

Of the two timing components, the resistors are the easiest to select, because there is such a great number of standard values from which to choose, particularly

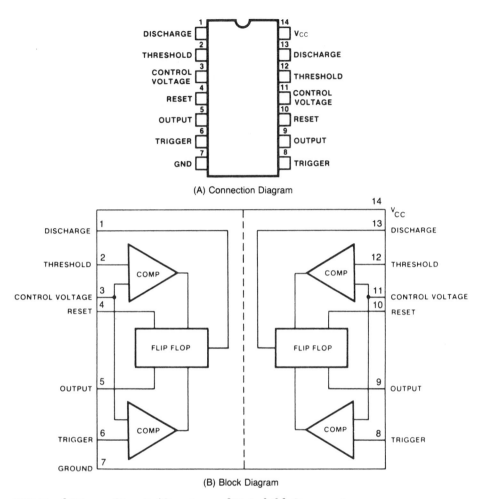

Figure 11–4.   556 Dual Timer Circuit (Courtesy of Fairchild Camera &
Instrument Corporation, Linear Division, © copyright 1982.)

for values up to 10 MΩ. For values above 10 MΩ, it may be necessary to procure
special components at a higher cost.

The most suitable resistor for precision timers is the metal-film type, available in a broad range of values from a large number of manufacturers. These
resistors cost more than carbon resistors, but their higher performance makes
the cost negligible.

Normally, potentiometers in the timing circuit should be avoided, because
the reliability and accuracy of the circuit may be compromised. In applications
demanding adjustable timing resistance, such as panel-mounted controls, only
the highest quality potentiometers should be used.

Capacitor Selection

Capacitor selection creates more problems than resistor selection for the designer. There are many different types of capacitors, but few types have tolerances below 5 percent. In general, capacitors are imprecise components with varying performance characteristics, and the range of available values is limited.

Some desirable characteristics for timing circuit capacitors are *low leakage, low dielectric absorption,* and *low temperature coefficient.* For example, if a capacitor is to be charged from a 1 μA source, it must have a leakage much lower than that, and the leakage must remain low over varying conditions of voltage, temperature, and time. Dielectric absorption means that when a capacitor is charged and then discharged by short-circuiting, all the energy that was stored during charge is not given up by the dielectric, and a residual voltage is retained across the capacitor. This is a serious problem because the timing principle depends on starting from zero volts. Therefore, high dielectric absorption capacitor types such as paper, ceramic, and some mica ones should not be used. The plastic-film capacitor family, which includes polystyrenes, polycarbonates, parylenes, and teflon, exhibits low dielectric absorption characteristics and is suitable for use in timing circuits.

A polystyrene capacitor is one of the best timing circuit choices in terms of cost and performance, but it is limited because it cannot be used above 85°C, and it is available in values only up to about 1 μF. The next best choice is a polycarbonate capacitor, which is available from many suppliers at reasonable cost, in values ranging up to about 100 μF, many with close tolerances.

Another possible problem with capacitors is their relatively large size, particularly for the higher values. In general, the higher the value, the larger the physical size and the higher the cost. Therefore, it is wise to minimize the size of the capacitor by reducing its value and increasing the value of the resistor associated with it in the timing circuit.

Except in very low-performance applications, electrolytic capacitors are not suitable for timing capacitor use, because of their characteristics of loose tolerances, high leakages, and poor stability over temperature changes.

11.5 Monostable Circuits

The most basic mode of operation of timer devices is the *triggered monostable* circuit shown in Figure 11–5. Note the simplicity of the circuit; it consists of only the two timing components, R_A and C, the timer itself, and bypass capacitor C_1, which is used only for noise immunity. When a negative-going input pulse applied to pin 2 reaches a level less than 1/3 V_{CC}, the timer is triggered and begins its timing cycle. The output goes HIGH and the voltage across C begins

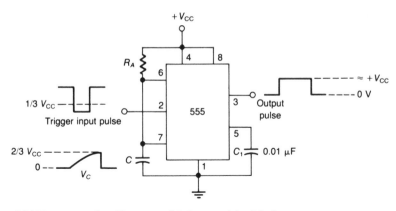

Figure 11–5. 555 Timer in the Triggered Monostable Mode

to charge towards V_{CC} at a rate determined by the RC time constant of R_A and C. When the capacitor voltage reaches 2/3 V_{CC}, the upper comparator (Figure 11–1) causes the output to go LOW, ending the timing cycle. The input and output voltage waveforms are shown in Figure 11–5.

The output pulse width PW, expressed in seconds, is defined as

$$PW = 1.1R_AC \tag{11.1}$$

There are relatively few restrictions on pulse width, and the timing components can have a wide range of values. Theoretically, there is no upper limit on PW, but practical limits are set by R and C values. The lower limit is set at 10 μs. It should be remembered that timer circuits can be cascaded to extend the pulse width to virtually any length, even periods of days, weeks, or months.

Example 11.1

Assume the values of the timing components in Figure 11–5 are $R_A = 10$ kΩ and $C = 0.001$ μF. Calculate the pulse width of the output pulse.

Solution

Use Equation 11.1:

$$PW = 1.1R_AC = 1.1 \times 10 \text{ k}\Omega \times 0.001 \text{ μF} = 11 \text{ μs}$$

The practical lower value for R_A is around 10 kΩ, and the practical lower value for C is around 0.001 μF. For capacitances below this value, stray capacitance affects accuracy and predictability of the timer circuit. In practice, it is best to select C first to minimize size and cost, then select R_A.

Example
11.2

Assume a desired output pulse width of 100 ms. Select the values for the timing resistor if $C = 1.0 \ \mu F$.

Solution

Rearrange Equation 11.1 to solve for R_A:

$$R_A = \frac{PW}{1.1C} = \frac{100 \text{ ms}}{1.1 \times 1.0 \ \mu F} = 90.9 \text{ k}\Omega$$

Example
11.3

If $R_A = 10 \text{ M}\Omega$, calculate C for an output pulse width of 100 s.

Solution

Rearrange Equation 11.1 to solve for C:

$$C = \frac{PW}{1.1R_A} = \frac{100 \text{ s}}{1.1 \times 10 \text{ M}\Omega} = 9.1 \ \mu F$$

Examples 11.2 and 11.3 demonstrate the wide range of values that may be used for the timing components. Figure 11–6 is a quick-reference design aid for selecting components and determining output pulse width.

Trigger Input Conditioning

Note that Figure 11–5 shows no trigger input conditioning components, which implies that the driving source must be capable of satisfying the trigger voltage requirements. In general, the only restriction on the input pulse is that it have a width less than the output pulse. Because of the internal latching mechanism (the flip-flop in Figure 11–1), the timer will always time out once triggered, even if subsequent noise pulses, such as bounce, are present on the trigger input. This feature makes the 555 an excellent device to use when connecting to noisy sources.

To increase stability and accuracy, the input conditioning network shown in Figure 11–7 is added to the circuit. Input resistor R_{in} insures that the output is LOW and that input capacitor C_{in} is charged to V_{CC} until the negative-going edge of the trigger pulse occurs. The time constant of R_{in} and C_{in} should be short compared with the output pulse width, established by the time constant of R_A and C. Diode D_1 prevents the timer from triggering on the positive-going edge of the input pulse.

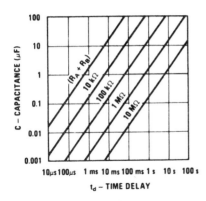

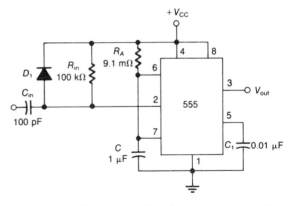

Figure 11–6. Quick-Reference Design Aid for Determining Output Pulse Width (Courtesy of National Semiconductor Corporation, © copyright 1980.)

Figure 11–7. Input Conditioning Network for the 555 Timer

Example 11.4

Calculate (a) the output pulse width (PW) and (b) the time constant (τ) of the input circuit in Figure 11–7.

Solutions

    a.   Use Equation 11.1:
$$PW = 1.1 R_A C = 1.1 \times 9.1 \ \text{M}\Omega \times 1 \ \mu\text{F} = 10 \ \text{s}$$
    b. $\tau = R_{in} C_{in} = 100 \ \text{k}\Omega \times 100 \ \text{pF} = 10 \ \mu\text{s}$

False Reset Prevention

Note that reset pin 4 is tied to V_{CC} in both Figure 11–5 and Figure 11–7. This is to prevent false resetting, for if pin 4 is grounded at any time, both output pin 3 and discharge pin 7 go to ground potential, removing any accumulated charge on C and pulling the output LOW. Those conditions will remain as long as the reset pin is grounded.

11.6 Astable Circuits

The 555 timer connected in the astable, or free-running, mode is shown in Figure 11–8A. This configuration uses three timing components—R_A, R_B, and C—the timer, and bypass capacitor C_1 for noise immunity. When voltage is first applied, the voltage across C is LOW, causing triggering at pins 2 and 6, which drives output pin 3 HIGH and opens pin 7. Capacitor C charges through resistors R_A

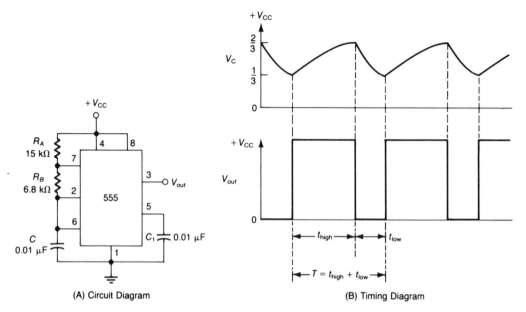

(A) Circuit Diagram

(B) Timing Diagram

Figure 11–8.   555 Timer in the Astable Mode

and R_B until V_C reaches the upper threshold level of 2/3 V_{CC}, at which time output pin 3 goes LOW. Discharge pin 7 also goes LOW, providing a discharge path for C through resistor R_B. When V_C drops to the lower threshold level of 1/3 V_{CC}, the timer is triggered again, starting a new cycle.

The output remains HIGH for the time interval (t) that C charges from 1/3 V_{CC} to 2/3 V_{CC}, as shown in Figure 11–8B. This time interval is

$$t_{high} = 0.695(R_A + R_B)C \tag{11.2}$$

For the time interval that C discharges from 2/3 V_{CC} to 1/3 V_{CC}, the output is LOW and is determined as follows:

$$t_{low} = 0.695R_BC \tag{11.3}$$

The total period of oscillation (T) is the sum of the HIGH and LOW time intervals:

$$T = t_{high} + t_{low} \tag{11.4a}$$

or

$$T = 0.695(R_A + 2R_B)C \tag{11.4b}$$

The frequency of oscillation of the free-running oscillator configuration is

$$f_{op} = \frac{1}{T} \tag{11.5a}$$

or

$$f_{op} = \frac{1.44}{(R_A + 2R_B)C} \tag{11.5b}$$

Example 11.5

Calculate (a) t_{high}, (b) t_{low}, (c) T, and (d) f_{op} for the astable timer circuit in Figure 11–8A.

Solutions

a. Use Equation 11.2:
$$t_{\text{high}} = 0.695(R_A + R_B)C = 0.695(15 \text{ k}\Omega + 6.8 \text{ k}\Omega)0.01 \text{ μF} \approx 152 \text{ μs}$$

b. Use Equation 11.3:
$$t_{\text{low}} = 0.695R_BC = 0.695 \times 6.8 \text{ k}\Omega \times 0.01 \text{ μF} \approx 47 \text{ μs}$$

c. Use Equation 11.4a:
$$T = t_{\text{high}} + t_{\text{low}} = 152 \text{ μs} + 47 \text{ μs} = 199 \text{ μs}$$
or use Equation 11.4b:
$$T = 0.695(R_A + 2R_B)C$$
$$= 0.695[15 \text{ k}\Omega + (2 \times 6.8 \text{ k}\Omega)]0.01 \text{ μF} \approx 199 \text{ μs}$$

d. Use Equation 11.5a:
$$f_{\text{op}} = \frac{1}{T} = \frac{1}{199 \text{ μs}} = 5 \text{ kHz}$$
or use Equation 11.5b:
$$f_{\text{op}} = \frac{1.44}{(R_A + 2R_B)C} = \frac{1.44}{[15 \text{ k}\Omega + (2 \times 6.8 \text{ k}\Omega)]0.01 \text{ μF}} = 5 \text{ kHz}$$

Duty Cycle

The duty cycle (D) is the ratio of the time interval of the LOW output (t_{low}) to the total period of oscillation (T), and is expressed as a percentage:

$$D = \frac{t_{\text{low}}}{T} \times 100\% \tag{11.6a}$$

or

$$D = \frac{R_B}{R_A + 2R_B} \times 100\% \tag{11.6b}$$

Within limits, the duty cycle can be programmed by the ratios of R_A and R_B. As you can see by observing Equation 11.6b, by making R_B large with respect to R_A, the duty cycle can approach 50 percent, or a square wave. On the other hand, if R_A is large with respect to R_B, the duty cycle approaches zero. However, since R_B cannot be allowed to reach zero, the duty cycle likewise can never reach zero. Therefore, the practical range for the duty cycle is from around 1 percent to 50 percent. Later we will see how to design a variable duty cycle circuit that can range from 1 percent to 99 percent.

Example 11.6

Determine the duty cycle for the circuit in Figure 11–8A.

Solution

Use Equation 11.6b:

$$D = \frac{R_B}{R_A + 2R_B} \times 100\% = \frac{6.8 \text{ k}\Omega}{15 \text{ k}\Omega + (2 \times 6.8 \text{ k}\Omega)} \times 100\% = 23.8\%$$

Example 11.7

Assume R_B is 1000 times the value of R_A in Figure 11–8A. Calculate the duty cycle.

Solution

Since $R_A = 15 \text{ k}\Omega$, $R_B = 1000 \times 15 \text{ k}\Omega = 15 \text{ M}\Omega$. Then,

$$D = \frac{15 \text{ M}\Omega}{15 \text{ k}\Omega + 30 \text{ M}\Omega} \times 100\% = 49.9\%$$

Example 11.8

Assume $R_A = 100R_B$ in Figure 11–8A. Calculate the duty cycle.

Solution

Since $R_B = 6.8 \text{ k}\Omega$, $R_A = 100 \times 6.8 \text{ k}\Omega = 680 \text{ k}\Omega$. Then,

$$D = \frac{6.8 \text{ k}\Omega}{680 \text{ k}\Omega + 13.6 \text{ k}\Omega} \times 100\% = 0.998\%$$

Fifty Percent Duty Cycle

A simple method of producing a 50 percent duty cycle is shown in Figure 11–9. In this network, R_A is equal to R_B, and a diode is connected in parallel with R_B. The capacitor C now charges through R_A and the diode, but discharges through R_B, and the output waveform times are determined as follows:

$$t_{\text{high}} = 0.695R_AC \tag{11.7a}$$
$$t_{\text{low}} = 0.695R_BC \tag{11.7b}$$
$$T = 0.695(R_A + R_B)C \tag{11.7c}$$

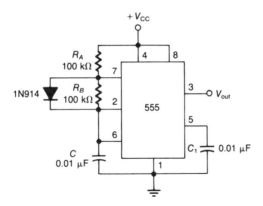

Figure 11–9. Circuit for the 50 Percent Duty Cycle

11.7 2240 Programmable Timer/Counter

The 2240 programmable timer/counter shown in Figure 11–10 is a monolithic controller capable of producing accurate delays from microseconds to five days. By cascading two of these timers, delays up to three years can easily be generated. The timer consists of a time-base oscillator, a programmable eight-bit counter, and a control flip-flop. An external RC network sets the oscillator frequency and allows delay times from $1RC$ to $255RC$ to be selected. In the astable mode of operation, 255 frequencies or pulse patterns can be generated from a single RC network. These frequencies or pulse patterns can also easily be synchronized to an external signal. The trigger, reset, and outputs are all TTL and DTL compatible for easy interface with digital systems. The timer's high accuracy and versatility in producing a wide range of time delays makes it ideal as a direct replacement for mechanical or electromechanical devices.

Operation

When power is applied to the 2240 with no trigger or reset inputs, the circuit starts with all outputs HIGH. The timing cycle is initiated with a positive-going trigger pulse at pin 11. This trigger pulse activates the time-base oscillator, enables the counter section, and sets the counter outputs LOW. The time-base oscillator generates timing pulses with a period of $T = 1RC$. These clock pulses are counted by the binary counter section. The timing sequence is completed when a positive-going reset pulse is applied to pin 10, which returns all outputs to HIGH.

Once triggered, the circuit is immune from additional trigger pulses until the timing cycle is completed or a reset input is applied. If both the reset and trigger are activated at the same time, the trigger takes precedence.

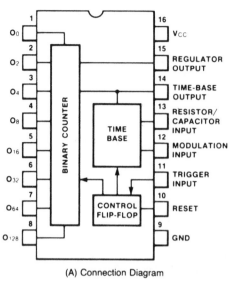

(A) Connection Diagram

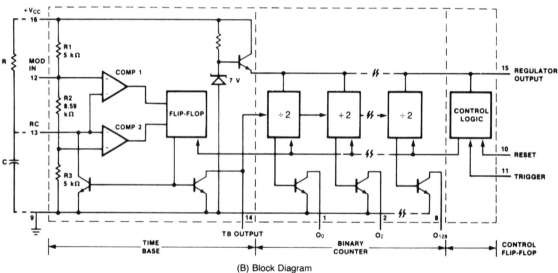

(B) Block Diagram

Figure 11–10.   2240 Programmable Timer/Counter (Courtesy of Fairchild Camera & Instrument Corporation, Linear Division, © copyright 1982.)

Basic Circuit

The basic circuit connection for timing applications is shown in Figure 11–11. With switch S_1 closed, the device is operating in the monostable mode. With S_1 open, the device is operating in the astable mode.

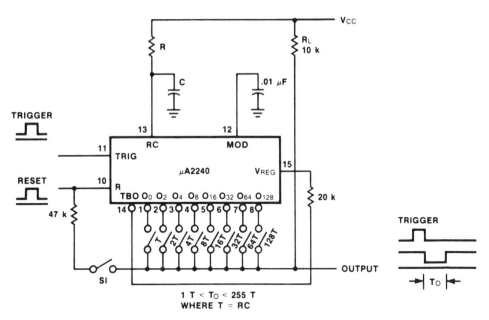

Figure 11–11. Basic Circuit Connection for Timing Applications (Courtesy of Fairchild Camera & Instrument Corporation, Linear Division, © copyright 1982.)

Monostable Operation

In precision timing applications the 2240 is used in its monostable mode. The output is normally HIGH and goes LOW following a trigger input. It remains in the LOW state for the time duration, T_{out}, and then returns to the HIGH state. The duration of the timing cycle is

$$T_{out} = N\tau \tag{11.8}$$

where

$\tau = RC$ = time base as set by the choice of timing components at RC pin 13

N = an integer between 1 and 255 as determined by the combination of counter outputs $0_0 \ldots 0_{128}$, pins 1 through 8, connected to the output bus.

The binary counter outputs, $0_0 \ldots 0_{128}$, pins 1 through 8, are open-collector type stages and can be shorted together to a common pull-up resistor to form a wired-OR connection; the combined output will be LOW as long as any one of the outputs is LOW. The time delays associated with each counter output can be added together by simply shorting the outputs together to form a common bus, as shown in Figure 11–11. For example, if only pin 6 is connected to the output and the rest are left open, the total timing duration T_{out} is 32τ. However, if pins 1, 5, and 6 are shorted to the output bus, the total time delay is $T_{out} = (1 + 16 + 32)\tau = 49\tau$. In this manner, by proper choice of counter terminals connected to the output bus, the timing cycle can be programmed to be any duration between 1τ and 255τ.

Cascading Units

Two 2240 units can be cascaded, as shown in Figure 11–12, to generate extremely long time delays. The total timing cycle of two cascaded units can be programmed from $T_{out} = 256\tau$ to $T_{out} = 65,536\tau$ in 256 discrete steps by selectively shorting one or more of the counter outputs from unit 2 to the output bus. In this application the reset and the trigger input terminals of both units are tied together and the unit 2 time base is disabled. Normally, the output is HIGH when the system is reset. On triggering, the output goes LOW where it remains for a total of $(256)^2$, or 65,536, cycles of the time-base oscillator.

Example 11.9	

Assume counter output pins 2, 6, and 8 in Figure 11–11 are shorted to the output bus. Determine the timing duration if $R = 2$ MΩ and $C = 1$ μF.

Solution

Use Equation 11.8:

$$T_{out} = N\tau = (2 + 32 + 128)(2 \times 10^6 \ \Omega \times 1 \times 10^{-6} \ F) = (162)(2)$$
$$= 324 \ s$$

Example 11.10	

Assume all counter output pins of unit 1 and pins 1, 3, 5, and 7 of unit 2 in Figure 11–12 are shorted to the output bus. $R = 10$ MΩ and $C = 100$ μF. Determine T_{out}.

Solution

Use Equation 11.8:

$$T_{out} = N\tau = (255 + 1 + 4 + 16 + 64)(10 \times 10^6 \ \Omega \times 100 \times 10^{-6} \ F)$$
$$= (340)(1000 \ s)$$
$$= 340,000 \ s = 5667 \ min = 94.4 \ hr = 3 \ days, \ 22 \ hr, \ 27 \ min$$

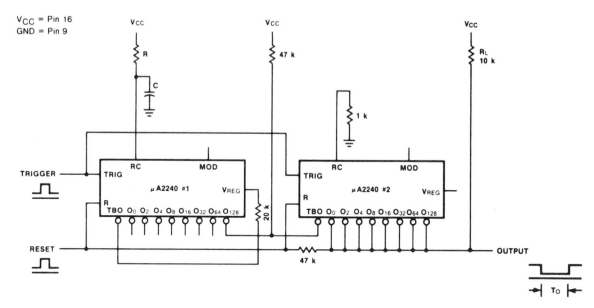

Figure 11-12. Circuit for a Long Time Delay (Courtesy of Fairchild Camera & Instrument Corporation, Linear Division, © copyright 1982.)

Example 11.11

Design a long time delay system using two 2240 units with all outputs tied together, to provide a timing cycle, T_{out}, of one year (365 days).

Solution

First, calculate the timing cycle in seconds:

$$T_{out} = \frac{60 \text{ s}}{1 \text{ min}} \times \frac{60 \text{ min}}{1 \text{ hr}} \times \frac{24 \text{ hr}}{1 \text{ day}} \times 365 \text{ days} = 31.536 \times 10^6 \text{ s}$$

Next solve for τ by rearranging Equation 11.8:

$$\tau = \frac{T_{out}}{N} = \frac{31.536 \times 10^6 \text{ s}}{65,536} = 481.2 \text{ s}$$

Now select a suitable value for C, say 100 μF, and solve for R:

$$\tau = RC = 481.2 \text{ s}$$
and
$$R = \frac{\tau}{C} = \frac{481.2 \text{ s}}{100 \times 10^{-6} \text{ F}} = 4.8 \text{ M}\Omega$$

Astable Operation

The 2240 can be operated in the astable mode by disconnecting reset terminal pin 10 from the counter outputs. The circuit in Figure 11–13 operates in its free-running mode with external trigger and reset signals. It starts counting and timing following a trigger input, continuing until an external reset pulse is applied. When a positive-going signal is applied to pin 10, the circuit reverts to its reset state, or HIGH outputs. This circuit is essentially the same as that in Figure 11–11 with switch S_1 open.

Continuous Operation

The circuit in Figure 11–14 is designed for continuous operation. It self-triggers automatically when the power supply is turned on, and continues to operate in its free-running mode indefinitely.

Synchronous Operation

When operating in the astable, or free-running, mode, the counter outputs can be used individually as synchronized oscillators, or they can be interconnected to generate complex pulse patterns, such as those shown in Figure 11–15. The pulse pattern repeats itself at a rate equal to the period of the highest

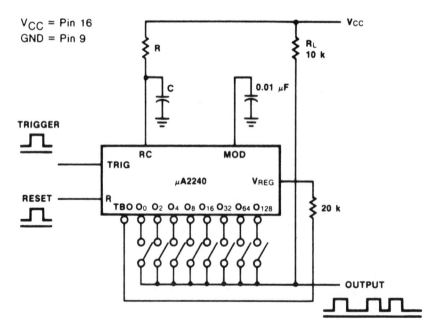

Figure 11–13. Astable Operation with External Trigger and Reset Inputs (Courtesy of Fairchild Camera & Instrument Corporation, Linear Division, © copyright 1982.)

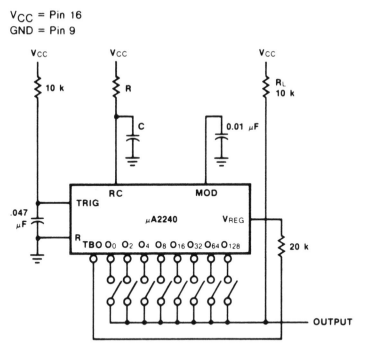

V_{CC} = Pin 16
GND = Pin 9

Figure 11–14. Free-Running Continuous Operation (Courtesy of Fairchild Camera & Instrument Corporation, Linear Division, © copyright 1982.)

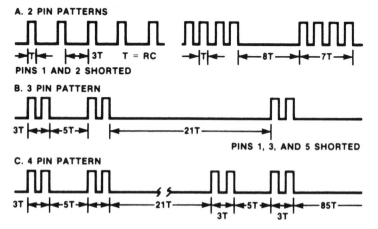

Figure 11–15. Binary Pulse Patterns (Courtesy of Fairchild Camera & Instrument Corporation, Linear Division, © copyright 1982.)

counter bit connected to the common output bus. The minimum pulse width contained in the pulse train is determined by the lowest counter bit connected to the output.

External Clock Operation

The 2240 can be operated with an external clock or time base when connected as shown in Figure 11–16. The internal time-base oscillator is disabled by connecting a 1 kΩ resistor from RC pin 13 to ground. The counters are triggered on the negative-going edges of the external clock pulse applied to the time-base oscillator (TBO) pin 14. For proper operation, a minimum clock pulse amplitude of 3 V and a clock pulse width of more than 1 μs is required.

Low Power Operation

For low power operation with supply voltages of 6 V or less, the internal time-base section can be powered down by connecting V_{CC} to pin 15 and leaving pin 16 open. In this configuration the internal time base does not draw any current, and the overall current drain is reduced by approximately 3 mA.

11.8   LM322 Precision Timer

The LM322 shown in Figure 11–17 is a precision timer of great versatility and high accuracy. It operates with unregulated supplies of 4.5–40 V while maintaining constant timing periods from microseconds to hours. Internal logic and

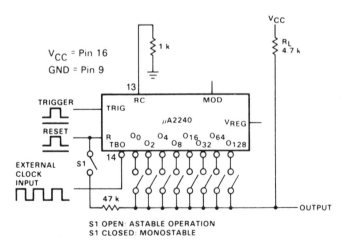

Figure 11–16.   Operation with an External Clock (Courtesy of Fairchild Camera & Instrument Corporation, Linear Division, © copyright 1982.)

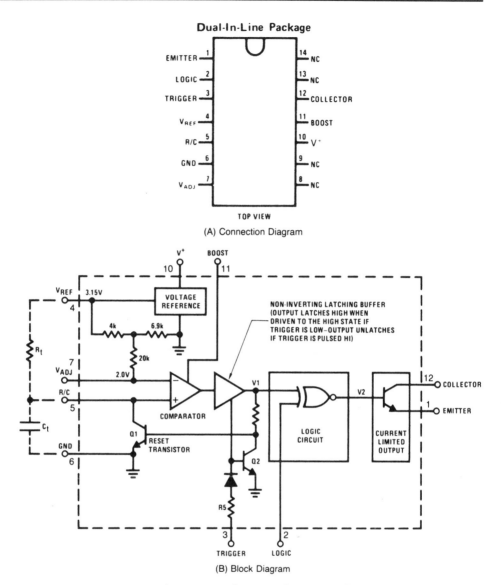

Dual-In-Line Package

EMITTER — 1   14 — NC
LOGIC — 2   13 — NC
TRIGGER — 3   12 — COLLECTOR
V_{REF} — 4   11 — BOOST
R/C — 5   10 — V^+
GND — 6   9 — NC
V_{ADJ} — 7   8 — NC

TOP VIEW

(A) Connection Diagram

(B) Block Diagram

Figure 11–17. LM322 Precision Timer (Courtesy of National Semiconductor Corporation, © copyright 1980.)

regulator circuits complement the basic timing function and enable the 322 to operate in many different applications with a minimum of external components.

Operation

The output of the timer is a floating transistor with built-in current limiting. It can drive either ground-referred or supply-referred loads up to 40 V and 50 mA.

The floating nature of this output makes it ideal for interfacing, driving lamps and relays, and signal conditioning where an open collector or emitter is required. You can program a "logic reverse" circuit to make the output transistor either ON or OFF during the timing period.

Trigger input pin 3 has a threshold of 1.6 V independent of supply voltage, but it is fully protected against inputs as high as ± 40 V, even when using a 5 V power supply. The circuitry reacts only to the rising edge of the trigger pulse, and is immune to any trigger voltage during the timing periods.

V_{ref} pin 4 is the output of an internal 3.15 V regulator referenced to ground pin 6. Up to 5.0 mA can be drawn from this regulator for driving external loads. An internal 2 V divider between the reference and ground sets the timing period to $1RC$. The timing period can be voltage-controlled by driving this divider with an external source through V_{ADJ} pin 7. Timing ratios of 50:1 can be easily achieved. In most applications the timing resistor is tied to V_{ref}, but it need not be if a more linear charging current is required. The regulated voltage is very useful in applications where the 322 is not used as a timer. Such applications are switching regulators, variable reference comparators, and temperature controllers.

The comparator used in the 322 utilizes high-gain PNP input transistors to achieve 300 pA typical input bias current over a common-mode range of 0–3 V. Boost terminal pin 11 allows you to increase comparator operating current for timing periods less than 1 ms. This allows the timer to operate over a 3 µs to multihour timing range with excellent repeatability.

The RC pin 5 is tied to the noninverting side of the comparator and to the collector of reset transistor Q_1. Timing ends when the voltage on this pin reaches 2.0 V ($1RC$ time constant referenced to the 3.15 V regulator). Transistor Q_1 turns ON only if the trigger voltage has dropped below threshold. In comparator and regulator applications of the timer, the trigger is held permanently HIGH and the RC pin acts just like the input to an ordinary comparator. The maximum voltages that can be applied to this pin are $+5.5$ V and -0.7 V. Gain of the comparator is 200,000 or more, depending on the state of logic reverse pin 2 and the connection of the output transistor.

Ground pin 6 of the 322 need not necessarily be tied to system ground. It can be connected to any positive or negative voltage as long as the supply is negative with respect to V^+ pin 10. However, level shifting may be necessary for the input trigger if the trigger voltage is referred to system ground. This can be done by capacitive coupling, or by actual resistive or active level shifting.

Output Variations

The emitter and collector outputs of the timer can be treated just as if they are an ordinary transistor with 40 V minimum collector-emitter breakdown voltage. Normally, the emitter is tied to ground and the signal is taken from the collector, or the collector is tied to V^+ and the signal is taken from the emitter. Variations of these basic connections are possible. The collector can be tied to any positive voltage up to 40 V when the signal is taken from the emitter. However, the emitter will not be pulled higher than the supply voltage on V^+ pin 10. Connecting the collector to a voltage less than V^+ is allowed, but the emitter should not be connected to a low impedance load other than that to which the ground pin is tied.

Logic pin 2 is used to reverse the signal that appears at the output transistor. An open or HIGH condition on pin 2 programs the output transistor to be OFF during the timing period and ON all other times. Grounding pin 2 reverses the sequence, making the transistor ON during the timing period. Minimum and maximum voltages that may appear on the logic pin are 0 V and +5.0 V, respectively.

Basic Circuits

A basic timer using the 322 is shown in Figure 11–18A. The output is taken from the collector. R_t and C_t set the time interval with R_L as the load. During the timing interval, the output may be either HIGH or LOW depending on the connection of the logic pin. In the timing waveforms shown, note that the trigger pulse can be either shorter or longer than the output pulse width.

Another basic timer using the 322 is shown in Figure 11–18B. In this circuit, the output is taken from the emitter of the output transistor. As with the collector output, the output taken from the emitter may be either HIGH or LOW during the timing interval. Also, the trigger input may be either longer or shorter than the output pulse width.

Self-Starting

The 322 can be made into a self-starting oscillator by feeding the output back to the trigger input through a capacitor, as shown in Figure 11–19A. The operating frequency is determined by

$$f_{op} = \frac{1}{(R_t + R_1)(C_t)} \tag{11.9}$$

The output is a narrow negative pulse whose width is approximately twice the product of resistor R_2 and feedback capacitor C_f:

$$PW = 2R_2C_f \tag{11.10}$$

For optimum frequency stability, C_f should be as small as possible. The minimum value is determined by the time required to discharge C_t through the internal discharge transistor. A conservative value for C_f can be chosen from the graph shown in Figure 11–19B. For frequencies below 1 kHz, the frequency error introduced by C_f is a few tenths of one percent or less for R_t greater than 500 kΩ.

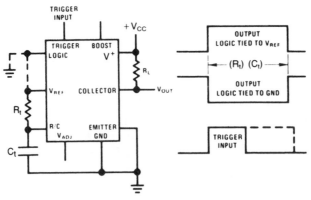

(A) Circuit with Collector Output

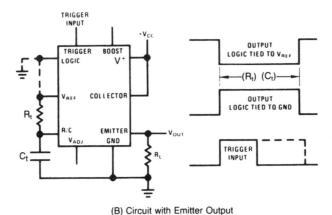

(B) Circuit with Emitter Output

Figure 11–18. Basic Timer Circuits (Courtesy of National Semiconductor Corporation, © copyright 1980.)

Example 11.12

Determine (a) the operating frequency of the self-starting oscillator in Figure 11–19A and (b) the pulse width of the output. Let $C_t = 0.015\ \mu F$.

Solutions

a. Use Equation 11.9:

$$f_{op} = \frac{1}{(R_t + R_1)(C_t)} = \frac{1}{(100\ k\Omega + 3\ k\Omega)(0.015\ \mu F)} = 647\ Hz$$

b. First determine C_f from Figure 11–19B, which indicates a value of approximately 0.1 μF for $C_t = 0.015\ \mu F$. Then use Equation 11.10 to solve for *PW*:

$$PW = 2R_2C_f = 2 \times 3\ k\Omega \times 0.1\ \mu F = 600\ \mu s$$

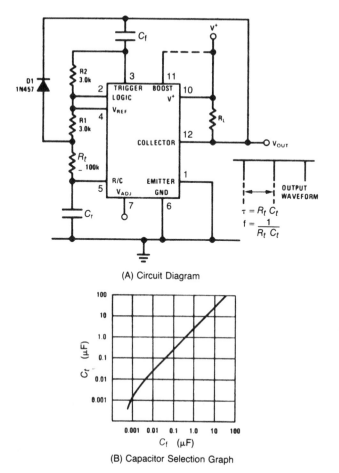

(A) Circuit Diagram

(B) Capacitor Selection Graph

Figure 11–19. An Astable Oscillator (Courtesy of National Semiconductor Corporation, © copyright 1980.)

11.9 Some Typical Timer Applications

The list of practical applications for IC timers is almost limitless, far too great for the scope of this book. In this section we will present but a few typical applications—a divider, a pulse width modulator, a pulse width detector, a one-hour timer, a staircase generator, a frequency-to-voltage converter, a digital interface, a variable duty-cycle pulse generator, a digital S & H, and an eight-bit ADC.

Divider

The circuit in Figure 11–20 is a divide-by-three circuit that uses a 555 timer in the monostable mode of operation. The frequency divider is designed by making

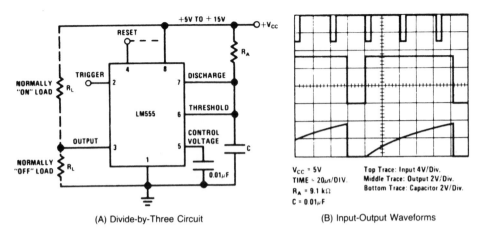

(A) Divide-by-Three Circuit

(B) Input-Output Waveforms

Figure 11–20. Frequency Divider Circuit (Courtesy of National Semiconductor Corporation, © copyright 1980.)

the timing interval longer than the period of the input signal, in this case, three times as long. The one-shot is triggered by the first negative-going pulse of the input signal, but the output will still be HIGH when the next two negative-going pulses occur. The one-shot will be retriggered on the fourth negative-going pulse. There is one output pulse for every three input pulses, therefore the input is divided by three.

Pulse Width Modulator

When the timer is connected in the monostable mode and triggered with a continuous pulse train, the output pulse width can be modulated by a signal applied to pin 5. The 555 in Figure 11–21A is connected as a pulse width modulator, and Figure 11–21B shows some waveform examples.

Pulse Width Detector

A simple, accurate pulse width detector using the LM322 is shown in Figure 11–22. In this application the logic terminal is normally held HIGH by resistor R_3. When a trigger pulse is received, transistor Q_1 is turned ON, driving the logic terminal to ground. The result of triggering the timer and reversing the logic at the same time is that the output does not change from its initial LOW condition. The only time the output will change states is when the trigger input stays HIGH longer than one time period set by R_t and C_t. The output pulse width is equal to the input trigger width minus R_tC_t. C_2 insures no output pulse for short trigger pulses (those less than RC) by prematurely resetting the timing capacitor when the trigger pulse drops. C_L filters the narrow spikes that would occur at the output due to propagation delays during switching.

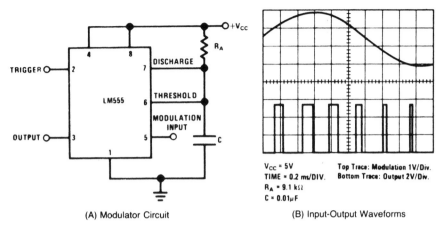

(A) Modulator Circuit

(B) Input-Output Waveforms

Figure 11–21. Pulse Width Modulator (Courtesy of National Semiconductor Corporation, © copyright 1980.)

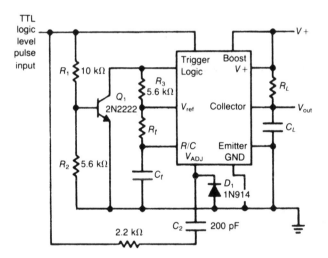

Figure 11–22. Pulse Width Detector (Courtesy of National Semiconductor Corporation, © copyright 1980.)

One-Hour Timer

The LM322 connected as a one-hour timer with manual controls for start, reset, and cycle end is illustrated in Figure 11–23. Switch S_1 starts timing, but has no effect after timing has started. Switch S_2 is a center-OFF switch that can either end the cycle prematurely with the appropriate change in output state and

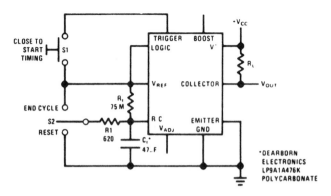

Figure 11–23. One-Hour Timer with Reset and Manual Cycle End (Courtesy of National Semiconductor Corporation, © copyright 1980.)

discharging of C_t, or cause C_t to be reset to 0 V without a change in output. In the latter case, a new timing period starts as soon as S_2 is released. The average charging current through R_t is about 30 nA, so some attention must be paid to parts layout to prevent stray leakage paths. The suggested timing capacitor has a typical self time constant of 300 hours and a guaranteed minimum of 25 hours at +25°C. Other capacitor types may be used if sufficient information is available on their leakage characteristics.

Staircase Generator

The 2240 timer/counter can be interconnected with an external op amp and a precision resistor ladder to form a staircase generator, as shown in Figure 11–24. Under reset conditions, the output is LOW. When a trigger is applied, the op amp output goes HIGH and generates a negative-going staircase of 256 equal steps. The time duration of each step is equal to the time-base period T. The staircase can be stopped at any level by applying a disable signal to pin 14 through a steering diode. The count is stopped when pin 14 is clamped at a voltage level not greater than 1.0 V.

Frequency-to-Voltage Converter

An accurate frequency-to-voltage converter can be made with the LM322 by averaging output pulses with a simple filter, as illustrated in Figure 11–25. Pulse width is adjusted with R_2 to provide initial calibration at 10 kHz. The collector of the output transistor is tied to V_{ref}, giving constant amplitude pulses equal to

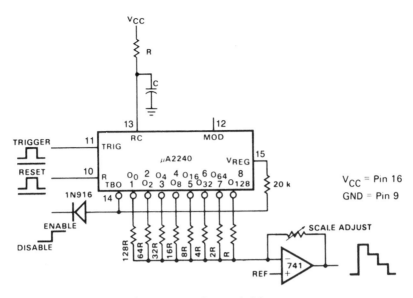

Figure 11–24. Staircase Generator (Courtesy of Fairchild Camera & Instrument Corporation, Linear Division, © copyright 1982.)

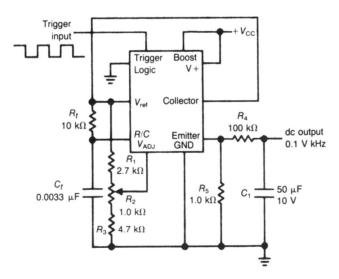

Figure 11–25. Frequency-to-Voltage Converter (Courtesy of National Semiconductor Corporation, © copyright 1980.)

V_{ref} at the emitter output. R_4 and C_1 filter the pulses to give a dc output equal to

$$V_{out(dc)} = R_t C_t V_{ref} f_{op} \qquad (11.11)$$

Linearity is about 0.2 percent for a 0–1 V output. If better linearity is desired, R_5 can be tied to the summing node of an op amp that has the filter in the feedback path. If a low output impedance is desired, a unity gain buffer can be tied to the output. An analog meter can be driven directly by placing it in series with R_5 to ground. A series RC network across the meter to provide damping will improve response at very low frequencies.

Digital Interfacing

The ability of the timer to interface to digital logic when operating off a high supply voltage is demonstrated in Figure 11–26. V_{out} swings between +5 V and ground with a minimum fanout of 5 for medium speed TTL. If the logic is sensitive to rise/fall time of the trailing edge of the output pulse, the trigger pin should be LOW at that time.

Variable Duty-Cycle Pulse Generator

A variable duty-cycle pulse generator can be constructed using the 555 timer, as illustrated in Figure 11–27. Independent charge and discharge paths for ca-

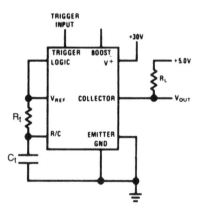

Figure 11–26. 30 V Supply Interface with 5 V Logic (Courtesy of National Semiconductor Corporation, © copyright 1980.)

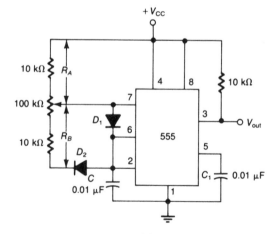

Figure 11–27. Variable Duty-Cycle Pulse Generator

pacitor C are established by diodes D_1 and D_2. The charge path for C is from V_{CC} through R_A and D_1. The discharge path for C is through D_2, R_B, and pin 7. (Note that R_A and R_B are made up of fixed resistors plus a portion of a potentiometer.) The charge and discharge times, t_{high} and t_{low}, are given by Equations 11.2 and 11.3, respectively. The period, T, and duty cycle, D, for this circuit are as follows:

$$T = 0.7(R_A + R_B)C \tag{11.12}$$

and

$$D = \frac{R_B}{R_A + R_B} \times 100\% \tag{11.13}$$

With this circuit configuration, duty cycles from 1 percent to 99 percent can be achieved.

Example 11.13

In Figure 11–27 assume the potentiometer to be set so that $R_A = 20$ kΩ. Calculate (a) the period T and (b) the duty cycle.

Solutions

a. Use Equation 11.12:
$$T = 0.7(R_A + R_B)C = 0.7(20 \text{ k}\Omega + 100 \text{ k}\Omega)0.01 \text{ }\mu\text{F} = 840 \text{ }\mu\text{s}$$
b. Use Equation 11.13:
$$D = \frac{R_B}{R_A + R_B} \times 100\% = \frac{100 \text{ k}\Omega}{20 \text{ k}\Omega + 100 \text{ k}\Omega} \times 100\% = 83.3\%$$

Example 11.14

In Figure 11–27 assume the potentiometer to be set so that $R_A = 90$ kΩ. Calculate (a) the period T and (b) the duty cycle.

Solutions

a. Use Equation 11.12:
$$T = 0.7(90 \text{ k}\Omega + 30 \text{ k}\Omega)0.01 \text{ }\mu\text{F} = 840 \text{ }\mu\text{s}$$
b. Use Equation 11.13:
$$D = \frac{30 \text{ k}\Omega}{90 \text{ k}\Omega + 30 \text{ k}\Omega} \times 100\% = 25\%$$

Close observation of Examples 11.13 and 11.14 will reveal that $R_A + R_B$ is constant, that T is constant, and that placement of the wiper arm on the potentiometer determines the duty cycle, or the percentage of the total period (T) that the output signal is LOW.

Digital Sample-and-Hold Circuit

A digital S & H circuit that uses the 2240 is shown in Figure 11–28. Circuit operation is similar to that of the staircase generator discussed earlier. When a strobe input is applied, the *RC* low-pass network between the reset and the trigger inputs resets the timer, then triggers it. This strobe input also sets the output of the bistable latch to a HIGH state and activates the counter.

The circuit generates a staircase voltage at the op amp output. When the level of the staircase reaches that of the analog input to be sampled, the comparator changes state, activates the bistable latch, and stops the count. At this point, the voltage level at the op amp output corresponds to the sampled analog input. Once the input is sampled, it is held until the next strobe signal. Minimum recycle time of the system is approximately 6 ms.

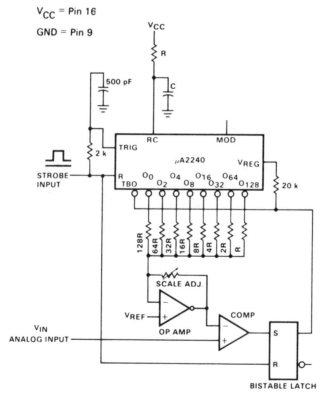

Figure 11–28.   Digital Sample-and-Hold Circuit (Courtesy of Fairchild Camera & Instrument Corporation, Linear Division, © copyright 1982.)

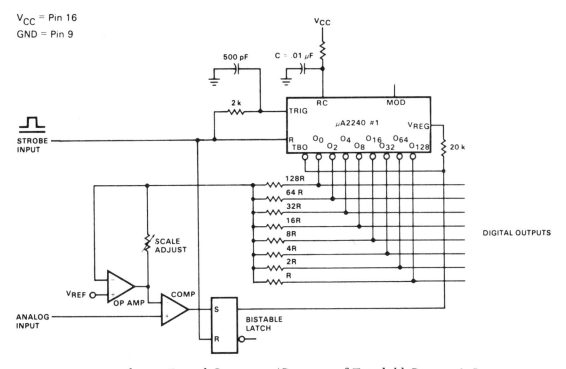

V_{CC} = Pin 16
GND = Pin 9

Figure 11–29.   Analog-to-Digital Converter (Courtesy of Fairchild Camera & Instrument Corporation, Linear Division, © copyright 1982.)

Eight-Bit ADC

A simple eight-bit analog-to-digital converter (ADC) system using the 2240 is illustrated in Figure 11–29. Circuit operation is very similar to that of the digital S & H circuit in Figure 11–28. In the case of the ADC, the digital output is obtained in parallel format from the binary-counter outputs with the output at pin 8 corresponding to the most significant bit (MSB). Recycle time for this circuit is approximately 6 ms.

There are a great number of other applications for the timers discussed in this chapter, as well as for other available timers. Reference to manufacturers' data sheets and application notes is recommended for those interested in further investigation of timers and timer circuits.

11.10 Summary

1. Timers are sophisticated, versatile linear IC devices that can be used in a multitude of applications.

2. In general, IC timers require only a minimum of external components to provide the desired results.

3. IC timer circuits can be connected in such a manner as to provide timing cycles from microseconds to days, weeks, months, or years.

4. IC timers operate in two basic multivibrator modes: monostable, or one-shot, and astable, or free-running.

5. Frequency, duty cycle, and pulse width are all determined by externally connected components.

6. Most linear IC timers will operate accurately when supplied with a wide range of supply voltages.

7. One of the most important considerations when designing timer circuits is the careful selection of timing components. Capacitors are the most critical components because of the wide variance in their characteristics, while resistors are more easily selected.

8. The 555 IC timer is an ideal timer for applications where noisy sources are used. The timer will always time out once triggered, providing immunity to noise pulses.

9. The 2240 programmable timer/counter IC can deliver extremely long time delays when cascaded. Programming the outputs of the device is a simple matter of selecting the number of outputs to be tied to the output bus.

10. The LM322 precision timer can operate with unregulated supply voltages over the range from 4.5 V to 40 V while maintaining accuracy in output timing periods from microseconds to hours.

11. The 322 can provide outputs that are normally ON or normally OFF by use of a "logic reverse" circuit.

12. Some of the applications of timers are as follows: frequency dividers, frequency multipliers, pulse width modulators, pulse width detectors, missing pulse detectors, window detectors, square-wave generators, staircase generators, digital sample-and-hold circuits, analog-to-digital converters, frequency-to-voltage converters, voltage-to-frequency converters, dc-dc converters, and variable duty-cycle pulse generators. Many other applications are available through manufacturers' data sheets and application notes.

11.11 Questions and Problems

11.1 What are the basic operating modes of IC timers?

11.2 List the functional blocks inside a 555 IC timer.

11.3 Explain the most basic mode of operation of the 555 timer.

11.4 How many external components are required with a 555 timer operating in the time delay mode? in the free-running mode?

11.5 Describe the characteristics of the 555 output pin 3.

11.6 Is there such a thing as an override function on the 555 timer? If so, describe its operation.

11.7 What is the purpose of a bypass capacitor in timer circuits?

11.8 What is one of the most important considerations in timer circuit design?

11.9 What is the best choice of resistor types for timer applications?

11.10 Describe some of the desirable characteristics of timing capacitors.

11.11 What is a good choice of timing capacitor types?

11.12 Are electrolytic capacitors an acceptable choice for timing circuit applications? Explain your answer.

11.13 Refer to Figure 11–5. Assume values of $R_A = 22$ kΩ and $C = 0.01$ μF. Calculate the pulse width of the output pulse.

11.14 A circuit design calls for a timing capacitor value of 10 μF and a desired output pulse width of 100 seconds. Calculate the value for the timing resistor.

11.15 A circuit design shows a timing resistor value of 2 MΩ and a desired output pulse width of 1 minute. Calculate the timing capacitor value.

11.16 Refer to Figure 11–6. Determine the timing component values for an output signal width of 100 ms. (NOTE: There are several. List them all.)

11.17 What is the purpose of an input-conditioning network?

11.18 What function does the input resistor of the input-conditioning network serve?

11.19 Refer to Figure 11–7. Assume $R_A = 10$ MΩ and $C = 0.01$ μF. Calculate the output pulse width.

11.20 Refer to Figure 11–7. Assume $R_{in} = 10$ kΩ and $C_{in} = 1000$ pF. Calculate the input network time constant.

11.21 Refer to Figure 11–30. Calculate (a) t_{high}, (b) t_{low}, (c) T, and (d) f_{op}.

11.22 Calculate the duty cycle for the circuit in Figure 11–30.

11.23 Refer to Figure 11–11. Assume $R = 20$ MΩ and $C = 10$ μF. Determine the timing duration for the following conditions: (a) output pins 2, 4, 6, and 8 are shorted to the output bus; (b) output pins 1, 3, 5, and 7 are shorted to the output bus; and (c) all pins are shorted to the output bus.

11.24 Refer to Figure 11–12. Assume $R = 100$

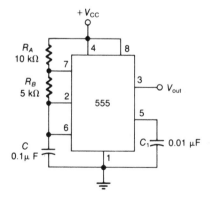

Figure 11–30. Circuit for Problem 11.21 and Problem 11.22

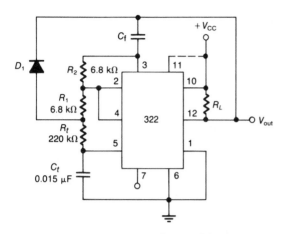

Figure 11–31.   Circuit for Problem 11.26

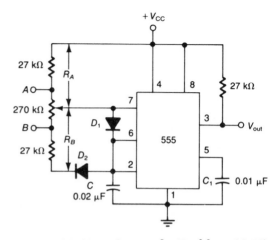

Figure 11–32.   Circuit for Problem 11.27

MΩ, $C = 100$ μF, and the desired time delay is exactly 50 days. Determine which pins must be shorted to the output bus.

11.25 Design a long time delay system with two 2240 units that will provide a timing cycle of 3 years. (HINT: Short all output pins to the output bus. This helps to keep the component values lower.)

11.26 Refer to Figure 11–31. Determine the operating frequency and the pulse width of the output signal. (HINT: Refer to Figure 11–19B to determine the value of C_f.)

11.27 Refer to Figure 11–32. Determine (a) the period T and the duty cycle when the potentiometer is set at point A, and (b) the period T and the duty cycle when the potentiometer is set at point B.

Chapter 12

Special Analog Circuits and Devices

12.1 Introduction

The wide range of linear IC devices that we have discussed up to this point is but a small sample of those available to the electronics technician. The versatility of such devices, and the applications for which they may be used, are limited only by the imagination of the designer. In this chapter we will present some special-purpose devices, as well as practical applications that use many circuits with which you are familiar. Some applications use individual ICs and some use a combination of ICs and discrete components.

12.2 Objectives

When you complete this chapter, you should be able to:

☐ Describe a variety of consumer-type linear IC applications.

☐ Design a variety of useful circuits using op amps, monolithic ICs, discrete devices, and external components.

12.3 Audio, Radio, and Television Circuits

In this section we will present a wide selection of consumer-type applications of linear IC devices. Many of the circuits discussed here can be constructed using either op amps with external components or monolithic ICs with the major circuit components contained within a single package.

Dual Power Amplifier

The LM377 is a dual power amplifier that offers high-quality performance for stereo phonographs, tape players, recorders, and AM-FM stereo receivers. It will deliver two watts per channel into 8 Ω or 16 Ω loads. The device contains an internal bias regulator to bias each amplifier and contains both internal current limit and thermal shutdown for device protection.

The LM377 operates from a 10–26 V supply, and provides a typical open-loop output gain of 90 dB, with 70 dB channel separation and 70 dB ripple rejection. The connection diagram is shown in Figure 12–1.

A minimum of external components is required for this device, as shown in the simple stereo amplifier circuit of Figure 12–2. Addition of an *RC* network in the feedback path between the speakers and the inverting input of the two amplifiers, as shown in Figure 12–3, provides bass boost to the basic stereo amplifier.

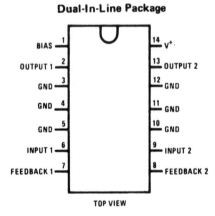

Dual-In-Line Package

Pin	Signal
1	BIAS
2	OUTPUT 1
3	GND
4	GND
5	GND
6	INPUT 1
7	FEEDBACK 1
14	V⁺
13	OUTPUT 2
12	GND
11	GND
10	GND
9	INPUT 2
8	FEEDBACK 2

TOP VIEW

Figure 12–1. Connection Diagram for LM377 Dual 2-Watt Audio Amplifier (Courtesy of National Semiconductor Corporation, © copyright 1980.)

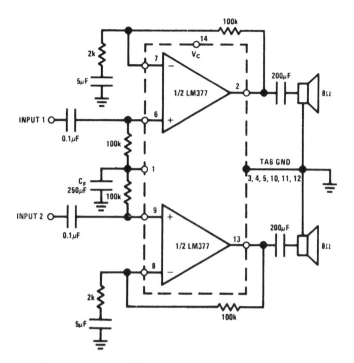

Figure 12–2. Simple Stereo Amplifier Using LM377 (Courtesy of National Semiconductor Corporation, © copyright 1980.)

Stereo Amplifiers

The LM1877 is a pin-for-pin replacement for the LM377. This device operates with a minimum of external components while still providing flexibility for use in stereo phonographs, tape recorders, and AM-FM stereo receivers. Other applications are servo amplifiers, intercom systems, and automotive products.

The LM1877 operates for a supply range of 6–24 V, has very low cross-over distortion, internal current limiting, short circuit protection, and thermal shutdown. It provides 70 dB channel separation and 65 dB ripple rejection. Each amplifier is biased from a common internal regulator to provide high power supply rejection and output Q point centering.

A stereo phonograph amplifier with bass tone control is shown in Figure 12–4A. The frequency response of the bass tone control of this circuit is essentially flat from 100 Hz to 20 kHz. The circuit voltage gain, A_v, equals 50 (34 dB) when feeding an 8 Ω load.

The circuit in Figure 12–4B is a stereo amplifier with output voltage gain A_v of 200, which is about 46 dB, into an 8 Ω load.

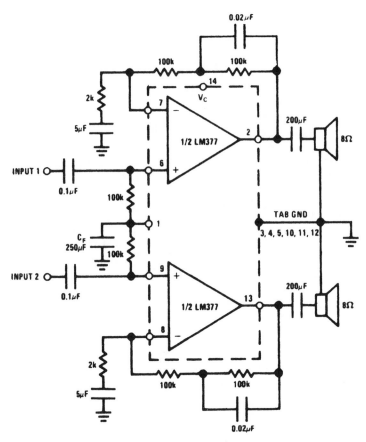

Figure 12–3. Stereo Amplifier with Bass Boost (Courtesy of National Semiconductor Corporation, © copyright 1980.)

Power Audio Amplifier

The LM380, illustrated in Figure 12–5, is a power audio amplifier designed for low-cost consumer application. Gain is internally fixed at 34 dB, the unique configuration of the input stage allows inputs to be ground referenced, and the output is automatically self-centering to one-half the supply voltage. The output is short circuit-proof with internal thermal limiting.

Uses of the LM380 include simple phonograph amplifiers (Figure 12–6A), phase-shift oscillators (Figure 12–6B), intercoms (Figure 12–6C), and bridge amplifiers (Figure 12–6D). Other uses are as line drivers, teaching machine outputs, alarms, ultrasonic drivers, TV sound systems, AM-FM radios, small servo drivers, and power converters.

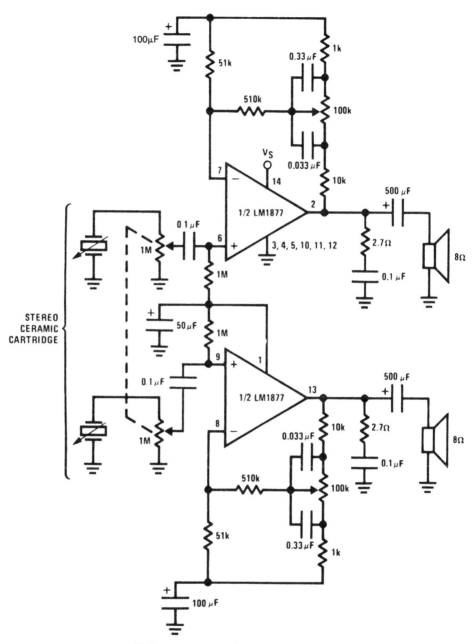

(A) Stereo Phono Amplifier with Bass Tone Control

Figure 12–4. Stereo Amplifiers (Courtesy of National Semiconductor Corporation, © copyright 1980.) (Continued)

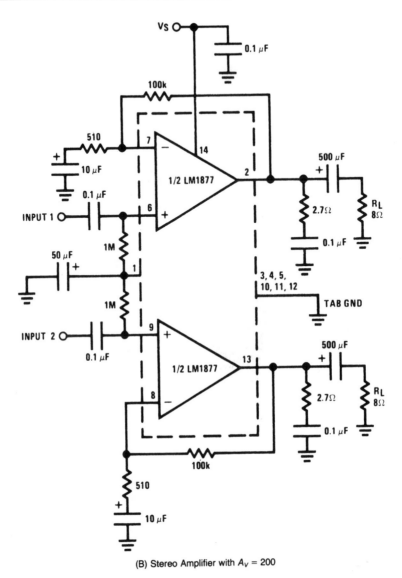

(B) Stereo Amplifier with $A_V = 200$

Figure 12–4.   *Continued*

Battery-Operated Low-Voltage Audio Amplifier

A battery-operated device is the LM389 low-voltage audio power amplifier with NPN transistor array, shown in Figure 12–7. The amplifier inputs are ground referenced, and the output is automatically biased to one-half the supply voltage.

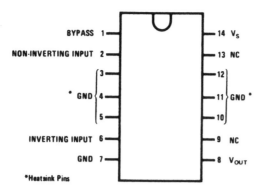

BYPASS 1 — — 14 V_S

NON-INVERTING INPUT 2 — — 13 NC

— 12

* GND { 3 — — 11 } GND *

4 —

5 —

— 10

INVERTING INPUT 6 — — 9 NC

GND 7 — — 8 V_{OUT}

*Heatsink Pins

Figure 12–5. Connection Diagram for LM380 Audio Power Amplifier (Courtesy of National Semiconductor Corporation, © copyright 1980.)

The three transistors have high gain and excellent matching characteristics. They are well suited to a wide variety of applications in dc through VHF systems. The transistors are general-purpose devices that can be used in the same manner as other small-signal transistors. As long as the currents and voltages are kept within the absolute maximum limitations, and the collectors are never allowed to go to ground potential with respect to pin 17, there is no limit to the way they can be used.

To make the LM389 a more versatile amplifier, two pins (4 and 12) are provided for gain control. With pins 4 and 12 open, the 1.35 kΩ resistor sets the gain at 20. However, bypassing the 1.35 kΩ resistor with a capacitor between pins 4 and 12 increases the gain to 200. If a resistor is placed in series with the bypass capacitor, gain can be set at any level between 20 and 200.

Additional external components can be placed in parallel with the internal feedback resistors to tailor the gain and frequency response for individual applications. For example, we can compensate for poor speaker bass response by placing a series RC network from pin 1 to pin 12, thus paralleling the internal 15 kΩ resistor.

The schematic of Figure 12–7B shows that both inputs are biased to ground with a 50 kΩ resistor. The base current of the input transistors is about 250 nA, so the inputs are at about 12.5 mV when left open. If the dc source resistance driving the LM389 is higher than 250 kΩ it will contribute very little additional offset (about 2.5 mV at the input, 50 mV at the output). If the dc source resistance is less than 10 kΩ, then shorting the unused input to ground will keep the offset low (about 2.5 mV at the input, 50 mV at the output). For dc source resistances between these values we can eliminate excess offset by putting a resistor of equal value to the dc source resistance, from the unused input to ground. Elimination

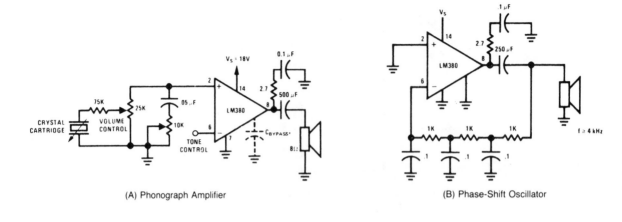

(A) Phonograph Amplifier

(B) Phase-Shift Oscillator

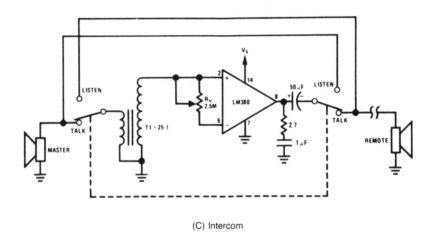

(C) Intercom

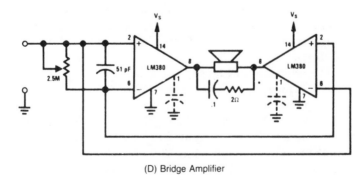

(D) Bridge Amplifier

Figure 12–6. LM380 Applications (Courtesy of National Semiconductor Corporation, © copyright 1980.)

Dual-In-Line Package

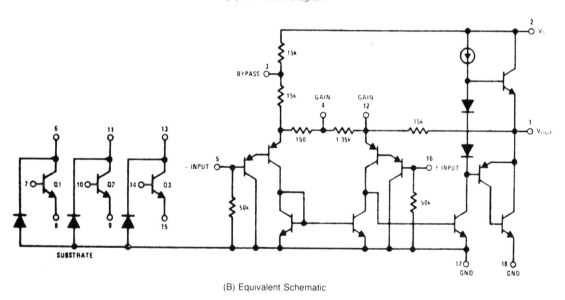

(A) Connection Diagram

(B) Equivalent Schematic

Figure 12–7. LM389 Low-Voltage Audio Power Amplifier with NPN Transistor Array (Courtesy of National Semiconductor Corporation, © copyright 1980.)

of all offset problems can be accomplished by capacitively coupling the inputs.

When using the LM389 with higher gains (bypassing the internal 1.35 kΩ resistor), it is necessary to bypass the unused input to prevent gain degradation and possible instability. This is done with a 0.1 μF capacitor or a short to ground, depending on the dc source resistance.

Figure 12–8 shows four typical applications for the LM389.

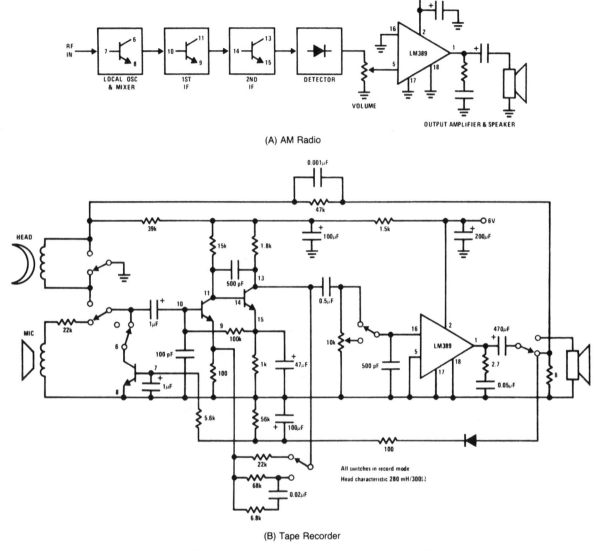

(A) AM Radio

(B) Tape Recorder

Figure 12–8. LM389 Applications (Courtesy of National Semiconductor Corporation, © copyright 1980.)

TV Modulator Circuit

The MC1374 shown in Figure 12–9 is a TV modulator circuit that includes an FM audio modulator, sound carrier oscillator, RF oscillator, and RF dual input modulator. It is designed to generate a TV signal from audio and video inputs.

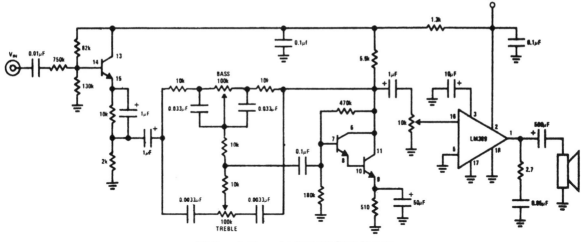

(C) Ceramic Phono Amplifier with Tone Controls

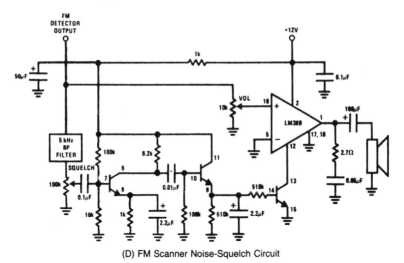

(D) FM Scanner Noise-Squelch Circuit

Figure 12–8.   *Continued*

The wide dynamic range and low distortion audio of the device make it particularly well suited for applications such as video tape recorders (VTR) and video
disc players (VDP). It operates off a single supply, 5–12 V, and is designed for
channel 3 or 4 operation.

The oscillator components shown in the typical application circuit of Figure
12–10 are selected to have a parallel resonance at the carrier frequency of the
desired TV channel. The values of C_2 (56 pF) and L_1 (0.1 μH) were chosen for
a channel 4 carrier frequency of 67.25 MHz. For channel 3 operation, the resonant
frequency is 61.25 MHz, so the values would be $C_2 = 67.5$ pF and $L_1 = 0.1$

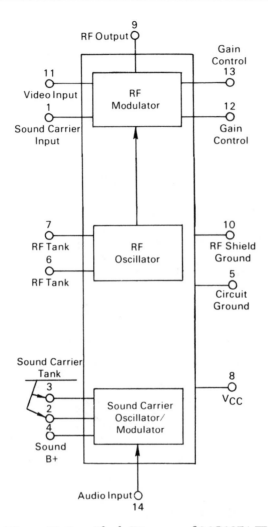

Figure 12–9. Block Diagram of MC1374 TV
Modulator Circuit (Courtesy
of Motorola Semiconductor
Products, Inc., © copyright.)

μH. Resistors R_2 and R_3 are chosen to provide an adequate amplitude of switching voltage, while R_1 is used to lower the maximum dc level of switching voltage below V_{CC}, thus preventing saturation within the IC.

The video modulator is a balanced modulator. Sound-carrier and video information are applied to pins 1 and 11. The other modulator inputs are internally connected to the RF oscillator. The modulator output appears at pin 9.

In a typical application where the composite video information is dc coupled to pin 11, the bias on pin 1 is set to give the desired modulation characteristics.

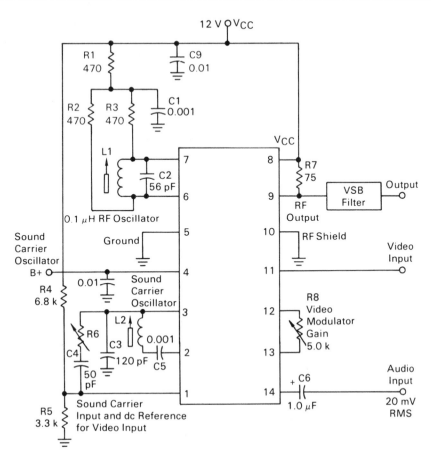

Figure 12–10. MC1374 Channel 4 Application, Stereo Application Circuit (Left Channel Shown), $V_s =$ is V (Courtesy of Motorola Semiconductor Products, Inc., © copyright.)

Minimum carrier occurs when the voltage on pins 1 and 11 is equal, a desirable trait. The minimum permissible voltage on either input is 1.6 V. The maximum voltage should be 1.5 V below the dc voltage on pins 6 and 7. The value for gain-setting resistor R_8, between pins 12 and 13, is selected to give the proper modulation depth for the available composite video amplitude.

The modulated RF signal is presented as a current at the RF output pin 9. Since this pin represents a current source, any load impedance may be selected for matching purposes and gain selections, as long as the voltage on pin 9 is high enough to prevent output devices from reaching saturation. Lowering the dc voltage on pins 6 and 7 gives increased RF output capability at the expense of video input range.

The sound-carrier oscillator and audio modulator have internally set bias. A separate $B+$ is supplied to the oscillator through pin 4, so the oscillator may be easily disabled while tuning the RF tank. The sound-carrier frequency is determined by L_2 and C_3. The oscillator feedback is fed to L_2 through dc blocking capacitor C_5, and 4.5 MHz appears at the input to the oscillator, pin 3.

The sound carrier is coupled to the modulator input, pin 1, through variable resistor R_6 and capacitor C_4. The value for R_6 is chosen to give the desired sound-carrier amplitude and depends upon the Q of L_2 and the values for R_4 and R_5, and the RF modulator gain is set by R_8.

Baseband audio is fed to the audio modulator on pin 14, where it directly modulates the sound-carrier oscillator for a flat characteristic and very low distortion. The input impedance on pin 14 is nominally 6 kΩ. If the audio available is much greater than necessary for proper deviation, a series resistor may be added to allow a low value for coupling capacitor C_6.

When the application calls for tight frequency stability, the sound carrier may be frequency controlled by supplying a dc current to pin 14 from a suitable automatic frequency control (AFC) circuit. The nominal voltage at pin 14 is approximately 3 V. Supplying current to pin 14 increases the frequency, and pulling current out of pin 14 reduces the frequency.

Two-channel operation is possible by switching in a second capacitor to tune the lower channel by means of a PIN diode. Figure 12–11 shows the circuit for channel 3 or 4 operation.

High-Gain, Low-Power FM-IF Circuit

The MC3359 illustrated in Figure 12–12 is a high-gain, low-power FM-IF circuit that includes oscillator, mixer, limiting amplifier, AFC, quadrature discriminator, op amp, squelch, scan control, and mute switch. It is designed primarily for use in voice communication scanning receivers.

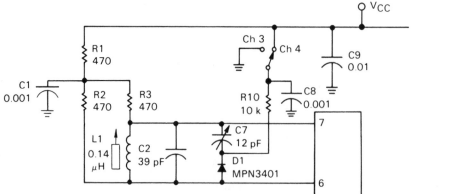

Figure 12–11. Oscillator Components Circuit for Channel 3 and Channel 4 Operation (Courtesy of Motorola Semiconductor Products, Inc., © copyright.)

PIN CONNECTIONS

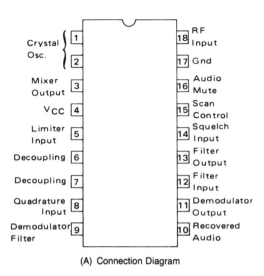

(A) Connection Diagram

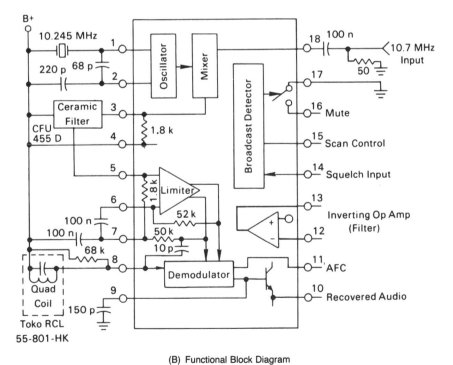

(B) Functional Block Diagram

Figure 12–12. MC3359 High-Gain, Low-Power FM-IF Circuit (Courtesy of Motorola Semiconductor Products, Inc., © copyright.)

The mixer-oscillator combination converts the 10.7 MHz input frequency down to 455 kHz, where, after external bandpass filtering (ceramic filter at pin 3), most of the amplification is done. The audio is recovered using a conventional quadrature FM detector (at pin 8). The absence of an input signal is indicated by the presence of noise after the desired audio frequencies. This "noise band" is monitored by an active filter and a detector. A squelch-trigger circuit (pin 14) indicates the presence of noise (or a tone) by an output (pin 15) that can be used to control scanning. At the same time, an internal switch (pins 16 and 17) is operated. This switch can be used to mute the audio.

The oscillator is an internally-biased Colpitts with the collector, base, and emitter connections at pins 4, 1, and 2, respectively. A 10.245 MHz crystal is used in place of the usual coil.

The mixer is doubly balanced to reduce spurious responses. The input impedance at pin 16 is set by a 3.6 kΩ internal biasing resistor and has low capacitance, which allows the circuit to be preceded by a crystal filter. The mixer output at pin 3 has a 1.8 kΩ impedance to match the external ceramic filter.

After bandpass filtering, the signal goes to the input of a six-stage limiter at pin 5 whose impedance is again 1.8 kΩ. The output of the limiter drives a multiplier, both directly and through the quadrature coil, to detect the FM.

The external capacitor at pin 9 can combine with the internal 50 kΩ resistor to form a low-pass filter for the audio.

The audio is delivered through the emitter follower to pin 10, which may require an external resistor-to-ground to prevent the signal from rectifying with some capacitive loads.

Pin 11 provides AFC. If AFC is not required, pin 11 should be grounded, or it can be tied to pin 9 to double the recovered audio amplitude.

A simple inverting op amp is provided, with an output at pin 13 providing dc bias externally to the input at pin 12, which is referred internally to 2.3 V. A filter can be made with external impedance elements to discriminate between frequencies. With an external AM detector, the filtered audio signal can be checked for the presence of either noise above the normal audio band or a tone signal. The result is applied to pin 14.

An external negative bias to pin 14 sets up the squelch-trigger circuit such that pin 15 is HIGH, at an impedance level of about 2.5 kΩ, and the audio mute (pin 16) is open-circuit. If pin 14 is raised to 0.7 V by the noise or tone detector, pin 15 will go open-circuit and pin 16 is internally short-circuited to ground. There is no hysteresis. Audio muting is accomplished by connecting pin 16 to a high-impedance ground-reference point in the audio path between pin 10 and the audio amplifier.

VCO/Modulator

The MC1376 shown in Figure 12–13 is a voltage-controlled oscillator/modulator that is ideally suited to cordless telephone and television intercarrier applications. It operates over a supply range of 5–12 V dc, has a useful frequency range of

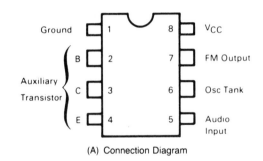

(A) Connection Diagram

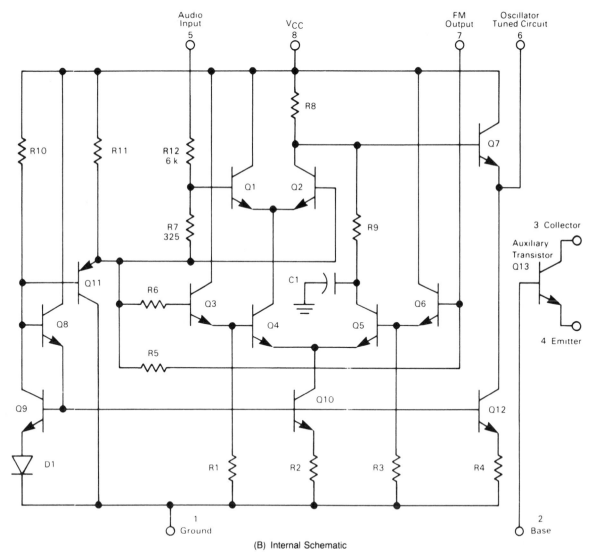

(B) Internal Schematic

Figure 12–13. MC1376 FM Modulator Circuit (Courtesy of Motorola Semiconductor Products, Inc., © copyright.)

1.4–14 MHz, has less than 1 percent distortion, and offers excellent oscillator stability.

This device was originally designed for the base station of a cordless telephone. It includes a separate, or auxiliary, transistor (pins 2, 3, and 4) suitable for service as an output buffer or amplifier for up to 50 mA. Though the oscillator contains internal phase-shift components that are not accessible, the device still has an operating range of 1.4–14 MHz. This range makes the MC1376 a good companion to other devices, such as the MC1372 and MC1373, as a 4.5 MHz or 5.5 MHz intercarrier sound modulator for television signal generation. Also, the device can be used as a low-cost FM-IF (10.7 MHz) signal source. The modulator section of this device is identical to the FM portion of the MC1374 TV modulator discussed earlier.

Cordless Telephone Base Station

A 1.76 MHz cordless telephone base station transmitter is shown in Figure 12–14. The oscillator center frequency is approximately the resonance of the inductor (pin 6 to pin 7) and the total capacitance from pin 7 to ground. If the internal capacitance of about 6 pF is included, the circuit strays in the resonant frequency calculations for the higher frequency applications. For overall oscillator stability,

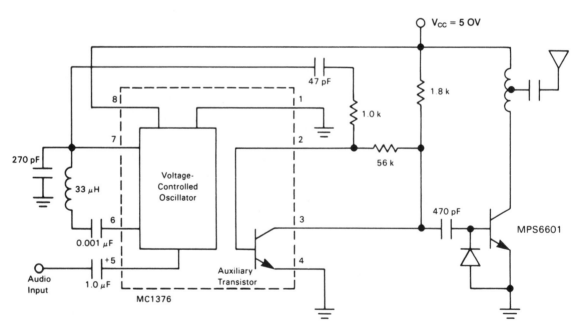

Figure 12–14.   1.76 MHz Cordless Telephone Base Station Transmitter (Courtesy of Motorola Semiconductor Products, Inc., © copyright.)

it is best to keep X_L (inductive reactance) and X_C (capacitive reactance) in the range of 300 Ω to 1 kΩ.

Most applications will require no dc connection at the audio input, pin 5. However, some performance improvements can be achieved by the addition of biasing circuitry. The unaided device will usually establish its own pin 5 bias at 2.9–3.0 V. This bias is a little high for optimum modulation linearity and results in some modulation distortion. This distortion can be significantly reduced by pulling the pin bias down to 2.6–2.7 V. Temperature and supply voltage factors must also be considered when determining biasing.

Temperature stability can be improved by pulling pin 5 down to 2.6 V through a 27 kΩ resistor. If V_{CC} is well regulated, then a simple 180 kΩ/30 kΩ resistor divider, as shown in Figure 12–15, provides optimum distortion-and-frequency stability versus temperature.

The FM output at pin 7 is usually about 600 mV$_{p-p}$ and has low harmonic content and high (2 kΩ) output impedance. The oscillator behavior is relatively unaffected by loading above 1.0 kΩ. If lower impedance must be driven, the capacitive divider, shown in Figure 12–16, can be used, or the auxiliary transistor can be used as a buffer.

Electronically Switched Audio Tape System

The LM1818 electronically switched audio tape system shown in Figure 12–17 is a linear IC that contains all of the active electronics necessary for building a tape recorder deck (excluding the bias oscillator). The electronic functions on the chip include a microphone and playback preamplifier, record and playback

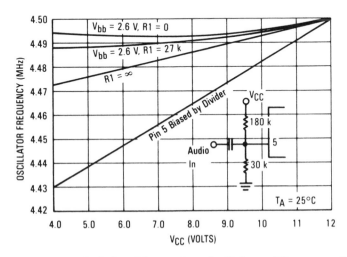

Figure 12–15. Frequency Stability Versus Supply Voltage (Courtesy of Motorola Semiconductor Products, Inc., © copyright.)

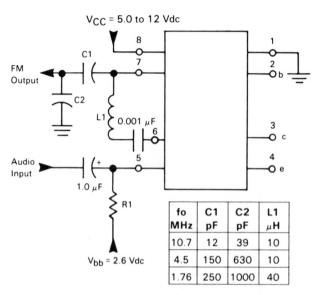

V$_{CC}$ = 5.0 to 12 Vdc

V$_{bb}$ = 2.6 Vdc

fo MHz	C1 pF	C2 pF	L1 μH
10.7	12	39	10
4.5	150	630	10
1.76	250	1000	40

Figure 12–16. Test Circuit for Capacitive Divider Output (Courtesy of Motorola Semiconductor Products, Inc., © copyright.)

amplifiers, a meter driving circuit, and an automatic input level control circuit. The IC features complete internal electronic switching between the record and playback modes of operation.

A monaural application circuit using the LM1818 is shown in Figure 12–18. Table 12–1 lists the external components and their functions.

No-Holds TV Circuit

The LM1880 no-holds vertical/horizontal circuit shown in Figure 12–19 uses compatible linear/I^{2}L (integrated injection logic) technology to produce a TV vertical and horizontal processing system that completely eliminates the hold controls. The heart of the system is a precision 32-times horizontal frequency VCO that is designed to use a low-cost resonator as a tuning element.

The VCO signal is divided in the horizontal section to produce a predriver output that is locked to negative synchronization (sync) by means of an on-chip phase detector. The vertical output ramp is injection-locked by vertical sync that is subject to a sync window derived from the vertical countdown section. A gate pulse centered on the chroma burst is also provided.

A typical application circuit is illustrated in Figure 12–20. Since the LM1880 uses a counter to derive the horizontal frequency, care must be taken to prevent extraneous signals from the horizontal driver and output stages from feeding back

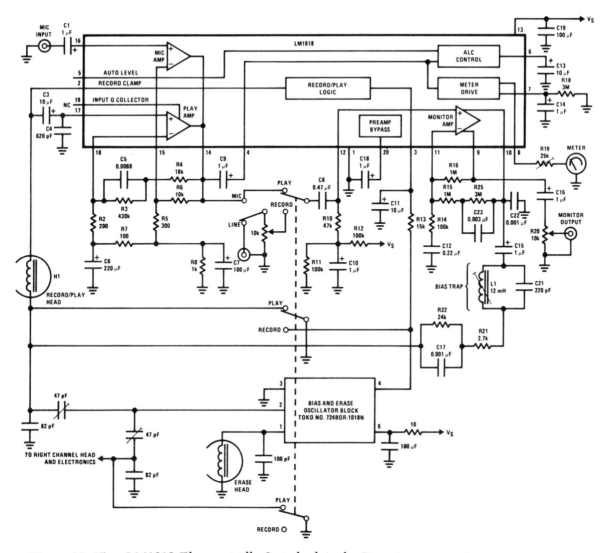

Figure 12–17. LM1818 Electronically Switched Audio Tape System Application (Courtesy of National Semiconductor Corporation, © copyright 1980.)

to the VCO where they could cause false counts and consequent severe phase jitter. To prevent this problem from occurring, keep the VCO feedback capacitor, C_L, as close as possible to device pins 6 and 7, and limit the lead length on the horizontal output pin 8. If a long line is required to the driver base, isolate it with a small series resistor (200–300 Ω) next to pin 8.

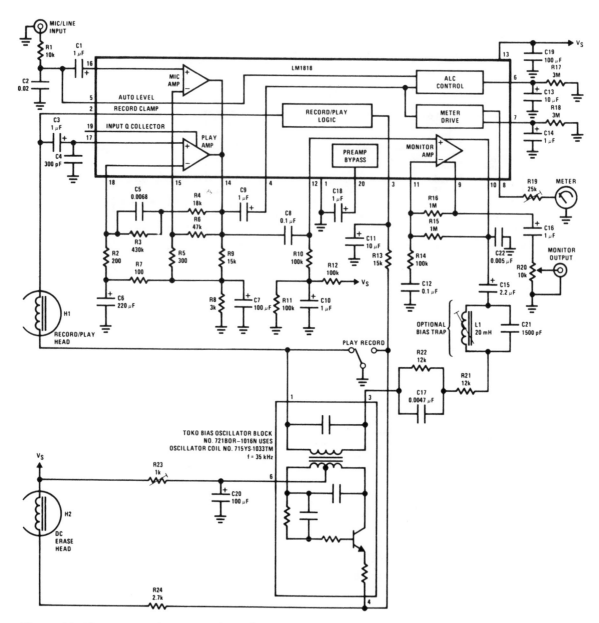

Figure 12–18. LM1818 Monaural Application Circuit (Courtesy of National Semiconductor Corporation, © copyright 1980.)

Table 12–1	External Components for Monaural Application Circuit, Figure 12–19	
COMPONENT	**EXTERNAL COMPONENT FUNCTION**	**NORMAL RANGE OF VALUE**
R1	Used in conjunction with varying impedance of pin 5, forming a resistor divider network to reduce input level in automatic level control circuit	500 Ω – 20 kΩ
C2	Forms a noise reduction system by varying bandwidth as a function of the changing impedance on pin 5. With a small input signal, the bandwidth is reduced by R1 and C2. As the input level increases, so does the bandwidth.	0.01 μF – 0.5 μF
C1, C3	Coupling capacitors. Because these are part of the source impedance, it is important to use the larger values to keep low frequency source impedance at a minimum.	0.5 μF – 10 μF
C4	Radio frequency interference roll-off capacitor	100 pF – 300 pF
R2 R3 R4 C5	Playback response equalization. C5 and R3 form a pole in the amplifier response at 50 Hz. C5 and R4 form a zero in the response at 1.3 kHz for 120 μs equalization and 2.3 kHz for 70 μs equalization.	50 Ω – 200 Ω 47 kΩ – 3.3 MΩ 2 kΩ – 200 kΩ
R5 R6	Microphone preamplifier gain equalization	50 Ω – 200 Ω 5 kΩ – 200 kΩ
R7 R8 R9 C6 C7	DC feedback path. Provides a low impedance path to the negative input in order to sink the 50 μA negative input amplifier current. C6, R9, R7 and C7 provide isolation from the output so that adequate gain can be obtained at 20 Hz. This 2-pole technique also provides fast turn-ON settling time.	0 – 2 kΩ 200 Ω – 5 kΩ 1 kΩ – 30 kΩ 200 μF – 1000 μF 0 – 100 μF
C8	Preamplifier output to monitor amplifier input coupling	0.05 μF – 1 μF
C9	ALC coupling capacitor. Note that ALC input impedance is 2 kΩ	0.1 μF – 5 μF
R10 R11 R12 C10	These components bias the monitor amplifier output to half supply since the amplifier is unity gain at DC. This allows for maximum output swing on a varying supply.	10 kΩ – 100 kΩ 10 kΩ – 100 kΩ 10 kΩ – 100 kΩ 1 μF – 100 μF
C11 R13	Exponentially falling or rising signal on pin 3 determines sequencing, time delay, and operational mode of the record/play anti-pop circuitry. See anti-pop diagram.	0 – 10 μF 0 – 50 kΩ
R14 R15 R16 C12	R16, R14 and C12 determine monitor amplifier response in the play mode. R15, R14 and C12 determine monitor amplifier response in the record mode.	1k – 100k 30 kΩ – 3 MΩ 30 kΩ – 3 MΩ 0.1 μF – 20 μF
C13 R17	Determines decay response on ALC characteristic and reduces amplifier pop	5 μF – 20 μF 100k – ∞
C14 R18	Determines time constant of meter driving circuitry	0.1 μF – 10 μF 100k – ∞
R19	Meter sensitivity adjust	10 kΩ – 100 kΩ
C15	Record output DC blocking capacitor	1 μF – 10 μF
C16	Play output DC blocking capacitor	0.1 μF – 10 μF
C17 R21 R22	Changes record output response to approximate a constant current output in conjunction with record head impedance resulting in proper recording equalization	500 pF – 0.1 μF 5 kΩ – 100 kΩ 5 kΩ – 100 kΩ
C18	Preamplifier supply decoupling capacitor. Note that large value capacitor will increase turn-ON time	0.1 μF – 500 μF
C19	Supply decoupling capacitor	100 μF – 1000 μF

Table
12–1   *(continued)*

COMPONENT	EXTERNAL COMPONENT FUNCTION	NORMAL RANGE OF VALUE
C20	Decouples bias oscillator supply	$10\,\mu F - 500\,\mu F$
R23	Allows bias level adjustment	$0 - 1\,k\Omega$
R24	Adjusts DC erase current in DC erase machines (for AC erase, "Stereo Application Hook-up")	
L1 C21	Optional bias trap	$1\,mH - 30\,mH$ $100\,pF - 2000\,pF$
C22	Bias Roll-Off	$0.001\,\mu F - 0.01\,\mu F$
H1	Record/play head	$100\,\Omega - 500\,\Omega$; $70\,mH - 300\,mH$
H2	Erase head (DC type, AC optional)	$10\,\Omega - 300\,\Omega$

Source: (Courtesy of National Semiconductor Corporation, © copyright 1980.)

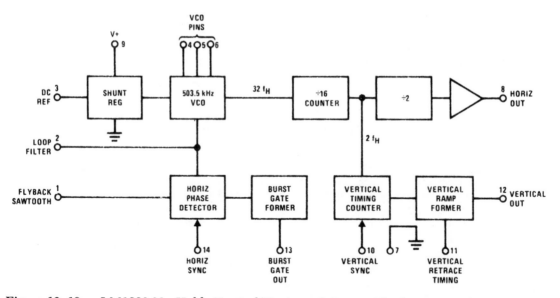

Figure 12–19. LM1880 No-Holds Vertical/Horizontal Circuit Block Diagram (Courtesy of National Semiconductor Corporation, © copyright 1980.)

TV Signal Processing Circuit

The TBA950-2 TV signal processing circuit of Figure 12–21 is designed for pulse separation and line synchronization in TV receivers with transistor output stages. It consists of a sync separator with noise suppression, a frame pulse integrator, a phase comparator, a switching stage for automatic changeover of noise im-

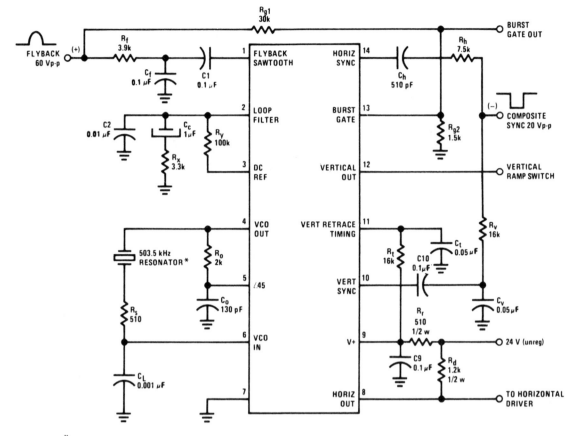

*MuRata Corporation of America, Part No. FX-1028, Vernitron Corp. VTFA3-01-503.5

Figure 12–20. LM1880 Application Circuit (Courtesy of National Semiconductor Corporation, © copyright 1980.)

munity, a line oscillator with frequency range limiter, a phase control circuit, and an output stage. It delivers prepared frame sync pulses for triggering the frame oscillator. The phase comparator may be switched for video recording operation. Because of large scale integration, few external components are needed.

The sync separator separates the synchronizing pulses from the composite video signal. The noise inverter circuit, which needs no external components, in connection with an integrating and differentiating network frees the synchronizing signal from distortion and noise.

The frame sync pulse is obtained by multiple integration and limitation of the synchronizing signal and is available at pin 7. It is recommended to use the leading edge of the frame sync pulse for triggering, because of possible pulse duration differences in production of the sync pulses.

The frequency of the line oscillator is determined by a 10 nF polystyrene

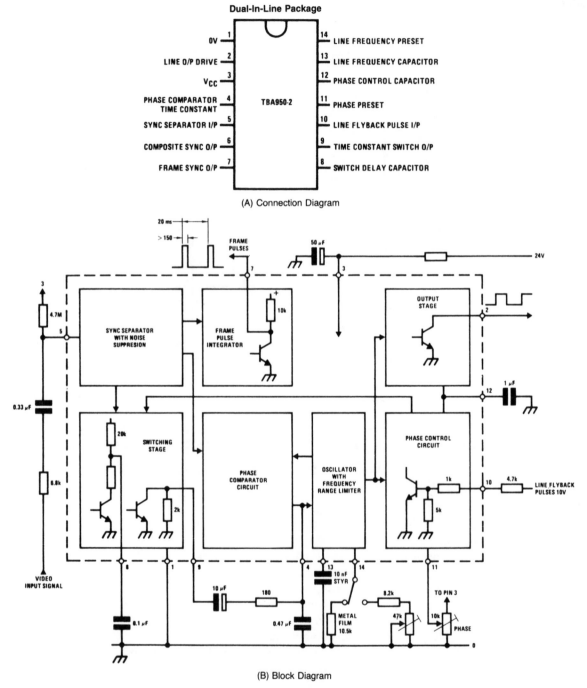

Figure 12–21. TBA950-2 Television Signal Processing Circuit (Courtesy of National Semiconductor Corporation, © copyright 1980.)

capacitor at pin 13 that is charged and discharged periodically by two internal current sources. The external resistor at pin 14 defines the charging current and, consequently, in conjunction with the oscillator capacitor, the line frequency.

The phase comparator compares the sawtooth voltage of the oscillator with the line sync pulses. Simultaneously, an AFC voltage is generated that influences the oscillator frequency. A frequency range limiter restricts the frequency holding range.

The oscillator sawtooth voltage, which is in a fixed ratio to the line sync pulses, is compared with the flyback pulse in the phase control circuit to compensate all drift of delay times in the driver and line output stage. The correct phase position and, hence, the horizontal position of the picture, can be adjusted by the 10 kΩ potentiometer connected to pin 11. Within the adjustable range, the output pulse duration (pin 2) is constant. Any larger displacements of the picture (for example, that due to a nonsymmetrical picture tube) should not be corrected by the phase potentiometer, since in all cases the flyback pulse must overlap the sync pulses on both edges.

The switching stage has an auxiliary function. When the two signals supplied by the sync separator and the phase control circuit, respectively, are synchronized, a saturated transistor is in parallel with the integrated 2 kΩ resistor at pin 9. Thus the time constant of the filter network at pin 4 increases and, consequently, reduces the pull-in range of the phase comparator circuit for the synchronized state to approximately 50 Hz. This arrangement ensures disturbance-free operation.

For video recording operation, this automatic switchover can be blocked by a positive current fed into pin 8 through a resistor connected to pin 3. It may also be useful to connect a resistor of about 680 Ω or 1 kΩ between pin 9 and ground. The capacitor at pin 4 may be lowered in value to 0.1 μF. These alterations do not significantly influence the normal operation of the IC and thus do not need to be switched.

At pin 2 the output stage delivers output pulses of duration and polarity suitable for driving the line driver stage. If the supply voltage goes down (for example, when power is turned off), a built-in protection circuit ensures defined line frequency pulses down to $V_3 = 4$ V and shuts OFF when V_3 falls below 4 V, thus preventing pulses of undefined duration and frequency. Conversely, if the supply voltage rises, pulses defined in duration and frequency will appear at the output pin as soon as V_3 reaches 4.5 V. In the range between $V_3 = 4.5$ V and full supply voltage, the shape and frequency of the output pulses are practically constant.

12.4 Collection of Practical Circuits

A variety of simple, easy to construct, practical circuits is covered in this section. Amplifiers are very important in all aspects of electronics, so we will begin the discussion with amplifiers.

Balanced Bridge Circuit

A balanced circuit where the input resistances are equal is the bridge circuit shown in Figure 12–22. This circuit uses a transducer in the feedback path as a sensing element. The transducer converts an environmental change to a resistive change. With all resistances in the circuit equal, the bridge is balanced and the output voltage is zero. If the resistance of the transducer changes, an output voltage is present. The transducer used may be a strain gauge, thermistor, or photodetector. Almost any type of op amp may be used. The output voltage V_{out} is determined as follows:

$$V_{\text{out}} = -V_{\text{ref}}\left(\frac{R_T + \Delta R_T}{R_1 + R_T}\right) \qquad (12.1)$$

where

R_T = transducer resistance
ΔR_T = a change in the transducer resistance

Current Amplifier

An easily assembled current amplifier is shown in Figure 12–23A. A small current generated by the solar cell is amplified by the op amp, providing enough current to cause a visible indication from the LED. Taking this simple circuit one step further, as shown in Figure 12–23B, an optoisolator can be connected to the amplifier, with a resultant higher voltage output.

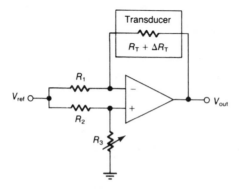

Figure 12–22.   Bridge Amplifier with
Transducer Sensing Element

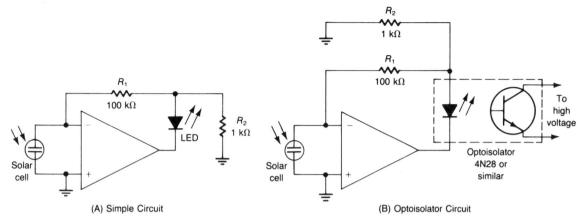

Figure 12–23. Current Amplifiers (Part B: Courtesy of National Semiconductor Corporation, © copyright 1980.)

dc and ac Controllers

An important very effective amplifier circuit for controlling dc servomotors is illustrated in Figure 12–24. The speed and direction of the motor are determined by the amplitude and polarity of the voltage at the input to the circuit.

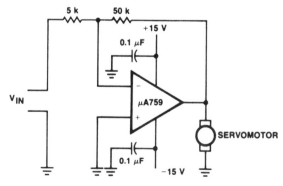

Figure 12–24. dc Servo Amplifier (Courtesy of Fairchild Camera & Instrument Corporation, Linear Division, © copyright 1982.)

Using two op amps, as shown in Figure 12–25, produces an extremely effective ac servo amplifier. The speed and direction of the motor are again determined by the input voltage. The noninverting inputs are held at a predetermined dc reference voltage, usually slightly less than the level of the power supply for the op amps.

Absolute Value Amplifier

An absolute value amplifier, illustrated in Figure 12–26, produces a positive voltage at the output, regardless of the polarity of the input signal. A positive input signal causes diode D_2 to conduct, and the circuit acts as a noninverting amplifier to produce a positive output voltage. A negative input signal causes diode D_1 to conduct, and the circuit acts as an inverting amplifier, again producing a positive output voltage. Input signals greater than 1 V provide the best accuracy for the circuit.

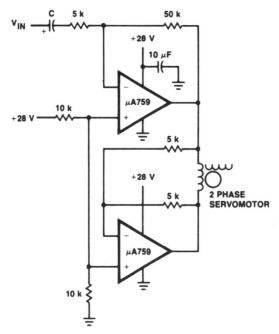

Figure 12–25. ac Servo Amplifier (Courtesy of Fairchild Camera & Instrument Corporation, Linear Division, © copyright 1982.)

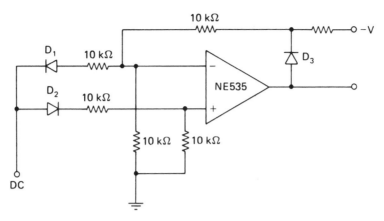

Figure 12–26. Absolute Value Amplifier (Courtesy of Signetics, a subsidiary of U.S. Philips Corporation, © copyright 1980.)

More Timer Circuits

We have already established that IC timers are extremely versatile. We will now discuss some practical and easily constructed timer circuits.

Tone-Burst Generator

A tone-burst generator constructed with one 556 dual timer is shown in Figure 12–27. With switch S_1 set to the *continuous* position, timer B functions

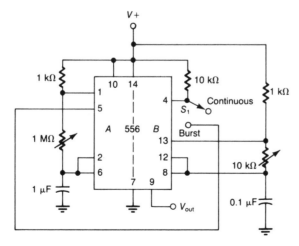

Figure 12–27. Tone-Burst Oscillator Using a Single 556 Timer

as a free-running multivibrator. The oscillating frequency can be varied from about 1.3 kHz to 14 kHz by the 10 kΩ potentiometer. If the potentiometer is replaced by a thermistor or photoconductive cell, the oscillating frequency will be proportional to temperature or light intensity, respectively.

When S_1 is set to the *burst* position, timer A output pin 5 alternately places a LOW or HIGH voltage on reset pin 4 of timer B. When pin 4 is LOW, timer B cannot oscillate; but when pin 4 is HIGH, timer B does oscillate. The result of the alternately high and low voltage, then, is that timer B oscillates in bursts. The output of the tone-burst generator is taken from pin 9 of timer B. This output can be used to drive either an audio amplifier or a stepdown transformer directly to a speaker.

Water-Level Fill Control

A simple water-level fill control circuit that uses a 555 timer is shown in Figure 12–28. When switch S_1 is closed, the output of the timer is LOW. When S_1 is opened, the output goes HIGH and starts the pump. The time interval for the HIGH output sets the height of the water level, and is determined as follows:

$$t_{high} = 1.1R_A C \qquad \text{(in seconds)} \qquad (12.2)$$

To prevent overflow, at some predetermined water level, the overfill switch

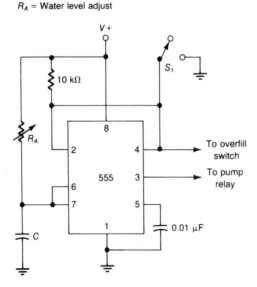

Figure 12–28.   Water-Level Fill Control
Using a 555 Timer

will place a LOW on reset pin 4, causing the output of the timer to go LOW and stop the pump.

Missing-Pulse Detector

Adding a transistor to a 555 one-shot multivibrator, as illustrated in Figure 12–29, produces a missing-pulse detector. When $V_{in} = 0$ V, the emitter diode of Q_1 clamps capacitor voltage V_C to a few tenths of a volt above ground, forcing the 555 timer into its idle state with a HIGH output voltage at pin 3. When V_{in} goes HIGH, Q_1 is cut off and capacitor C begins to charge towards V_{CC}. If V_{in} goes LOW before the timer completes its timing cycle, the voltage across capacitor C is reset to near 0 V. However, if V_{in} does not go LOW before the timer completes its timing cycle, the timer will enter its normal state, and output pin 3 will go LOW. This is what happens if the timing interval is slightly longer than the input pulse at V_{in} and V_{in} misses a pulse. The missing-pulse detector is useful in medical electronics, where it can be used to detect a missing heartbeat.

Remote Temperature Sensor

The LM134 shown in Figure 12–30 is a three-terminal adjustable current source. Current is established with one external resistor (R at pin 1) and no other components are needed. The device makes an ideal remote temperature sensor because its current-mode operation does not lose accuracy over long wire runs. Figure 12–31 shows two possible methods for constructing a low output impedance thermometer using the LM134 current source. Output current (I_{out}) is

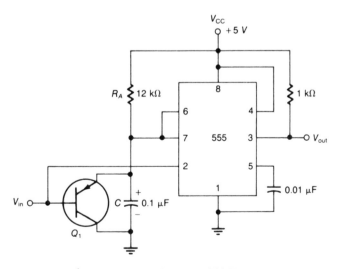

Figure 12–29. Missing-Pulse Detector Using a 555 Timer

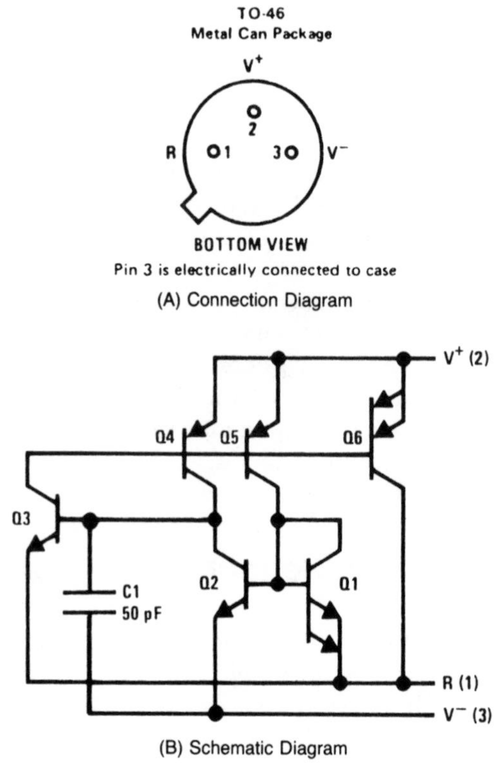
TO-46
Metal Can Package
V+
2
R
1
3
V−
BOTTOM VIEW
Pin 3 is electrically connected to case
(A) Connection Diagram
V+ (2)
Q4
Q5
Q6
Q3
C1
50 pF
Q2
Q1
R (1)
V− (3)
(B) Schematic Diagram

Figure 12–30. LM134 Three-Terminal Adjustable Current Source (Courtesy of National Semiconductor Corporation, © copyright 1980.)

directly proportional to absolute temperature (T) in degrees Kelvin (K) and is determined as follows:

$$I_{out} = \frac{(227\ \mu V/K)(T)}{R} \tag{12.3}$$

where

R = the external resistor connected between the V^- and R pins

To maintain high accuracy, a low temperature coefficient resistor must be used for R.

Fahrenheit Thermometer

A ground-referred Fahrenheit thermometer that uses the LM134 device is shown in Figure 12–32. In this circuit, the value for R_3 is selected so that

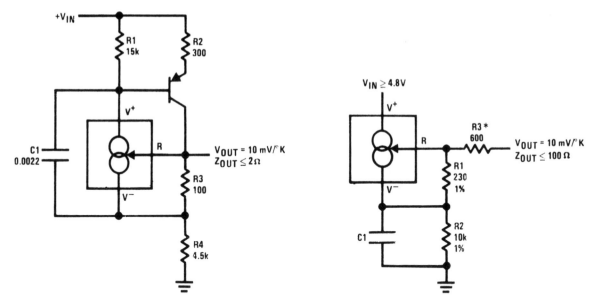

Figure 12–31.   Two Methods for Constructing Low Output Impedance Thermometers
(Courtesy of National Semiconductor Corporation, © copyright 1980.)

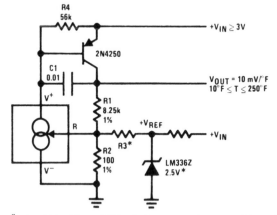

*Select R3 = V_{REF}/583 μA. V_{REF} may be any stable positive voltage ≥ 2V ·
Trim R3 to calibrate

Figure 12–32.   Ground-Referred Fahrenheit
Thermometer (Courtesy of
National Semiconductor
Corporation, © copyright
1980.)

$$R_3 = \frac{V_{ref}}{583 \ \mu A} \tag{12.4}$$

and V_{ref} may be any stable positive value greater than 2 V. To calibrate the circuit, trim R_3.

Precision Temperature Sensor

The LM135 is a precision, easily calibrated, IC temperature sensor. Applications for this device include almost any type of temperature sensing over a range from $-55°C$ to $+150°C$. The device has low impedance and linear output, making it especially easy to interface with readout or control circuitry. Its linear output is unlike other temperature sensors.

Included on the LM135 chip is an easy method for calibrating the device for higher accuracies. A potentiometer connected across the device with the arm tied to the adjustment terminal allows a single-point calibration of the sensor that corrects for inaccuracies over the full temperature range.

The output of the device (calibrated and uncalibrated) is expressed as

$$V_{out(T)} = V_{out(T_O)} \times \frac{T}{T_O} \tag{12.5}$$

where

T = unknown temperature expressed in degrees Kelvin (K)
T_O = reference temperature, also expressed in K

By calibrating the output to read correctly at one temperature, the output at all temperatures is correct. Nominally, the output is calibrated at 10 mV/K.

A wide range of applications for this temperature sensor is shown in the data sheets in Appendix H.

Smoke Detectors, Gas Detectors, Intrusion Alarms

The LM1801 shown in Figure 12–33A is an ionization type smoke detector designed to operate off a 9 V alkaline battery, although provisions are made for operation off supplies up to 14 V and for line operation.

Low battery threshold, alarm threshold, hysteresis (for noise immunity), and stand-by current drain are externally programmed by resistors. The device includes a power transistor capable of directly driving a typical 85 dB horn. The ionization chamber requires an external FET buffer.

A parallel alarm output is provided to enable up to eight similar detectors to be connected in parallel. In this mode of operation, a fault in the line cannot prevent local operation. The low battery alarm signal is confined to the local unit.

Applications for the LM1801 include domestic smoke detectors (Figure 12–

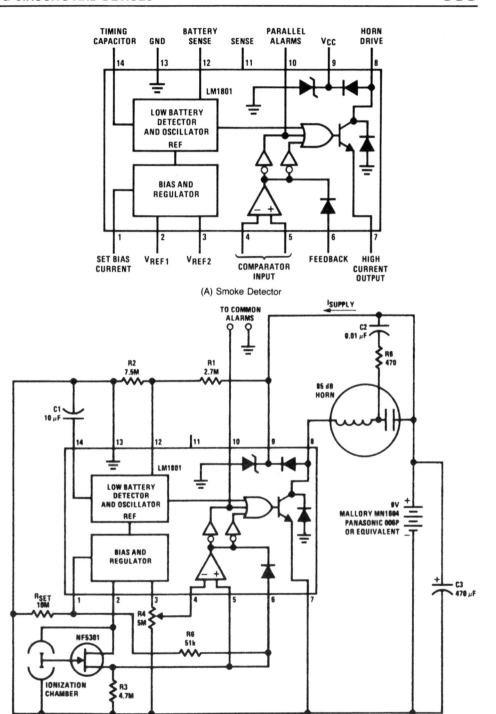

(A) Smoke Detector

(B) 9 V Battery-Operated Ionization Type Smoke Detector

Figure 12–33. LM1801 Applications (Courtesy of National Semiconductor Corporation, © copyright 1980.)

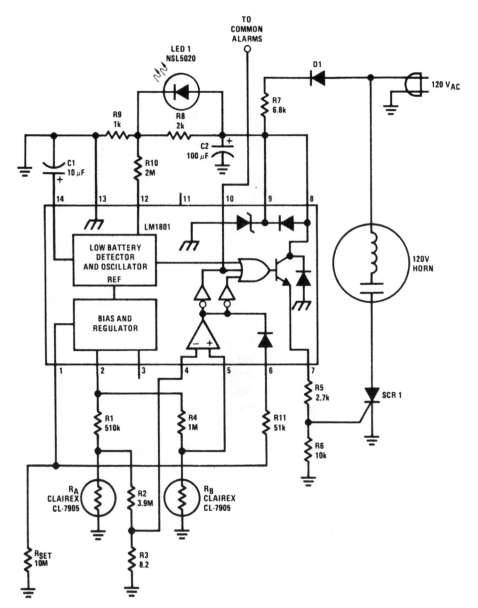

(C) Line-Operated Photoelectric Smoke Alarm Using Light-Sensitive Resistor
(Includes detection of open-circuited LED)

Figure 12–33. *Continued*

33B), line-operated smoke detectors (Figure 12–33C), gas detectors, and intrusion alarms.

In Figure 12–33B low battery threshold is set by resistors R_1 and R_2. The value for these resistors is selected so that the voltage at pin 12 is equal to the oscillator trip voltage when the battery voltage is at the low limit at which the low battery alarm is to operate. The values shown provide a warning at about 8.2 V.

Parallel operation of two or more units is easily achieved with a pair of wires connecting pin 10 of each unit and ground. In this mode, every alarm will sound if any single unit detects smoke.

Fluid Detection System

An IC designed for use in fluid detection systems is the LM1830 shown in Figure 12–34. Applications include beverage dispensers, water softeners, irrigation pumps, sump pumps, radiators, reservoirs, and boilers. In typical applications, the output can be used to drive an LED, loud speaker, or a low-current relay.

An ac signal is passed through two probes within the fluid. A detector determines the presence or absence of the fluid by comparing the resistance of the fluid between the probes with the resistance internal to the IC. A pin is available for connecting an external resistance in cases where the fluid impedance

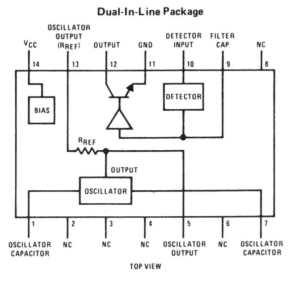

Figure 12–34. LM1830 Fluid Detector
(Courtesy of National
Semiconductor Corporation,
© copyright 1980.)

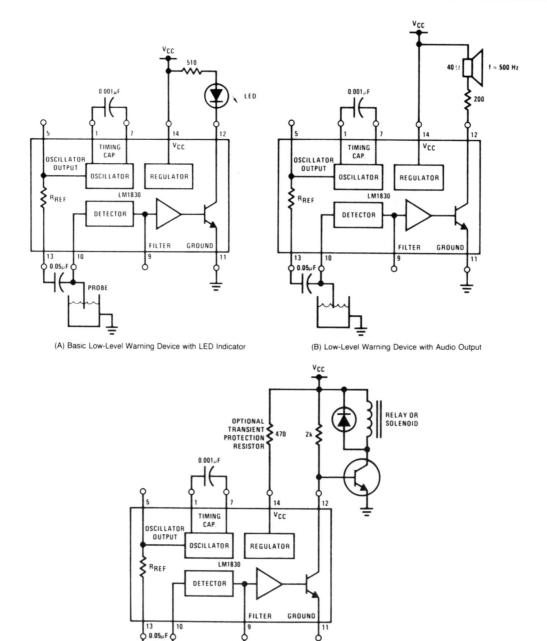

(A) Basic Low-Level Warning Device with LED Indicator

(B) Low-Level Warning Device with Audio Output

(C) High-Level Warning Device with Relay Output

Figure 12–35. LM1830 Applications (Courtesy of National Semiconductor Corporation, © copyright 1980.)

is of a different magnitude than that of the internal resistance. When the probe resistance increases above the preset value, the oscillator signal is coupled to the base of the open-collector output transistor, which drives the external alarm indicator.

Low- and High-Level Warning Devices

Typical circuits that use the LM1830 are the basic low-level warning device with an LED indicator in Figure 12–35A, the low-level warning device with audio output in Figure 12–35B, and the high-level warning device in Figure 12–35C, which is suitable for driving a sump pump or opening a drain valve. In all cases, note the simplicity of the circuit.

LED Flasher/Oscillator

An extremely useful device is the LM3909 LED flasher/oscillator shown in Figure 12–36, designed to provide low power drain and operation from weak batteries so that continuous operation life exceeds that expected from battery rating. By using the timing capacitor for voltage boost, the device delivers pulses of 2 V or more to the LED while operating on a supply of 1.5 V or less.

Application is made simple by inclusion of internal timing resistors and an internal LED current limit resistor. Timing capacitors will generally be of the electrolytic type, and a small 3 V rated part will be suitable for any LED flasher that uses a supply up to 6 V. The circuit is inherently self-starting and requires the addition of only a battery and a capacitor to function as an LED flasher.

Typical applications of the LM3909 IC are the flashlight finder in Figure 12–37A, the emergency lantern/flasher in Figure 12–37B, and the 1 kHz square wave oscillator in Figure 12–37C. Other typical applications include devices for locating boat mooring floats in the dark, sales and advertising gimmicks, emergency locators for such things as fire extinguishers, and electronic applications such as trigger and sawtooth generators.

12.5 Summary

1. Consumer applications offer the largest single use for linear ICs.

2. Single-chip ICs have been developed that contain all the necessary circuitry for complete radio systems, except for the antenna and speaker.

3. Many single-chip ICs that replace entire stages in radio, television, and two-way communications systems are readily available.

4. Linear ICs are the basic active devices in such life-saving systems as missing-pulse detectors, which are used by many hospitals for monitoring patients with heart problems.

5. Smoke detectors and intrusion alarms are easily constructed using simple ICs.

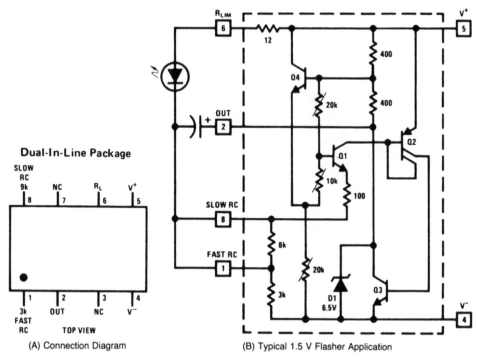

Dual-In-Line Package

(A) Connection Diagram

(B) Typical 1.5 V Flasher Application

Figure 12–36.   LM3909 LED Flasher/Oscillator (Courtesy of National Semiconductor Corporation, © copyright 1980.)

6. There are a great number of useful and practical applications for ICs available to the electronics technician. Research of manufacturer's literature is extremely helpful for finding the IC best suited to a particular application.

12.6 **Questions and Problems**

12.1   Of the devices discussed in this chapter, select one that would best suit the requirements for a stereo phono amplifier with bass tone control that operates on a 12 V system and provides at least 60 dB channel selection and 60 dB ripple rejection. You desire $A_v = 50$, and you are using 8 Ω speakers.

12.2   You plan to build a simple intercom system with ground-referenced input having $A_v = 50$. The output will be centered to one-half the supply voltage. What will be a low-cost device to use for this application?

12.3   You are using a battery-operated LM389 audio amplifier as an AM radio. You desire to increase the gain from 20 to 200. What modification to the circuit will you make to accomplish this increase?

12.4   You are in an area in which TV channel 3 provides full-length feature movies. You wish to record these movies on a video

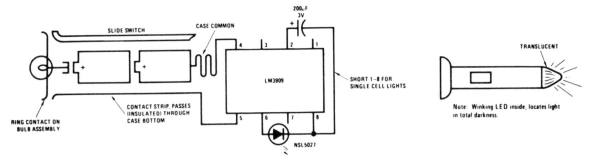

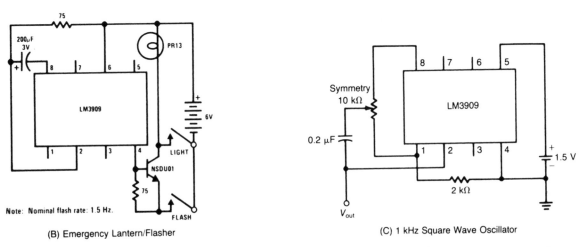

Figure 12–37. LM3909 Applications (Courtesy of National Semiconductor Corporation, © copyright 1980.)

tape recorder (VTR). What device can you use for this purpose, and what oscillator frequency will you need to design? What are the component values for this frequency?

12.5 If you also want to design the circuit in Problem 12.4 so that both channel 3 and channel 4 are available for recording, what frequencies must be used, and what are the component values for those frequencies?

12.6 You are using an MC3359 FM-IF circuit in a voice communications system. You find that AFC is not required, but you desire to increase the audio output level. What action will you take to accomplish this increased output?

12.7 You find that the circuit of Problem 12.6 sometimes rectifies signals when using capacitive loads. How can you correct this problem?

12.8 You find that the cordless base station

transmitter you operate is showing some modulation distortion. What action can you take to minimize this problem? What values are the components used to correct the problem?

12.9 You are using a no-holds vertical/horizontal circuit to eliminate the hold controls on your TV receiver. State the actions necessary to prevent false counts and consequent severe phase jitter.

12.10 A TBA950-2 TV signal processing circuit is to be used for video recording. What must be done to block the automatic switchover feature of the device?

12.11 Explain the operation of a balanced bridge circuit using a transducer in the feedback path.

12.12 What determines the speed and direction of a motor in a dc servo amplifier? in an ac servo amplifier?

12.13 You desire to have a positive output signal regardless of the polarity of the input signal. What type of circuit satisfies this requirement?

12.14 Explain the operation of the burst portion of a tone-burst generator constructed with a 556 timer.

12.15 In a water-level fill control circuit that uses a 555 timer, calculate the value needed for C if $R_A = 10\ \mathrm{M\Omega}$ and the HIGH time interval is 10 seconds.

12.16 What value would you need for R_A in the water-level fill control circuit if the HIGH time interval is 2 minutes and $C = 10\ \mu\mathrm{F}$?

12.17 What would the HIGH time interval be if $R_A = 2\ \mathrm{M\Omega}$ and $C = 0.1\ \mu\mathrm{F}$?

12.18 How might you use a 555 timer to detect a missing heartbeat in medical electronics?

12.19 You have the task of monitoring the temperature in an environmentally controlled room that is located several hundred feet from your work station. What device can you use that will allow you to perform the monitoring task without leaving your work area?

12.20 You have the task of designing a smoke alarm system for a large home. The customer wants all alarms to be activated if any one of them detects smoke, and wants each alarm to operate independently if anything goes wrong with any one in the system. What type of device can you select to satisfy this requirement?

12.21 You are a technician with an industrial firm that uses large boilers in the plant. You are instructed to design a system that will monitor the water in the boilers and automatically cut off the burner for a boiler that loses its water. What is one type of device that can be used for this task? What type of load would be logically driven by the output of the device for this task?

12.22 You are hired by a marina to design a system for marking boat mooring floats and fire extinguishers so that they may be located easily in the dark. The fire extinguishers should be identified with a red signal and the floats by green. What is a simple device selection for this problem? What type of output indicators will you use?

Troubleshooting

13.1 Introduction

One of the major problems that you face as a technician is to find the defect in a faulty system or circuit, correct it, and make the system or circuit perform as designed. When faced with such a problem, you must make use of any technique available to determine the fault, keeping in mind that there is no single approach to solving every problem. It takes a combination of experience, knowledge of the system, sound analytical ability, and persistence to reach any solution.

In this chapter we will look at several basic troubleshooting techniques. While these are standard procedures, you will, as a technician, develop your own method.

13.2 Objectives

When you complete this chapter, you should be able to:

☐ List several troubleshooting techniques.

☐ Realize the limitations of the quick-check method of troubleshooting.

☐ Recognize that a different approach is needed for troubleshooting circuits with feedback.

☐ List several precautions that must be taken when troubleshooting circuits containing ICs.

☐ Recognize the value of the information contained within systems manuals.

☐ List the important troubleshooting information contained within systems manuals.

☐ Select the right piece of test equipment to use for the troubleshooting task at hand.

☐ Realize the value of data sheets for setting up test circuits.

☐ List some PCB troubles and the steps that can be taken to correct or minimize those problems.

☐ List several typical linear IC failure modes and steps that can be taken to minimize failures.

13.3 Troubleshooting Methods

It is not always possible to draw a fine line between good and bad troubleshooting procedures. An acceptable method of locating trouble in one system can result in the destruction of a device in another system. As more sophisticated devices and systems become available, troubleshooting methods have to be altered.

Shotgun Method

In the vacuum-tube circuit days, some technicians resorted to the shotgun approach; that is, they replaced all the parts in the circuit without taking the time to determine which of the individual parts was faulty. This was considered to be a very poor practice, since it overlooked the value of logically determining the fault. In modern systems, however, with their integrated circuits and packaged subassemblies, or *modules,* it is an accepted practice to replace the IC or module after isolating the fault to that stage, because it is impossible to locate and replace the faulty component within such a device.

Troubleshooting Procedures

When troubleshooting an electronic system, a good procedure is to start with general facts about the system's behavior and move to the specific device at fault, as outlined in the block diagram in Figure 13–1.

Defining the Problem

The first step is to define the problem and record the failure mode. The operator of the system can often provide valuable information about its behavior, both prior to and after failure. Also, make a record of the failure mode to refer to when future failures of a similar nature occur in the same type of system. Such information should become a permanent part of the records for that particular system.

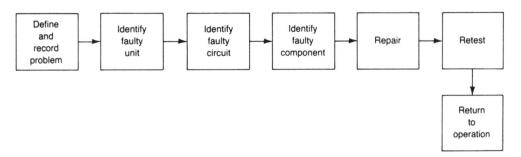

Figure 13–1.   Block Diagram for General Troubleshooting Procedures

Isolating the Faulty Unit

Many troubleshooting manuals for systems identify possible failure modes and possible causes. Such manuals also describe important checkpoints in the system, usually with waveforms and voltage levels. These manuals are extremely helpful for quickly isolating the faulty unit. If the faulty unit cannot be identified from the symptoms, *signal injection* or *signal tracing* can be used, as shown in Figure 13–2.

If the system has several units, the divide-and-conquer concept works well with signal injection. This method involves dividing the system units in half, injecting a signal at that halfway point, and checking for a normal output. If the

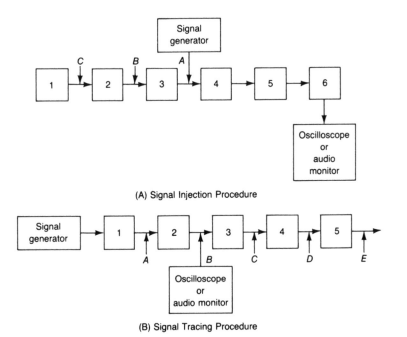

(A) Signal Injection Procedure

(B) Signal Tracing Procedure

Figure 13–2.   Two Standard Troubleshooting Techniques

output signal is normal, then you know that the last half of the system is working properly. The next step is to divide the remaining portion of the system in half and repeat the procedure as many times as necessary to locate the problem. In this manner, the number of tests required to isolate the faulty unit is reduced.

Locating the Faulty Circuit

Once the faulty unit is located, the next step is to locate the faulty circuit within that unit. The same techniques of signal injection and signal tracing can be used to locate faulty circuits within the unit. The divide-and-conquer concept is useful, also, if there are several stages within the unit.

Identifying the Faulty Component

After locating the faulty circuit, the next step is to identify the faulty device or component in that circuit. This is usually done by making voltage, current, and resistance checks.

Repairing and Retesting

Replacing the faulty device or component should complete the trouble-shooting procedures, but the system must be tested for proper operation before it is returned to its operator. *Retesting* is a step often overlooked by many technicians, yet it is as important as all the other procedures. It is advisable to let the system operate over an extended period of time to verify complete operational capability.

Quick-Check Methods

Quick-check methods are useful only if you understand their limitations and take proper precautions to protect the equipment under test. For example, a popular quick-check for bipolar transistor circuits is to cause a temporary short between the base and emitter while measuring the collector voltage, as illustrated in Figure 13–3. The short circuit should cause the collector voltage to rise to the V_{CC} voltage.

This procedure must be used with caution, because a momentary short between the base and collector instead of between the base and emitter can destroy the transistor. Also, if you are working with the direct-coupled amplifier circuit shown in Figure 13–4, a base-emitter short at Q_1 will destroy transistor Q_2. The following is a description of what happens. The collector current of Q_1 is conducted through resistor R_2. The base voltage of Q_2 is V_{CC} minus the voltage drop across R_2 caused by the collector current of Q_1. The base current of Q_2 also flows through R_2, but is so small that it can be ignored. Shorting the base-emitter

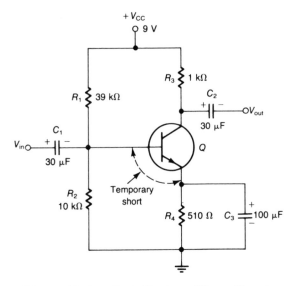

Figure 13–3. Base-Emitter Short Circuit
Quick Test

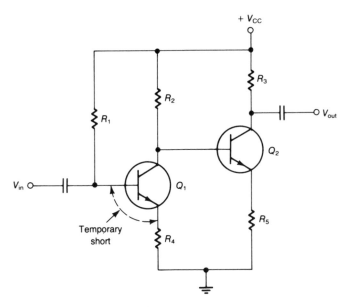

Figure 13–4. Direct-Coupled Amplifier

of Q_1 shuts off Q_1 and causes Q_1 collector current to stop flowing, forcing the base voltage of Q_2 to rise to the V_{CC} voltage and destroying transistor Q_2.

Bad-Habit Quick-Check

A quick-check that has been used by some technicians could prove to be dangerous to your health and well-being. That *bad-habit* quick-check is placing a finger on the grid, base, or gate of the amplifying device. A change in the output that results from the disturbance caused by the finger at the input generally indicates that the stage is operating correctly. However, this test is based on the false assumptions that the voltages present at these points are always too small to cause personal injury and that solid-state circuits always use low voltage supplies.

In some vacuum-tube circuits, direct coupling of amplifiers could place voltages as high as 300–500 V on the grids. Likewise, solid-state circuits may have voltages as high as 100 V or more, and accidentally placing your finger on the wrong terminal could cause severe shock.

A *safer* method for this type of quick-check is to use a screwdriver blade instead of your finger to disturb the input of the stage. However, this approach also must be used cautiously, because the screwdriver blade may cause an accidental short circuit and introduce problems into the circuit that did not exist before the tests began.

Other Quick-Checks

Quick-checks using ohmmeters in transistor or semiconductor diode circuits also must be made with care. Since an ohmmeter delivers a voltage, it can forward bias the device into conduction, resulting in faulty resistance readings. Reversing the ohmmeter leads may eliminate the forward biasing problem (if the semiconductor device is operating properly), but you still have to interpret the readings in relationship to the circuit under test.

Some ohmmeters are designed to provide either *high ohms* or *low ohms* switchable positions. In the low ohms position, the ohmmeter supplies such a low voltage to the circuit that it cannot forward bias a PN junction. Interpretation of readings in relationship to the circuit under test is still required.

Troubleshooting Closed-Loop Circuits

Troubleshooting in closed-loop circuitry requires special methods and techniques, because each section in the circuit depends upon the others for its operation.

A block diagram of a regulated power supply that uses closed-loop circuitry is shown in Figure 13–5. Let's assume that this device has incorrect output voltage. The problem might be due to defects in the series-pass power amplifier,

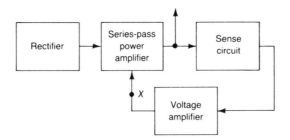

Figure 13–5. Closed-Loop Power Supply Regulator

the sense circuit, or the voltage amplifier. In the closed loop, the power amplifier gets its input voltage from the voltage amplifier, which receives its voltage from the sense circuit, which in turn receives its voltage from the power amplifier. A defect in any of these stages causes all stages in the closed loop to receive improper voltage and results in incorrect output voltage. The defective stage cannot be isolated by making voltage measurements around the loop.

The most popular method used to troubleshoot closed loops is to *open the feedback loop* at some convenient place, such as at point X in Figure 13–5, and insert a dc power supply to provide the control voltage for the series-pass power amplifier. Measurements can then be taken in each section to determine the faulty stage.

Another procedure that works well in modular systems where the circuits can be replaced without much soldering is to simply replace each unit, one at a time, until the output voltage is correct. Keep in mind that the power to the circuit must be turned off before making such replacements, otherwise transient voltage spikes may destroy the unit you are working with.

All or part of the feedback signal in some closed-loop systems may be ac rather than dc. An example is the motor-speed control system shown in Figure 13–6. The voltage-controlled oscillator (VCO) develops the basic motor signal, which is amplified by the voltage and power amplifier and delivered to the motor.

In the feedback loop the VCO frequency, f_1, is compared in the discriminator with the countdown frequency, f_2, from the crystal-controlled oscillator (CCO). The output of the discriminator is a dc voltage that is amplified and used to set the VCO frequency. The gain control at the dc amplifier input allows setting the VCO frequency to produce the desired motor speed.

Two possible approaches, each equally effective, can be used to isolate the faulty unit. One method is to open the circuit at point X and insert a dc power supply to set and control VCO frequency f_1. The other is to open the circuit at point Y and use a signal generator to inject frequency f_1 into the discriminator. After the loop is opened, using either method, measurements are made to determine the faulty stage of the circuit.

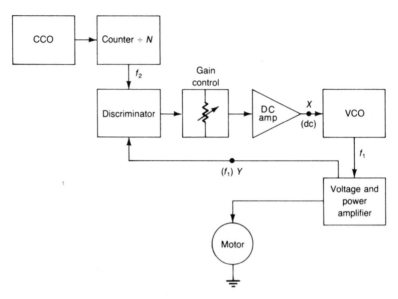

Figure 13–6. Closed-Loop Motor-Speed Control System

Precautions

When troubleshooting circuits that contain ICs, take the following precautions. Do not short pins by using large test probes, do not use excessive heat when unsoldering a component, and never remove or insert an IC into its socket without first making sure the power to the circuit has been removed. With power on, excessive surge currents can occur and destroy the IC. Whenever possible, use available aids such as IC insertion tools, test clips, and so on, when servicing IC units.

A logical approach to troubleshooting IC units is as follows:

1. Check the power supply input at the IC pins. If it is within its rated value, and the ripple level is low, proceed to the next step.
2. Check for the proper input signal at the IC pin indicated as input on the IC diagram.
3. Check the output pin of the IC for correct output signal.
4. Visually check, and measure with an ohmmeter, for any open or short circuits in the copper track to the IC.
5. If the IC must be removed from the circuit by desoldering, take the extra time necessary to remove the solder completely from each pin with a good desoldering tool, so that the IC can be lifted out without placing strain on the copper tracks of the printed circuit board (PCB).

13.4 Using the Systems Manual

Most equipment manufacturers provide a manual that usually includes a schematic diagram for the particular piece of equipment. A good schematic provides much valuable information for the technician, such as how the components are electrically connected, the physical layout of the unit, typical voltage values for a properly operating unit, and expected waveforms at key points. With this information, you can go through the circuit, make voltage measurements, check waveforms, and compare your measurements with those of a properly operating unit. When discrepancies are found in a section, closer examination of that section is called for.

Some manufacturer's manuals also provide a *trouble/cause table*. These tables are extremely helpful for speeding up the troubleshooting process. However, you should not rely completely on such aids, because the manufacturer cannot list all possible faults and their causes. It is important that you develop and use good troubleshooting techniques. Use what aids are available to you, but rely on your own good judgment.

13.5 Selecting the Right Test Equipment

In order to locate and correct faults in a system, test instruments of some kind are essential. Apart from any specialized instrument that may be required for complex systems, a large majority of system faults can be located using only the following three standard test instruments:

1. **Multirange meter (multimeter)**—either analog or digital, for voltage, current, and resistance measurements
2. **Oscilloscope**—single trace or dual trace, for checking waveforms
3. **Function generator**—to provide sine wave, square wave, triangle wave, or pulse inputs at various points in a system (signal injection)

Multimeter

The simplest and most useful of these instruments is the multimeter. The range and accuracy of the meter must be such that it can accurately measure all voltages, ac and dc, that you might be expected to encounter in the system under test. Perhaps the most important consideration is the input impedance of the instrument. It must be as high as possible to prevent loading effects on the circuit being measured. Most modern multimeters use FET input devices, which produce an input impedance of 10 MΩ or more.

As mentioned earlier in this chapter, the ohmmeter function of the multimeter is most valuable to you if it has the high-low ohms capability. The high ohms mode allows in-circuit resistance measurements when no active devices are connected to the resistance under test. This mode could cause a PN junction to be forward biased. The low ohms mode offers such a low voltage that it cannot forward bias a PN junction, so it can be used in circuits where active devices are connected to the resistance under test. However, you must still interpret the results of the measurements in relationship to the overall circuit. Also, remember that all power to the circuit under ohms test must be turned off.

Oscilloscope

The oscilloscope is a versatile and extremely useful test instrument. It is possible to measure both dc values and ac waveforms with the oscilloscope. It offers high sensitivity, typically 10 mV/division, and low loading effect, with typical input impedances greater than 1 MΩ. The frequency, shape, and time period of a single waveform can be determined, or waveforms can be displayed in time and phase relationship to one another. Since many of the ICs we have discussed involve time-varying signals, the oscilloscope will prove to be very useful in any fault-finding effort in these circuits. For best results, the oscilloscope used should have a *minimum* bandwidth of 5 MHz.

The accuracy of both the Y (vertical, or amplitude) input and the X (horizontal, or time) input is limited to about ±3 percent. At low frequencies, the voltage signal to be measured can be applied directly to the Y input through suitable wires or a coaxial cable. However, when measuring high frequencies, to prevent signal degradation, a fully *shielded* cable should be used. For example, when measuring a rapidly changing signal, such as in comparators, or determining a slew-rate, you should make sure that the probe used is *compensated* to account for the loading of the oscilloscope amplifier input impedance. This compensation consists of an *RC* parallel network in series with the input impedance of the oscilloscope. Figure 13–7 illustrates a typical compensated probe circuit, in which the actual input impedance components of the oscilloscope have typical values of 1 MΩ and 30 pF.

Such probes are commercially available. To verify full compensation, the probe is connected to a square wave generator (usually available on the oscilloscope as a calibration signal), and the waveform is observed on the cathode ray tube (CRT). The variable capacitor, normally a small screwdriver adjustment on the probe, is adjusted until the square wave seen on the CRT has no overshoot and shows sharp rising and falling edges.

Function Generator

Function generators are used in troubleshooting to inject a suitable test signal into the system. The resulting output signal is observed with an oscilloscope or voltmeter and then compared with the data listed in the manufacturer's manual.

The complexity and performance characteristics of this instrument are usu-

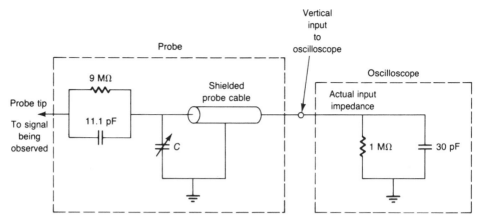

Figure 13–7. Equivalent Circuit for Compensated Oscilloscope Probe and Input Impedance

ally dictated by the system under test. Signal generators are available that provide sine, square, triangle, and pulse waveforms. In some sophisticated units, a sweep control is added, which allows you to automatically vary the frequency of the applied signal between two predetermined values. This feature allows examination of the frequency response of a device.

Signal Injector and Continuity Tester

A more simple troubleshooting device for use in analog systems is a small *handheld signal injector*. This is usually a fixed-frequency battery-powered oscillator that runs at about 1 kHz, with its output available from a metal prod and a lead with an alligator clip provided for a ground connection. An extension of this design is the *continuity tester,* a battery-powered 1 kHz oscillator with an audible output provided by a small loudspeaker. When the two output leads are shorted together, or connected by the low resistance of a cable wire, the output of the oscillator is fed to the loudspeaker. This type of tester is very useful for checking the continuity of cables, connecting wires, and PCB traces.

13.6 Setting Up Test Circuits

Most data sheets provide a test section and a diagram for testing the IC. This testing can be done before the IC is used in a circuit, or when it is removed from a circuit if it is suspected of being at fault. An example is the test circuit for the LM1830 fluid detector shown in Figure 13–8. In this test circuit, using the 0.001 μF capacitor between pins 1 and 7 produces an oscillator output

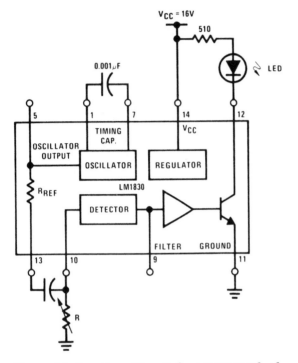

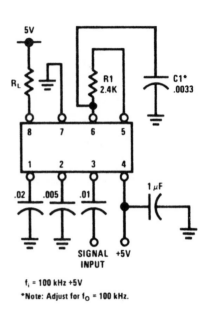

Figure 13–8. Test Circuit for LM1830 Fluid Detector (Courtesy of National Semiconductor Corporation, © copyright 1980.)

Figure 13–9. Test Circuit for LM567 Tone Decoder (Courtesy of National Semiconductor Corporation, © copyright 1980.)

frequency of approximately 6 kHz, available at pin 5. (In normal applications, the output is taken from pin 13 so that the internal 13 kΩ resistor, R_{ref}, can be used to compare with the probe resistance. Pin 13 is coupled to the probe by a blocking capacitor so that there is no net dc on the probe.)

The ac test circuit for the LM567 tone decoder is shown in Figure 13–9. In this circuit, a typical value for R_L is 20 kΩ. The input signal to pin 3 should be 100 kHz at ± 5 V, and capacitor C_1 should be adjusted for a center frequency, f_{op}, of 100 kHz.

Other examples of test circuits available in data sheets are shown in Figure 13–10, the LM1818 electronically switched audio tape system, and Figure 13–11, the LM3075 FM detector/limiter and audio preamplifier.

These test circuits are but a few examples of the valuable information available to you in the manufacturers' data sheets. The information provides test circuits and component values that enable you to test the relevant operating characteristics of the device. You should keep in mind, however, that the device may test good in the test circuit but may not perform properly when placed in the circuit of the system. This can be a result of feedback paths and faults in other parts of the system.

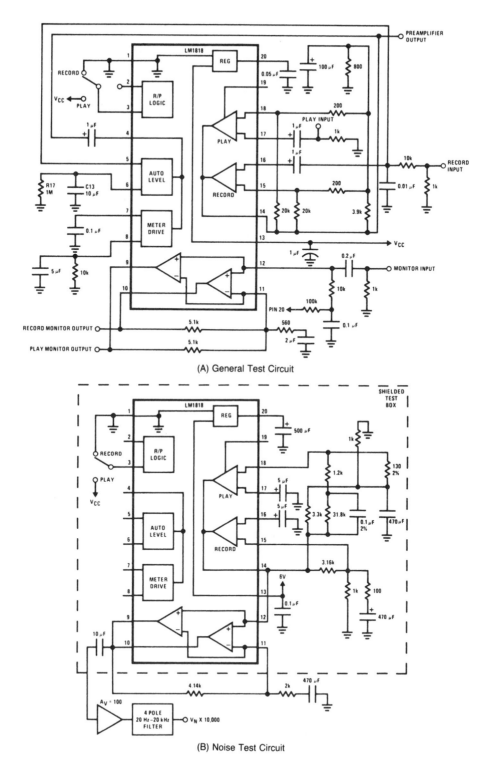

Figure 13–10. Test Circuits for LM1818 Electronically Switched Audio Tape System (Courtesy of National Semiconductor Corporation, © copyright 1980.)

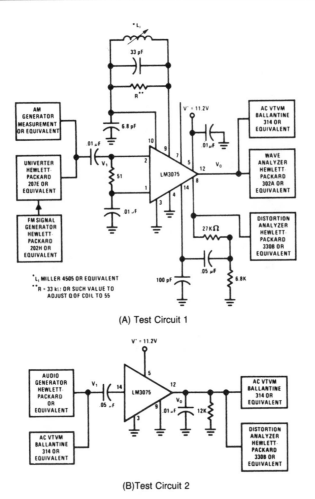

(A) Test Circuit 1

(B) Test Circuit 2

Figure 13–11. Test Circuits for LM3075 FM Detector/Limiter and Audio Preamplifier (Courtesy of National Semiconductor Corporation, © copyright 1980.)

13.7 Printed Circuit Boards

Printed circuit boards (PCBs) are used in almost all systems and circuits in which ICs are used. In some circuits the IC is soldered permanently to the board, and in others the IC is inserted into a mount soldered to the PCB.

Construction-Induced Problems

PCBs are constructed with strips of copper mounted to a rigid or flexible insulating material, which usually serves no other purpose than to support the copper strips and the components. Boards that are constructed with care and precision offer high reliability, but there are some problems that may arise in less well constructed boards. For example, if the original artwork or the etching process is poorly done, the copper strip may not have sufficient cross-sectional area, or it may be too narrow in spots. These deficiencies may cause high circuit resistances, or the copper strips may burn open due to heat generated at that portion of the board.

Another problem you may encounter is distortion of the shape of the PCB due to either improper insertion into the PCB holder or the heat generated in the operating unit. Such distortion may cause the copper strips to crack, resulting in open circuits or intermittent failures. This same problem may arise in PCBs used in systems that are subjected to serious vibrations.

Precautions

There are some precautions that you can take when working with PCBs. If you make your own boards, be painstaking in the artwork and in the etching process. The extra time spent in this phase will provide reliability and peace of mind when the board is installed in the system. When soldering components to the board, use only as much heat as is necessary to make a good solid connection. Excess heat can damage or destroy the component, or it can cause the copper strip to be lifted from the board. Be especially careful that excessive solder is not used, because it may flow between two strips and cause a bridge. After all soldering is completed, use a good solvent to clean the board of flux.

To prevent high resistance or open circuits caused by corrosion of the copper, hermetically seal the PCB. If this is not possible (or not desired), occasional cleaning of the board after it is put into operation (preventive maintenance) will provide longer life and longer trouble-free operation of the system.

13.8 Typical Linear IC Failures

Catastrophic Failure

In general, linear ICs are extremely reliable devices, but they can and do fail. Sudden, or catastrophic, failures occur when the device is overloaded. This overload condition can result from a shorted output or some condition that causes excessive current flow over a long period of time, or when excessive voltages or voltages of the wrong polarity are applied to the device. Transient power supply spikes on the voltage supply line can also cause a device to be destroyed. Identifying and eliminating the cause of the catastrophic failure is essential prior to replacing the faulty device. The most difficult problem to identify is the power supply spike, because this occurs so infrequently. However, if the replacement device fails again soon after installation, and you are sure that all other possible causes of the failure have been corrected, look to the power supply.

Gradual Failure

Long periods of usage may cause degradation of the IC's operating parameters. This is particularly true if the device has been operating in a high temperature or high humidity environment. The change in the device's specifications will show up as a gradual deterioration in system performance. Replacing the device and altering the environmental conditions that caused the gradual failure should eliminate this type of failure.

Poor Performance

Poor performance of a device in a circuit designed for a specific purpose may be the result of the external components or improper connections. If you assume that the device is being used properly, check that the external components are of the correct value. They could have changed value because of excessive heat used during soldering, leakage paths, or unwanted feedback paths. Solder bridges or leakage paths can very easily provide a path for current or signals to flow to the wrong location in the circuit, because the terminals of ICs are very close together. Care in mounting the IC and the external components, plus a thorough cleaning of the board should eliminate poor performance. If care and cleaning do not restore the IC to the proper condition, recheck the design specifications. You may decide to use a different design to achieve higher performance.

Other Considerations

The effects of offset voltage, offset current, and temperature drift are often overlooked when evaluating poor performance, particularly in circuits using op amps. If circuitry that allows you to null the output or compensate for offsets is not included in the circuit design, you should look at the effect these parameters have on the overall performance of the circuit. A check of the manufacturer's data sheet will generally provide information on null and compensation circuits. Serious consideration should be given to adding the extra circuitry needed to overcome the influence of offset and drift.

13.9 **Summary**

1. There are five basic methods used for troubleshooting: (a) shotgun, (b) signal injection, (c) signal tracing, (d) quick-check, and (e) divide-and-conquer.

2. The best procedure for troubleshooting a system is to work from general facts about the problem to the specific fault.

3. There are six basic steps in the troubleshooting procedure: (a) define the problem, (b) isolate the faulty unit, (c) locate the faulty circuit, (d) identify the faulty component, (e) repair, and (f) retest.

4. A bad-habit quick-check to avoid is using your finger to cause a disturbance in a circuit.

5. Using an ohmmeter in transistor or semiconductor diode circuits may forward bias the device into conduction.

6. Troubleshooting closed-loop circuits requires special considerations.

7. When troubleshooting circuits containing ICs, special care must be taken not to short-circuit the pins of the IC.

8. Always make sure that the power to the circuit has been turned off before removing or replacing an IC.

9. To assist in the troubleshooting process, use equipment manufacturers' manuals and schematics.

10. It is essential to learn to select and correctly use the right test instruments for good troubleshooting practices.

11. The three standard test instruments are (a) the multimeter, (b) the oscilloscope, and (c) the function generator.

12. Simple devices that are helpful for troubleshooting analog systems are the signal injector and the continuity tester.

13. Troubleshooting printed circuit boards requires special skills.

14. Some of the types of failure in ICs are (a) catastrophic failure, (b) gradual failure, (c) poor performance, and (d) the effects of offset voltage, offset current, and temperature drift.

15. The best methods and procedures for you to use as a troubleshooter will come with experience, for which there is no substitute.

13.10 **Questions and Problems**

13.1 List the steps that should be taken when troubleshooting a system or circuit.

13.2 What type of information might you expect to find in a systems manual?

13.3 Are quick-check methods useful in all circumstances? Explain your answer.

13.4 What is meant by *divide-and-conquer?*

13.5 Can the divide-and-conquer method be used for circuit isolation as well as unit isolation?

13.6 Explain signal tracing.

13.7 Explain signal injection.

13.8 List the principal troubleshooting instruments and explain how they are used.

13.9 Explain how to verify full compensation of a probe for measuring high frequencies.

13.10 Where can you obtain information for testing ICs out of the circuit in which they operate?

13.11 List some problems you may encounter with printed circuit boards.

13.12 List the precautions you may take when working with PCBs that may prevent troubles in future operations.

13.13 List some possible failures of linear ICs.

13.14 What is often overlooked when evaluating poor circuit performance in op amp circuits?

13.15 What can you do to overcome the effects caused by the problem identified in Problem 13.14?

Glossary

Absolute accuracy   In ADC/DAC operations, a measure of the deviation of the analog output level from the ideal value under any input combination.

Absolute maximum ratings   Electrical limitations that must not be exceeded; specified by the device manufacturer and listed on data sheets.

Absolute value   The output of an absolute value amplifier, which is always positive regardless of the polarity of the input signal.

ac controller   A two op amp amplifier circuit for controlling ac servo-motors.

Acquisition time   In sample-and-hold operations, the time required to acquire a new analog voltage with an output step of 10 V.

ac stability   The capability of a power supply device to reject ac interference; improved by use of bypass capacitors with short leads.

Active filter   A network of resistors and capacitors around an active device, usually an op amp.

Adder   An amplifier circuit whose output is the sum of individual inputs (*see* Summing amplifier).

Adjustable voltage regulator   A regulator circuit that provides adjustable output voltages; formed by adding a few external components to a standard fixed regulator.

Alphanumeric code   A system used by manufacturers to identify their products; consists of a combination of alphabetic characters and decimal numbers.

Ambient   Surrounding; for example, the temperature of the air about a component or a device.

Amplitude modulation (AM)   An intelligence signal that is varied in amplitude (strength) while the frequency is held constant.

Amplitude response   The gain of an active filter, represented by the relationship between the filter's output voltage and its input voltage at various frequencies, usually expressed in decibels.

Analog mathematical circuit A variation of the difference amplifier; input and feedback resistors are made to be of equal values.

Analog signal An ac or dc voltage or current that varies smoothly or continuously.

Analog-to-digital converter (ADC) A device used as an interface between analog and digital systems; converts an analog input signal to a digital output signal.

Aperture time In sample-and-hold operation, the delay required between "hold" command and an input analog transition, so that the transition does not affect the held output.

Astable multivibrator A free-running MV with a square wave output whose frequency is determined by the values of externally connected components; requires no input trigger pulse.

Attenuation The reduction of signal strength through a circuit or device.

Attenuator A circuit or device used to reduce the strength of a signal being passed through it.

Automatic fine tuning (AFT) A circuit, usually found in the RF tuner section of TV receivers, that automatically locks on to the constant frequency of a TV channel.

Automatic frequency control (AFC) A circuit used to vary the frequency of an FM amplifier in proportion to the input signal frequency so that its output remains at a constant frequency.

Automatic gain control (AGC) A circuit used to vary the gain of an AM amplifier in proportion to the input signal strength so that its output remains at a constant level.

Averaging amplifier A variation of the summing amplifier; all input resistors (n) are of equal value and the feedback resistor is equal to $1/n$ of that value.

Band gap reference A low temperature coefficient reference used for reference voltages in compandors.

Bandpass (BP) A filter that allows a band of frequencies between a designated high value and a designated low value to pass, rejecting frequencies above and below those values.

Band reject A filter that prevents a band of frequencies between two designated values from passing.

Bandwidth (BW) The difference between the upper cutoff frequency and the lower cutoff frequency of a filter network; the band of frequencies allowed for transmitting a modulated signal.

Barkhausen Criterion A condition that must be met for an oscillator to be self-starting and self-sustaining; occurs when the product of amplifier gain and a fractional feedback factor equals one, that is, $A_vB_v = 1$.

Blanking The method used in oscilloscopes and TV receiver screens to blank out, or make invisible, the sweep beam as it returns to its starting point at the left of the screen.

Break point One designation for the frequency at which the voltage gain of a filter circuit drops to 70.7 percent of its maximum value.

Buffer A device used to prevent interaction between stages; an isolation amplifier.

Butterworth response A filter response where the passband amplitude response is relatively constant, or flat, with no oscillations; the response of a highly damped filter.

Capture mode One of three operating modes for PLLs; occurs when control voltage from the dc amplifier starts to change the VCO frequency.

Capture range The bandwidth over which capture is possible; can never be wider than the lock range.

Carrier A modulated radio frequency wave used to carry a transmitted intelligence signal over great distances.

Cascading Connecting circuits such as filters or amplifiers in series to improve system operation; the output of one stage becomes the input to the next stage.

Catastrophic failure The sudden failure of a device, usually caused by overloading or transient power supply spikes.

Cathode ray tube (CRT) A vacuum tube in which electrons emitted from the cathode are shaped into a narrow beam and accelerated to a high velocity before striking a phosphor-coated viewing screen; used in oscilloscopes and TV receivers.

Center operating frequency (f_{op}) The highest point of a filter response curve, or the point at which maximum voltage gain is reached.

Chebyshev response A filter response where the passband amplitude response is not flat, but has oscillations; the response of a filter with low damping.

Closed loop gain (A_{CL}) The gain of an op amp when feedback is present.

Colpitts oscillator An inductor-capacitor oscillator whose identifying feature is a tapped capacitor arrangement.

Common-mode gain (A_{CM}) The ratio of a change in output voltage to a change in common-mode voltage.

Common-mode rejection (CMR) The ability of an op amp to reject common-mode signals while amplifying differential signals.

Common-mode rejection ratio (CMMR) The ratio of difference gain to common-mode gain, expressed in decibels.

Common-mode signal Signal voltages that are in phase, of equal amplitude and frequency, applied to both inputs of a differential amplifier.

Compandor A gain control device used for dynamic gain expansion or compression; name derived from the two functions, compressor and expander; primary use is in telephone subscriber and trunk carrier systems.

Comparator An op amp circuit used as a sensing device that compares an input voltage to a reference voltage; not used to amplify signals; typically operates in the open-loop mode.

Compliance In ADC/DAC operations, the maximum output voltage range that can be tolerated while maintaining the specified accuracy.

Constant-current regulator A three-terminal fixed voltage regulator arranged to provide a constant-current output.

Constant-current source An op amp circuit configuration that has a fixed, stable reference voltage and a predetermined value for the input resistor, which results in a constant-current flow regardless of the value of the load resistance.

Continuity tester A simple, easily constructed troubleshooting device for checking the continuity of cables, wires, and PCB traces.

Control transistor The transistor that controls the current flow to a regulator output; series-pass transistor.

Control voltage An external voltage applied to a timer circuit that can alter both threshold and trigger voltages; can modify the timing period and can be used to modulate the output waveform.

Crossover distortion A nonlinearity in the output signal that occurs during the signal crossover from positive to negative in a push-pull output circuit.

Crystal oscillator An oscillator circuit that uses a crystal that is cut to a precise thickness and has a natural frequency of vibration; used for increased oscillator frequency stability.

Current-compensating resistor   A resistor placed in series with the ($+$) input terminal of an op amp to correct for output offset voltage caused by input current imbalances.

Current-differencing amplifier   Another designation for the Norton amplifier; derived from circuit negative feedback that is used to keep the two input currents equal.

Current-driven device   The Norton amplifier; driven by input currents rather than input voltages.

Current limit   The most commonly used regulator protection scheme; guards the output series-pass transistor against excessive output currents or short circuit conditions.

Current-mirror (biasing)   The method of biasing of the Norton amplifier; sets the resting dc output voltage at one-half the V^+ supply level.

Current regulator   A regulator circuit that maintains a constant current in a load, independent of changes in that load.

Current sinking   Occurs in a timer circuit when the output terminal is LOW and causes a floating supply load to be ON, thus providing a path between the load supply and ground through the timer.

Current sourcing   Occurs in a timer circuit when the output terminal is HIGH and causes a grounded load to be ON, thus providing a path between the timer supply and ground through the load.

Current-to-voltage converter   An inverting amplifier without an input resistor; input current is applied directly to the inverting input terminal of an op amp.

Cutoff frequency (f_c)   The frequency at which the voltage gain of a filter circuit drops to 70.7 percent of its maximum value; also called 0.707 point, 3 dB point, break point, and half-power point.

Damping   The determining factor in the shaping of a filter's passband; represented by the symbol alpha α.

Darlington pairs   The second stage of an op amp; a very high gain voltage amplifier stage; cascaded Darlington pairs produce the extremely high open-loop gain of the op amp.

Data acquisition   The gathering, or acquiring, of data in process control systems; usually involves ADC/DAC operations.

Data communications   The transmission of data, such as computer information, from point to point.

dc beta (β_{dc})   The dc current gain of a transistor.

dc controller   An op amp circuit used to control dc servo-motors.

dc-dc converter   A circuit that converts a dc input to a higher or lower dc output voltage.

Decade   A tenfold increase or decrease in bandwidth or in frequency.

Decibel (dB)   One-tenth of a bel; a unit used to express the relative increase or decrease in power or the gain or loss in a circuit.

Degenerative feedback   Feedback that is 180° out of phase with the input signal so that it subtracts from input; negative feedback.

Demodulation   The process of stripping the carrier wave from the intelligence information and amplifying the original intelligence signal to a usable level.

Demodulator   The section of a radio or TV receiver that rectifies the incoming modulated signal and removes the intelligence; a detector; a discriminator.

Deposition   The process of depositing impurities, or dopants, onto the surface of a wafer.

Detector   A circuit used for demodulation; for AM signals, diode detector or envelope detector; for FM signals, ratio detector or quadrature detector.

Dielectric absorption   The characteristic of a charged and discharged capacitor that retains a residual voltage across the capacitor; all the energy stored in the dielectric is not given up when the capacitor is discharged.

Difference amplifier An op amp circuit that has input signals simultaneously applied to both input terminals; output voltage is the algebraic sum of the inverting and noninverting outputs.

Difference gain (A_D) In a differential amplifier, the ratio of a change in the output voltage to a change in the differential input voltage.

Differential amplifier The input stage of an op amp; a circuit that amplifies the difference between two inputs; a difference amplifier.

Differential nonlinearity In ADC/DAC operations, a measure of the deviation between the actual output level change and the ideal output level change for a one-bit change in input code.

Differentiator A wave-forming circuit with a short time constant, the output of which is proportional to the rate of change of the input.

Diffusion The process of building electrical characteristics into silicon wafers, step by step; diffusing or driving dopants to the desired depth in a wafer by exposing the wafer to very high temperatures.

Digital signal A series of pulses or rapidly changing voltage levels that vary in discrete steps or increments.

Digital-to-analog converter (DAC) A device used for converting a digital input signal to an analog output signal; an interface between digital and analog systems.

Diode detector A type of detector (demodulator) used for AM signals.

Diode-transistor logic (DTL) A diode logic gate to which a transistor has been added for amplification purposes; encapsulated in a digital IC.

Discharge A timer terminal used to discharge an external timing capacitor during the LOW output period.

Discrete component Any component not a physical part of an IC; resistor, capacitor, transistor, diode, inductor.

Discriminator A type of detector (demodulator) used for FM signals.

Dissipative voltage regulator circuit A series or shunt regulator circuit in which the control transistor operates in its active region, resulting in power loss, or dissipation, in the transistor.

Divide-and-conquer A troubleshooting method of dividing system units or system stages in half to evaluate one-half at a time, repeating the division and evaluation until the problem is located.

Dopant An impurity, carrying either negative or positive charges, deposited onto wafer surfaces to establish either N-type or P-type sections.

Double-ended limit detector A circuit designed to monitor an input voltage and indicate predetermined upper and lower limits; a window detector.

Drift The change in offset current and offset voltage caused by temperature change.

Droop In sample-and-hold circuits, the variation, or drift, caused by the charge leaking out of the holding capacitor through the amplifier input terminals and the switch; expressed in millivolts per second.

Dual in-line package (DIP) A packaging design to improve IC mounting; internal connections are brought out to external legs on either side of the package; can be plugged into a mounting socket.

Dual tracking regulator A regulator circuit that provides both plus and minus regulated voltages, with the voltages tracking (changing) proportionally with line changes.

Dual power supply A power source that provides equal positive and negative voltages; can be two separate supplies.

Duty cycle A specific interval of ON time compared to the period of the signal; a ratio, normally expressed in percent.

Dynamic sampling error In sample-and-hold circuits, the error introduced into the held output caused by a changing analog input at the time the "hold" command is given; expressed in millivolts.

Effective series resistance (ESR) The internal effective series resistance that must be considered in the design of a switching regulator.

Envelope detector A type of detector (demodulator) used for AM signals.

Equal component filter Any second-order VCVS filter circuit where all capacitors are equal in value and all input resistors are of equal value; the gain must be fixed at 1.59.

Etching The process of cutting openings or windows into oxide, metal, or glass surfaces of a wafer.

External nulling A method for cancelling dc offset voltage in an op amp; a network of resistors and a potentiometer at the input terminal.

Fabrication The process of manufacturing an IC; the step-by-step procedure from wafer to packaged device.

Falloff The rate of decrease (or increase) in amplitude with an increase (or decrease) of frequency in a filter circuit; rate determined by the order of filter.

Feedback element The resistor or capacitor used between the output of an op amp and its input terminal; the determining factor for op amp gain and bandwidth.

Filter A device that screens out certain frequencies or passes electric current of only certain frequencies; can be active or passive.

Fixed voltage regulator A small, three-terminal device that requires no external components; provides a single, fixed, well regulated voltage.

Flat-pack An early packaging style for ICs; leads extend from two or four sides and are on the same plane as the package; used where space is a problem.

Flip-flop (FF) A nonoscillating circuit that has two conditions of equilibrium; an ON-OFF device.

Floating load A load that is ungrounded; it has no common connection.

Flyback The blanked out portion of a wave; the retrace portion of a wave; used in oscilloscopes and TV receivers.

Foldback current limiting A method of power supply protection; designed to cause the power supply voltage to decrease if load impedance becomes too low and draws excessive current from the power supply.

Frame pulse A synchronizing pulse used to trigger a frame oscillator in a TV signal processing circuit; there are 30 frames per second in a TV system.

Free-running mode The operating mode of a VCO before lock-in occurs; present when the VCO output frequency is too far from the standard frequency.

Free-running multivibrator An astable MV.

Frequency compensation A method used to prevent high-frequency output signals from being fed back to the input of an op amp and causing undesired oscillations; can be internally or externally connected.

Frequency converter stage The combined oscillator and mixer stages of a radio receiver.

Frequency-determining network The *RC* or *LC* configuration of a circuit that sets the operating frequency of the circuit.

Frequency deviation   The maximum departure from center frequency at the peak of the modulating signal.

Frequency modulation (FM)   An intelligence signal that is modulated by varying the frequency of the RF carrier wave at an audio rate.

Frequency shift keying (FSK)   The process of varying the carrier frequency rather than keying a transmitter on and off.

Frequency swing   The total frequency swing from maximum to minimum; it is equal to twice the deviation.

Frequency synthesis   The process of putting together, or mixing, two frequencies to provide a desired output; an application for the PLL.

Frequency-to-voltage converter   A device that converts an input pulse rate into an output analog value.

Full-scale range (FSR)   In ADC/DAC operations, the maximum range of steps, number of bits, or maximum analog output.

Function generator   Any signal generator that has two or more different output waveforms; usually has square, triangle, and sine wave outputs.

Gain-bandwidth product (GBP)   Equal to the unity-gain frequency of an op amp; determined by multiplying the gain and bandwidth of any specific circuit; indicates the upper useful frequency of a circuit and provides a means of determining bandwidth.

Gain drift   In ADC/DAC operations, a measure of the change in full-scale analog output, with all bits ones, over the specified temperature range; expressed in parts per million of full-scale range per °C.

Gain error   In sample-and-hold circuits, the ratio of output voltage swing to input voltage swing in the sample mode; expressed as a percent difference.

Gradual failure   The slow deterioration in performance of a device or system; caused by long periods of use or operation in a high temperature or high humidity environment.

Grounded load   A load with a ground, or common, connection on one side.

Half-power point   *See* Cutoff frequency

Hartley oscillator   An inductor-capacitor oscillator whose identifying feature is a tapped inductor arrangement.

Heat sink   A device for or method of dissipating heat away from components in which current flow generates heat at the junction.

High pass (HP)   A filter circuit that rejects or attenuates lower frequencies and allows only frequencies above a given value to pass.

High Q   Quality factor of a narrow bandwidth, high selectivity, high output voltage bandpass filter circuit.

Hold settling time   The time required for the output of a S & H circuit to settle with 1 mV of final value after the "hold" logic command.

Hold step   The voltage step at the output of the S & H circuit when switching from sample mode to hold mode with a steady dc analog input voltage.

Hold voltage drift   *See* Droop

Hybrid IC   An IC made up of a combination of monolithic and thin-film, thick-film, or individual semiconductor component circuits in a single package.

Hysteresis voltage (V_H)   The voltage difference between the upper threshold voltage and the lower threshold voltage in a Schmitt trigger circuit.

Index of modulation (m)   The variation in an AM signal compared with the unmodulated carrier; modulation factor.

Input bias current (I_b)   The average of the two input current (+ and −) of an op amp, with no signal applied.

Input capacitance   An op amp electrical characteristic, an important factor to be considered when the op amp is to be operated at high frequencies.

Input conditioning   The process of adding an RC network to the TRIGGER input of a timer circuit; increases stability and accuracy of the device.

Input element   The passive device (resistor or capacitor) between an input signal source and the input terminal of an op amp.

Input impedance (Z_{in})   An op amp electrical characteristic, usually designated as resistance, typically higher than 1 MΩ.

Input offset current (I_{oi})   The absolute value of the difference between the two input currents (+ and −) of an op amp, for which the output will be driven to change states.

Input offset voltage (V_{oi})   The absolute value of the voltage between the input terminals of an op amp required to make the output voltage greater than or less than some specified value.

Input voltage range   The range of voltage on the input terminals (common-mode) over which the offset specifications apply.

Instrumentation amplifier (IA)   An amplifier circuit with high-impedance differential inputs and high common-mode rejection; gain is set by one or two resistors that do not connect to the input terminals.

Integrated circuit (IC)   A complete electronic circuit formed in semiconductor material; many types may be used with no externally connected discrete components.

Integrated-injection logic (I^2L, or IIL)   A relatively new logic family; has high-speed low-power dissipation and high reliability.

Integrator   A wave-forming circuit with a long time constant, the output of which represents the average energy content of the input signal; square wave input produces triangle wave output.

Intelligence   The original signal, such as voice communication or music, being modulated and transmitted by a radio frequency signal; information.

Intermediate frequency (IF)   The resultant frequency produced in a receiver by combining a local oscillator and a mixer to form a frequency-converter stage; 455 kHz for AM and 10.7 MHz for FM.

Internal nulling   Internally connected circuits in an op amp that are used to cancel dc output offset voltage; controlled by an externally connected potentiometer across offset null or balance terminals.

Intrinsic input impedance (Z_{iin})   The input impedance (resistance) of an op amp under open-loop conditions.

Intrinsic output impedance (Z_{iout})   The output impedance (resistance) of an op amp under open-loop conditions.

Inverter   A circuit that converts a dc input voltage to a higher or lower ac output voltage.

Inverting amplifier   An op amp used with the input signal applied to the (−) input terminal; the output signal is 180° out of phase with the input signal.

Inverting input terminal (−)   One of two input terminals of an op amp, designated by a minus (−) sign; the output signal will be 180° out of phase with the input signal at this terminal.

Isolation amplifier   An amplifier that is electrically isolated between input and output in order to be able to amplify a differential signal superimposed on a high common-mode voltage.

JAN part number Joint Army-Navy numbering system used by manufacturers to designate devices developed to meet military standards.

Junction temperature (T_J) The heat generated at the junction of any semiconductor device that carries current.

Latching The effect of latch-up in the safe area protection circuit of a regulator.

Latch-up Occurs under heavy and high input-output conditions of a regulator; a safe area protection circuit action occurring after a momentary short.

Lead-lag network The frequency-selective network in a Wien-bridge oscillator in which both degenerative and regenerative feedback exists.

Leakage Current flow over the surface or through a path of high insulation value; the undesirable loss of charge in a capacitor.

Least significant bit (LSB) The rightmost bit in a data converter code; the analog output level shift associated with this bit, which is the smallest possible analog output step.

Level detector An op amp circuit that compares an input signal to a reference voltage; the output swing indicates which input signal is higher.

Limiter An IF amplifier tuned to the 10.7 MHz IF in an FM receiver; provides constant output level.

Linear integrated circuit An IC used specifically for analog applications; some may be a combination of analog and digital circuits.

Linear signal Sometimes used interchangeably with analog signal.

Line driver An amplifier circuit for driving a signal over transmission lines; data communications line amplifier.

Loading effect In voltage regulators, the effect that occurs in the sensing divider circuit because of the low input impedance of the sensing transistor.

Load protection A circuit, normally built into a power supply, that automatically reduces the voltage or limits the current of the supply if anything goes wrong with the power supply.

Local oscillator An oscillator circuit in a receiver, the frequency of which is mixed with an incoming signal frequency to produce a difference frequency called the intermediate frequency (IF).

Local regulation A method of physically locating a small regulator, usually in the form of a single IC, at any point in a system where it is needed; sometimes referred to as on-board regulation.

Lock-in mode The phase-locked operating mode of a PLL; occurs when the VCO output frequency is exactly the same as the standard frequency.

Lock range The frequency range over which a PLL can follow the incoming signal.

Logic gates Digital devices with two or more inputs and a single output; used in many combined analog-digital systems.

Logic reverse A characteristic of a timer circuit that allows the user to choose either an ON or OFF condition of the output transistor during the timing period.

Logic threshold voltage The voltage at the output of a comparator at which the driven logic circuitry changes its digital state; relates to the operation of digital devices.

Long-term stability The ability of a power supply to provide a constant output voltage over a long period of operation under constant load, temperature, line voltage, and circuit output control adjustment conditions.

Loop gain (A_L) The ratio of the open-loop gain to the closed-loop gain in an op amp circuit.

Lower frequency (f_{low}) In a bandpass filter, the frequency below center operating frequency at which the voltage is 0.707 times maximum.

Low pass (LP)   A filter that allows frequencies below a given value to pass, rejecting frequencies above that value.

Low Q   Quality factor of a wide bandwidth, low selectivity, low output voltage bandpass filter circuit.

Masking   A series of steps that selectively cut openings or windows into the oxide, metal, or glass surfaces of a wafer.

Metal package   One of several packaging types for ICs; leads connected to the chip are brought out from the base of the package, like transistors.

Missing-pulse detector   A timing circuit in which the output will go low if the timing interval is longer than the input pulse; useful in medical electronics for monitoring heartbeats.

Mixer   A circuit in a receiver that mixes an incoming RF signal with a local oscillator to produce an intermediate frequency.

Mode control signal   A signal pulse that allows FETs to turn on; usually a clock pulse of predetermined frequency.

Modulation   The process of varying or shaping an intelligence signal for transmission by an RF carrier wave.

Modulation envelope   The outline around a modulated signal, usually symmetrical.

Modulation factor (m)   *See* Index of modulation

Monolithic IC   An IC built entirely on a single base of semiconductor material.

Monostable multivibrator   A one-shot MV with a single output pulse, the width of which is determined by the values of external components; circuit must be triggered.

Monotonicity   A characteristic of a DAC that requires a nonnegative output step for an increasing input digital code.

Most significant bit (MSB)   The digital bit that carries the highest numerical weight; an analog output level shift associated with this bit.

Multifunction waveform generator   A signal generator that provides multiple output waveforms; normally provides square, triangle, and sine waves.

Multiple feedback   A filter circuit that has more than one feedback path; gain must be unity for a Butterworth response.

Multivibrator (MV)   An IC circuit in which the output is either a square wave or a rectangular wave; *see also* Astable MV and Monostable MV.

Narrow bandpass   A high Q, high selectivity, high voltage output bandpass filter; has sharp rolloff from center operating frequency.

Natural frequency of vibration (f_n)   The frequency of operation of a crystal; determined by the thickness of the crystal.

Noninverting amplifier   An op amp used with the input signal applied to the (+) input terminal; the output signal is in phase with the input signal.

Noninverting input terminal (+)   One of two input terminals of an op amp, designated by a plus (+) sign; the output signal will be in phase with the input signal at this terminal.

Nonlinearity   A measure of the deviation of an analog output level from an ideal straight-line transfer curve drawn between zero and full scale; commonly referred to as endpoint linearity.

Nonsinusoidal wave   All waveforms other than sine waves (square, triangle, sawtooth, staircase, rectangular).

Normalized curve   A curve in which the maximum amplitude response is set to unity, or 0 dB.

Norton amplifier   An op amp designed for use in ac amplifier circuits, operating from a single power supply; a current-driven device.

Notch   A filter that prevents a band of frequencies between two designated values from passing; also band-reject or band-stop.

Null frequency (f_{null})   The frequency at which maximum attenuation or rejection is obtained in a notch filter circuit.

Octave   A doubling or halving of frequency; the interval between frequencies that have a two-to-one ratio.

Offset drift   A measure of the change in analog output, with all bits zero, over a specified temperature range.

Offset null   The cancelling of dc output offset voltage in an op amp used in dc operations.

One-shot multivibrator   A monostable MV; has one stable state; requires trigger signal and produces a single output pulse before returning to its stable state.

Open-loop gain (A_{OL})   The gain of an op amp at 0 Hz or dc input, without feedback; can be considered as infinite gain.

Operational amplifier (op amp)   An amplifier circuit with very high gain and differential inputs.

Optoisolator   A device that contains both an infrared LED and a photodetector; used for electrical isolation between two stages, such as electronic monitoring devices and a hospital patient.

Order   For a filter, it defines the rate of decrease or rolloff from the operating frequency; the higher the order, the sharper the rolloff.

Oscillator   A circuit that generates a continuously repetitive output signal; used for timing or synchronizing operations.

Oscillograph   An instrument that provides a permanent visual trace of a waveform or shape; a chart recorder.

Output impedance (Z_{out})   The output resistance of an op amp under loaded conditions; the ratio of the intrinsic output impedance of the op amp to the loop gain.

Output offset voltage (V_{os})   A small dc output voltage inherent in op amps with zero input; not present when operating with ac signals.

Overmodulation   A condition that exists when the modulating wave exceeds the amplitude of the continuous carrier wave, thereby reducing the carrier wave power to zero.

Overvoltage crowbar   A load-protection circuit, built into many power supplies, that uses an SCR to automatically reduce voltage to protect the load if something goes wrong with the power supply.

Overvoltage protection   A load protection system that uses zener diodes and current-limiting resistors to protect op amps from an overvoltage condition of a power supply.

Oxidation   The process of growing a very thin layer of silicon dioxide on the surface of a wafer during IC fabrication.

Packaging   The method used for encapsulating a chip after the circuit is formed, resulting in a completed IC.

Passband   The range of frequencies passing through a filter with maximum gain or minimum attenuation.

Passive filter   A filter constructed with resistors, capacitors, and inductors; a filter that contains no active devices.

Peak detector   A circuit that detects and remembers the peak value of an input signal.

Phase detector   In analog systems, a double-balanced mixer using two diodes in a balanced rectifier circuit; a comparator circuit.

Phase difference detection   An ability of a comparator that uses the common-mode rejection characteristic to detect phase differences of out-of-phase input signals.

Phase-locked loop (PLL)   An electronic feedback loop consisting of a phase detector, a low-pass filter, a dc amplifier, and a voltage-controlled oscillator; a closed-loop circuit.

Phase-lock mode   Mode of operation of a PLL when the VCO output frequency is exactly the same as the standard frequency; lock-in.

Phase modulation (PM)   A process of changing the instantaneous frequency of RF energy already generated by a constant frequency; a variation of FM.

Phase-shifting network   An RC or LC network in the feedback path of an oscillator; designed to shift the feedback signal by 180° to cause it to be in phase with input to the amplifier.

Phase-shift oscillator   An oscillator circuit that uses an RC feedback network (usually three sections) to produce the required 180° phase shift.

Photoconductive cell   A device containing materials that increase in conductivity when exposed to light.

Pierce oscillator   A variation of the Colpitts crystal-controlled oscillator; the crystal replaces the inductor of a standard Colpitts oscillator and appears as an inductor.

Piezoelectricity   The quality of a crystal to generate a difference of potential across its face when subjected to mechanical pressure, and to compress when a difference of potential is applied across its face.

Power amplifier (PA)   The amplifier in a transmitter that supplies the antenna with the amount of current needed for the desired power output; the higher the current, the stronger the radiated signal.

Power supply rejection ratio (PSRR)   The ability of an op amp to reject power-supply-induced noise and drift.

Power supply sensitivity   A measure of the effect of power supply changes on a DAC full-scale output.

Precision level amplifier (PLA)   An amplifier placed between each digital input and its corresponding input resistor to a summing amplifier DAC; provides precise output levels of 0 V or 5 V.

Precision reference   A monolithic IC developed to provide precise reference voltages or currents for signal processing and conditioning applications; is temperature stabilized and contains an active reference zener diode.

Prescalar   An electronic device that produces an output pulse for a specified number of input pulses; used in frequency counters.

Printed circuit board (PCB)   A circuit in which electrical conductors are printed on an insulating base; constructed for use of insertion and removal of ICs.

Process control   A method used to measure such physical quantities as temperature, fluid flow rates, speed, pressure, light intensity, strain, and vibration.

Pro-Electron   An organization located in Belgium that has attempted to standardize an identification system for linear ICs; an IC numbering system.

Protective circuits   Circuits added on regular chips to improve reliability and to protect the regulator against certain types of overloads.

Pull-in range   The range of frequencies over which the phase comparator circuit in a TV signal processing circuit will react for synchronization.

Pulse width (PW)   The time duration of a pulse; pulse may be either positive or negative.

Pulse width detector A timer circuit used to determine when the trigger input stays HIGH longer than one time period set by the timing components.

Pulse width modulator A timer circuit connected in the monostable mode, triggered with a continuous pulse train, and modulated by a signal applied to the CONTROL VOLTAGE terminal.

Push-pull amplifier Two transistors used to amplify a signal in such a manner that each transistor amplifies one-half cycle of the signal; the output stage of an op amp.

Quadrature detector An FM discriminator circuit; a difference amplifier used in modern ICs for FM systems.

Quadrature oscillator An oscillator circuit that produces two sine wave output signals, 180° out of phase with one another; a sine-cosine oscillator.

Quality factor (Q) The relationship between the bandwidth and the center operating frequency in a bandpass filter circuit.

Quantization uncertainty A direct consequence of the resolution of a DAC or ADC; can only be reduced by increasing resolution.

Quick-check A troubleshooting method involving a minimum of test equipment; should be used with caution to prevent further problems in a system or a circuit.

Quiescent At rest, or static condition; inactive.

Radiated signal The radio frequency signal transmitted (radiated) from a transmitter; includes the carrier and intelligence.

Radio frequency (RF) The frequency that is used as a carrier for the intelligence transmitted.

Ramp generator A triangle wave generator

Reference voltage (V_{ref}) A small voltage applied to one of the inputs of a comparator to compare with the input signal at the other input.

Regenerative feedback Positive feedback; a small feedback signal that provides an in-phase signal from the output to the input.

Reset A control that allows an operator to disable and override command signals of a timer circuit, driving the output LOW.

Resistor-transistor logic (RTL) An early type of IC logic; easy to interface with discrete component circuits.

Resolution An indication of the number of possible analog output levels a DAC will produce, normally expressed as the number of input bits.

Response time The interval between the application of an input step function and the time when the output crosses the logic threshold voltage; similar to propagation delay of standard op amps.

Retest A troubleshooting step; the final step to verify proper operation after repairs are made.

Retrace The period of time when a display sweep in a CRT returns to a starting point after each line is scanned; also called flyback.

Ripple rejection The ability of a regulator to reject ripple from a power supply.

Rolloff *See* Falloff

R-2R ladder A resistive network in ladder form used in summing amplifier DACs.

Safe area limit In regulators, the limits placed on input-output differential conditions.

Safe-area protection Protection built into IC regulators to protect the series-pass transistor against excessive input-output differential conditions.

Safe operating area (SOA) A protective operating range for regulators that prevents excessive output current when excessive input-output differential conditions occur.

Sallen & Key (S & K)   A voltage-controlled voltage-source (VCVS) second-order active filter.

Sample-and-hold (S & H)   A circuit that periodically samples a signal and then holds it constant; used as an input to an ADC.

Saturation voltage (V_{sat})   In comparator circuits, normally one or two volts less than the applied power supply voltage of the op amp.

Sawtooth wave   A nonsymmetrical triangle wave; the time period of the negative ramp is extremely short compared to that of the positive ramp.

Scaling   A method of design convenient for use in second-order active filters.

Schmitt trigger   A comparator circuit with feedback that provides a reference voltage dependent upon the output voltage level; helps eliminate false output switching caused by noise at the input.

Scribing   Cutting into individual chips the hundreds of identical circuits fabricated on a wafer.

Selectivity   The relative ability of the bandpass filter circuit to select the desired frequency while rejecting all others.

Semiconductor   A material with resistivity somewhere in the range between conductors and insulators.

Semiconductor substrate   The base material, such as silicon, into which other semiconductor materials are diffused to form an IC.

Sensing device   A device used to determine if a voltage is greater or less than a given reference voltage; a comparator.

Series pass transistor   The variable control element in a regulator; control transistor.

Series regulator   A type of regulator circuit in which the variable control element is in series with the load.

Servo amplifier   An amplifier circuit for controlling either dc or ac servo-motors; speed and direction of the motor is determined by the input voltage to the ICs.

Settling time   The total time measured from a digital input change to the time the analog output reaches its new value within a specified error band.

Shotgun approach   A troubleshooting method in which all parts in a circuit are replaced without verifying the faulty part.

Shunt regulator   A type of regulator circuit in which the variable control element is in parallel (shunt) with the load.

Signal conditioning   The process of shaping or changing a signal to be used as a control signal; signal processing.

Signal injection   A troubleshooting method in which a signal of known value is injected into a circuit or system while the output is monitored.

Signal processing   *See* Signal conditioning

Signal-to-noise ratio (S/N)   The ratio of signal level to noise level expressed in similar units; the higher the ratio, the less noise interference.

Signal tracing   A troubleshooting method in which a signal of known value is applied to the input of a system and the signal is traced through the system.

Sine-cosine oscillator   An oscillator circuit that produces two sine wave output signals, 180° out of phase with one another; a quadrature oscillator.

Sine wave   A wave form of a single frequency ac, constantly changing smoothly in both amplitude and direction.

Single-voltage power supply   A power supply that provides only a positive or a negative output voltage; care must be taken when using such a voltage source with op amps.

Sink current   Current received by the output of an op amp or timer for an ungrounded (floating) load.

Sinusoidal A wave varying in proportion to the sine of an angle.

Slew rate *(SR)* An indication of how fast the output voltage of an op amp can change, normally designated in volts per microsecond.

Snap action The effect of comparator output signal switching action when hysteresis voltage is present.

Source current Current supplied to a grounded load by the output of an op amp or a timer.

Source follower A circuit in which the output signal is an exact reproduction of the input (source) signal; a voltage follower, unity-gain amplifier, buffer amplifier, isolation amplifier.

Square wave A wave that alternatively assumes two fixed values for equal lengths of time; symmetrical pulses.

Squaring circuit A comparator in which the (+) terminal is grounded and the output signal swings to the opposite polarity every time the sine wave input signal crosses the zero point, resulting in a square wave output; a zero-crossing detector.

Squelch To automatically quiet a radio receiver by reducing its gain when its input signal is below a specified level; a muting system.

Staircase wave A multisegmented wave produced by the addition of a number of square or rectangular waves.

State-variable A multiple feedback filter that uses three or four op amps, providing three different outputs from a single input.

Step size The amount that output voltage of a DAC will change as the digital input goes from one step to the next; always equal to the weight of the LSB.

Stopband The band of frequencies that is rejected by a filter; all frequencies other than those in the passband.

Strain gauge A type of transducer that converts a physical change to a resistance change.

Strobing A method used to enable or to disable a comparator.

Subtractor A variation of the difference amplifier; the output voltage is the difference between the two input voltages; all resistors are of equal values.

Summing amplifier An amplifier circuit whose output voltage is the sum of the individual input voltages applied through two or more resistors to the (−) input terminal of the op amp.

Summing point The point at which all summing amplifier inputs and the feedback resistor join; the virtual ground point.

Superheterodyne A radio receiver in which the incoming signal is converted to a fixed intermediate frequency before detecting the audio signal component.

Switching regulator A regulator in which the control transistor is switched between the cutoff and saturation modes; results in low power dissipation and high circuit efficiency.

Synchronizing signal A signal transmitted with a TV signal to lock in the received image in the same manner as the original image.

Sync pulse The horizontal (15,750 Hz) and the vertical (60 Hz) pulses used in TV signals for synchronization.

Sync separator A circuit in a TV receiver used to separate the horizontal and vertical pulses from the synchronizing signal.

Temperature coefficient A measure of the change in voltage or resistance of a device with temperature.

Thermal limit The maximum allowable temperature limit of a regulator IC.

Thermal overload A condition of excessive temperature in a regulator IC that activates the thermal shutdown transistor to protect the regulator circuit.

Thermal resistance (θ_{JA})   The opposition to the flow of heat from the junction of a device to the surrounding air.

Thermal shutdown   The effect of removing the base drive from a regulator transistor to prevent chip damage from excessive temperatures.

Thermistor   A type of transducer that converts a temperature change to a resistance change.

Thick-film   A circuit formed with passive elements (resistors, capacitors, conductors) on an insulating substrate, using a silk-screen process.

Thin-film   A circuit with passive elements formed from thin layers of metals and oxides deposited on an insulating substrate.

3 dB point   *See* Frequency cutoff

Threshold   One of two terminals of a timer that determines the operating state of the device.

Threshold voltage (V_{th})   The input voltage to a device at which the device is turned on or off, or the output is switched to the opposite polarity.

Time-base oscillator (TBO)   One stage of a programmable timer/counter; generates timing pulses with a period of 1 *RC* time constant.

Time delay   The length of a timing cycle; may be extended by cascading timers.

Timing state   The unstable or nonnormal state of a monostable MV.

Tracking mode   The phase-locked or lock-in mode of a PLL.

Transducer   A device that produces an electrical output signal proportional to an applied physical stimulus, or that converts an environmental change to a resistance change.

Transfer function   The gain or amplitude response of an active filter, represented by the ratio of the filter's output voltage to its input voltage at various frequencies.

Transistor-transistor logic (TTL or T²L)   The most widely used IC logic; characterized by low propagation delay and good noise immunity.

Transmission   The result of sending (transmitting) an intelligence signal; may be voice transmission over a telephone line, or other intelligence via radio frequency signal.

Transmittance amplifier   Voltage-to-current converter circuits similar in form to inverting and noninverting amplifiers, used to drive relays and analog meters.

Triangle wave   A wave consisting of symmetrical positive- and negative-going ramp voltages.

Triggered monostable circuit   A timer circuit that requires only two external timing components and an input trigger pulse.

Triggering   Applying a pulse to the input of a monostable timer to cause the timing cycle to begin.

Trouble-cause table   A manufacturer-prepared table of possible causes of specific equipment troubles; found in operating and/or maintenance manuals.

Troubleshooting   The logical step-by-step process of locating and repairing faulty systems; may be down to component level.

Twin-T notch   An active filter that consists of a passive twin-T filter connected to an op amp voltage follower.

Unity gain   In filters, $V_{out} = V_{in}$, and the transfer function has a maximum amplitude of 0 dB.

Unity gain amplifier   *See* Source follower

Unity gain analog subtractor   *See* Subtractor

Unity gain frequency   The gain-bandwidth product of an op amp circuit; the upper useful frequency of a circuit; the maximum frequency of an op amp operating at unity gain.

Unity gain inverter An op amp inverting amplifier in which both input and feedback resistors are of equal value.

Upper frequency (f_{high}) In a bandpass filter, the frequency above center operating frequency at which the voltage is 0.707 times maximum.

Varactor A semiconductor diode that operates on the principle of a varying capacitance inversely proportional to the amount of reverse dc voltage applied; a varicap.

Variable gain cell (ΔG) A current-in–current-out device used in compandors.

Varicap *See* Varactor

Virtual ground A point whose voltage is zero with respect to ground, yet is isolated from ground; the $(-)$ input terminal of an op amp inverting amplifier circuit.

Voltage-controlled oscillator (VCO) An oscillator in which the frequency is controlled by a signal voltage; the amount of change in frequency is directly proportional to the input voltage level.

Voltage-controlled voltage source (VCVS) A second-order active filter; also called Sallen and Key (S & K).

Voltage follower *See* Source follower

Voltage level detector A comparator circuit in which a reference voltage is applied to one of the input terminals and the input voltage is applied to the other; the output indicates which voltage level is higher.

Voltage subtractor *See* Subtractor

Voltage-to-current converter *See* Transmittance amplifier

Voltage-to-frequency converter A reference sometimes used for a voltage-controlled oscillator (VCO).

Wafer The starting piece of semiconductor material, a few inches in diameter, into which many identical circuits are formed in some process.

Wideband amplifier An amplifier circuit capable of amplifying a wide range of frequencies, extending from dc to several hundred megahertz.

Wide bandpass A bandpass filter that, in general, has a bandwidth greater than 10 percent of its center operating frequency.

Wien-bridge oscillator An oscillator in which RC networks are part of a bridge circuit that provides both degenerative and regenerative feedback, applied to both inputs of an op amp.

Window detector An op amp circuit designed to monitor an input voltage and indicate predetermined upper and lower limits; a double-ended limit detector.

Zero-crossing detector A comparator circuit in which the output swings to the opposite polarity every time the sine wave input crosses the zero point; a squaring circuit; the $(+)$ input terminal is grounded.

0.707 point *See* Cutoff frequency

Appendixes

Appendix A **Package Outlines**

2-Pin Metal Package
Similar to JEDEC TO-3

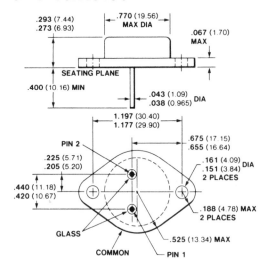

2-Pin Metal Package
Similar to JEDEC TO-39

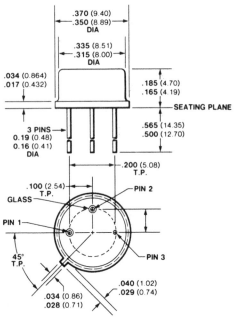

3-Pin Molded Package
Similar to JEDEC TO-220

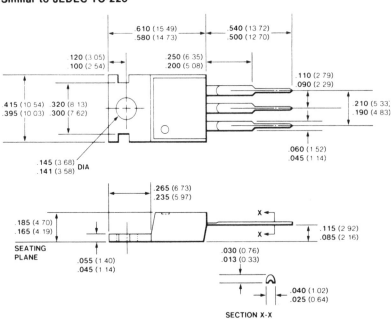

SECTION X-X

(Courtesy of Fairchild Camera & Instrument Corporation, Linear Division, © copyright 1982.)

3-Pin Molded Package
Similar to JEDEC TO-92

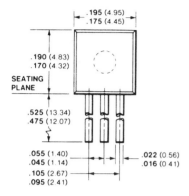

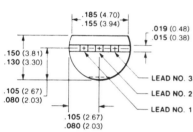

4-Pin Molded Single Wing

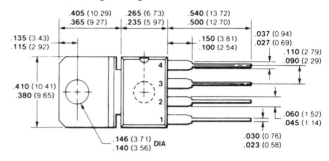

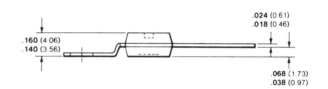

4-Pin Metal Package
Similar to JEDEC TO-3

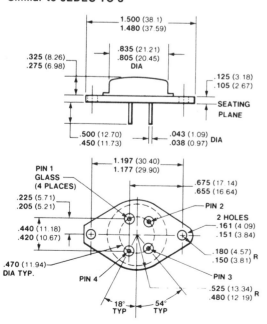

8-Pin Metal Package
Similar to JEDEC TO-3

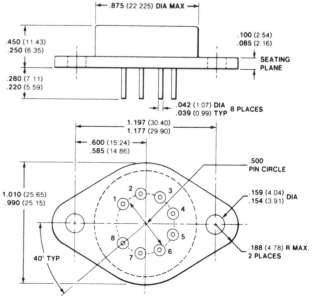

8-Pin Molded Dual In-Line

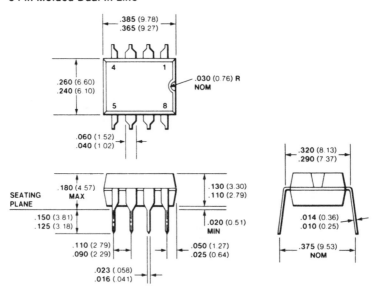

8-Pin Metal Package
In Accordance with JEDEC TO-99

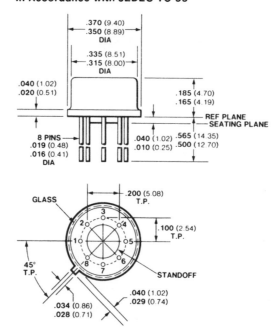

8-Pin Ceramic Dual In-Line

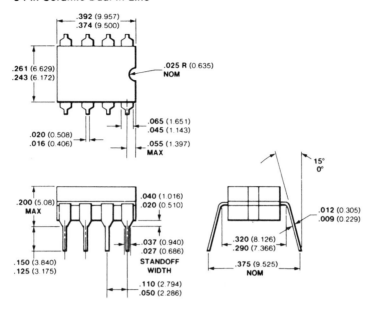

10-Pin Metal Package
In Accordance with JEDEC TO-100

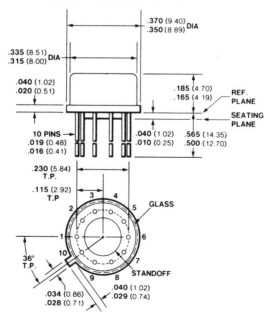

10-Pin Flatpak
In Accordance with JEDEC TO-91

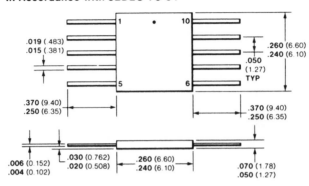

10-Pin Metal Package
Similar to JEDEC TO-3

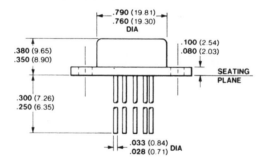

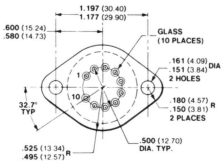

14-Pin Ceramic Dual In-Line
In Accordance with JEDEC TO-116

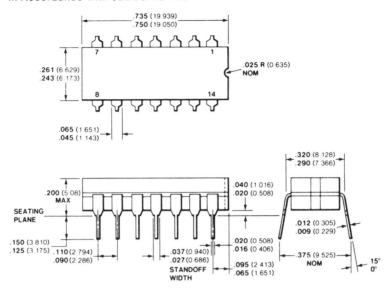

14-Pin Molded Dual In-Line
In Accordance with JEDEC TO-116

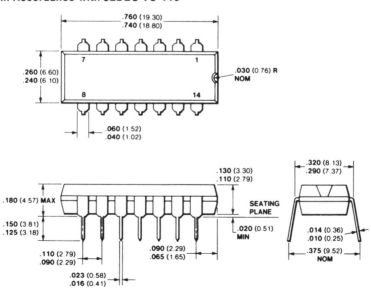

14-Pin Ceramic Dual In-Line

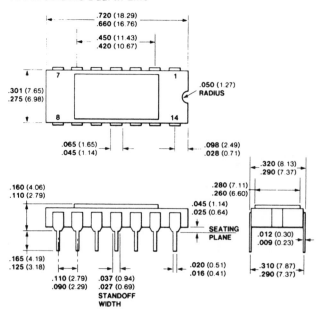

16-Pin Molded Dual In-Line

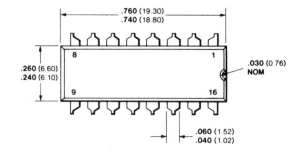

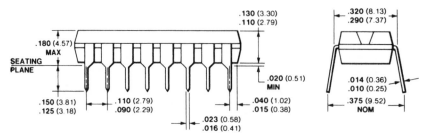

16-Pin Ceramic Dual In-Line

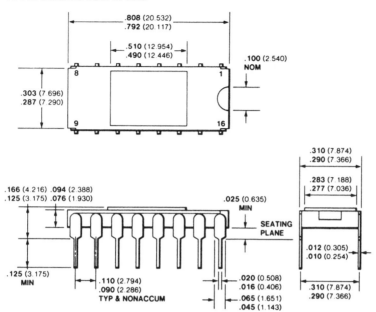

16-Pin Ceramic Dual In-Line

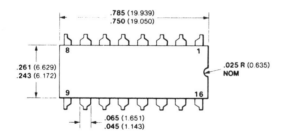

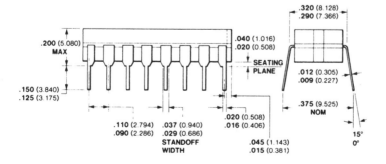

**18-Pin Ceramic DIP
Side-Brazed**

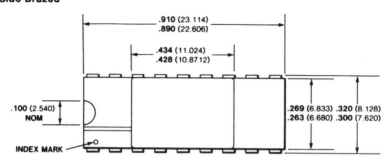

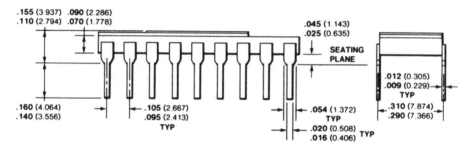

**24-Pin Ceramic Dual In-Line
Side Brazed**

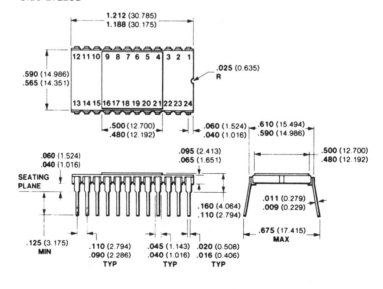

24-Pin Ceramic Dual In-Line

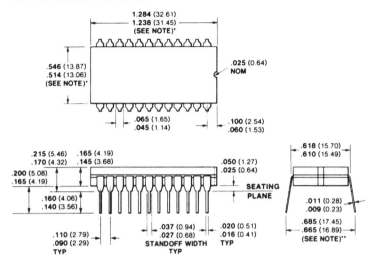

**40-Pin Ceramic Dual In-Line
Side Brazed**

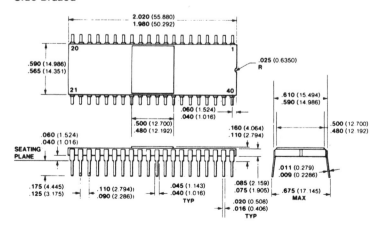

LM3900 Quad Amplifier

National Semiconductor

Operational Amplifiers/Buffers

LM2900/LM3900, LM3301, LM3401 Quad Amplifiers

General Description

The LM2900 series consists of four independent, dual input, internally compensated amplifiers which were designed specifically to operate off of a single power supply voltage and to provide a large output voltage swing. These amplifiers make use of a current mirror to achieve the non-inverting input function. Application areas include: ac amplifiers, RC active filters, low frequency triangle, squarewave and pulse waveform generation circuits, tachometers and low speed, high voltage digital logic gates.

Features

- Wide single supply voltage   4 V_{DC} to 36 V_{DC}
  range or dual supplies   ±2 V_{DC} to ±18 V_{DC}
- Supply current drain independent of supply voltage
- Low input biasing current   30 nA
- High open-loop gain   70 dB
- Wide bandwidth   2.5 MHz (Unity Gain)
- Large output voltage swing   (V^+ −1) Vp-p
- Internally frequency compensated for unity gain
- Output short-circuit protection

Schematic and Connection Diagrams

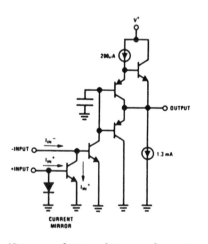

Order Number LM2900J
See NS Package J14A
Order Number LM2900N,
LM3900N, LM3301N
or LM3401N
See NS Package N14A

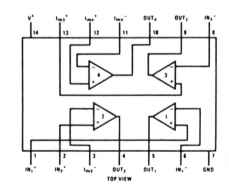

Dual-In-Line and Flat Package

(Courtesy of National Semiconductor Corporation, © copyright 1980.)

408

Typical Applications $(V^+ = 15\, V_{DC})$

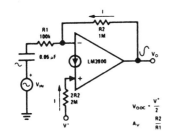

Inverting Amplifier

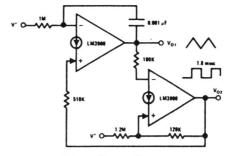

Triangle/Square Generator

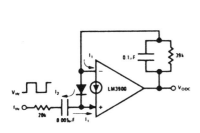

Frequency-Doubling Tachometer

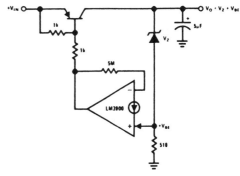

Low $V_{IN} - V_{OUT}$ Voltage Regulator

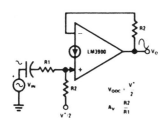

Non-Inverting Amplifier

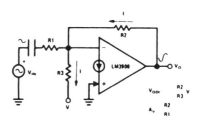

Negative Supply Biasing

Absolute Maximum Ratings

	LM3301	LM3401
Supply Voltage	28 V$_{DC}$	18 V$_{DC}$
	±14 V$_{DC}$	±9 V$_{DC}$
Power Dissipation (T$_A$ = 25°C) (Note 1)		
Cavity DIP		
Flat Pack		
Molded DIP	570 mW	570 mW
Input Currents, I$_{IN}$$^+$ or I$_{IN}$	20 mA$_{DC}$	20 mA$_{DC}$
Output Short-Circuit Duration — One Amplifier	Continuous	Continuous
T$_A$ = 25°C (See Application Hints)		
Operating Temperature Range		0°C to +75°C
LM2900	−40°C to +85°C	
LM3900		
Storage Temperature Range	−65°C to +150°C	−65°C to +150°C
Lead Temperature (Soldering, 10 seconds)	300°C	300°C

Electrical Characteristics (Note 6)

PARAMETER	CONDITIONS	LM2900 MIN	LM2900 TYP	LM2900 MAX	LM3900 MIN	LM3900 TYP	LM3900 MAX	LM3301 MIN	LM3301 TYP	LM3301 MAX	LM3401 MIN	LM3401 TYP	LM3401 MAX	UNITS
Open Loop														
Voltage Gain	T$_A$ = 25°C, f = 100 Hz										800			V/mV
Voltage Gain	T$_A$ = 25°C, Inverting Input	1.2	2.8		1.2	2.8		1.2	2.8		1.2	2.8		V/mV
Input Resistance			1			1			1		0.1	1		MΩ
Output Resistance			8			8			8			8		kΩ
Unity Gain Bandwidth	T$_A$ = 25°C, Inverting Input		2.5			2.5			2.5			2.5		MHz
Input Bias Current	T$_A$ = 25°C, Inverting Input		30	200		30	200		30	300		30	300	nA
	Inverting input												500	nA
Slew Rate	T$_A$ = 25°C, Positive Output Swing		0.5			0.5			0.5			0.5		V/μs
	T$_A$ = 25°C, Negative Output Swing		20			20			20			20		V/μs
Supply Current	T$_A$ = 25°C, R$_L$ = ∞ On All Amplifiers		6.2	10		6.2	10		6.2	10		6.2	10	mA$_{DC}$
Output Voltage Swing	T$_A$ = 25°C, R$_L$ = 2k, V$_{CC}$ = 15.0 V$_{DC}$													
V$_{OUT}$ High	I$_{IN}$$^-$ = 0	13.5			13.5			13.5			13.5			V$_{DC}$
V$_{OUT}$ Low	I$_{IN}$$^-$ = μA, I$_{IN}$$^+$ = 0		0.09	0.2		0.09	0.2		0.09	0.2		0.09	0.2	V$_{DC}$
V$_{OUT}$ High	I$_{IN}$$^-$ = 0 I$_{IN}$$^+$ = 0 R$_L$ = ∞, V$_{CC}$ = 30 V$_{DC}$		29.5			29.5			29.5			29.5		V$_{DC}$
Output Current Capability	T$_A$ = 25°C													
Source		6	18		6	10		5	18		5	10		mA$_{DC}$
Sink	(Note 2)	0.5	1.3		0.5	1.3		0.5	1.3		0.5	1.3		mA$_{DC}$
I$_{SINK}$	V$_{OL}$ = 1V, I$_{IN}$ = 5μA		5			5			5			5		mA$_{DC}$

Electrical Characteristics (Continued) (Note 6)

PARAMETER	CONDITIONS	LM2900			LM3900			LM3301			LM3401			UNITS
		MIN	TYP	MAX	MIN	TYP	MAX	MIN	TYP	MAX	MIN	TYP	MAX	
Power Supply Rejection	$T_A = 25°C$, f = 100 Hz		70			70			70			70		dB
Mirror Gain	@ 20µA (Note 3)	0.90	1.0	1.1	0.90	1.0	1.1	0.90	1	1.10	0.90	1	1.10	µA/µA
	@ 200µA (Note 3)	0.90	1.0	1.1	0.90	1.0	1.1	0.90	1	1.10	0.90	1	1.10	µA/µA
ΔMirror Gain	@ 20µA To 200µA (Note 3)		2	5		2	5		2	5		2	5	%
Mirror Current	(Note 4)		10	500		10	500		10	500		10	500	µADC
Negative Input Current	$T_A = 25°C$ (Note 5)		1.0			1.0			1.0			1.0		mADC
Voltage Gain	f = 100 Hz													V/mV
Input Bias Current	Inverting Input													nA

Note 1: For operating at high temperatures, the device must be derated based on a 125°C maximum junction temperature and a thermal resistance of 175°C/W which applies for the device soldered in a printed circuit board, operating in a still air ambient.

Note 2: The output current sink capability can be increased for large signal conditions by overdriving the inverting input. This is shown in the section on Typical Characteristics.

Note 3: This spec indicates the current gain of the current mirror which is used as the non-inverting input.

Note 4: Input V_{BE} match between the non-inverting and the inverting inputs occurs for a mirror current (non-inverting input current) of approximately 10µA. This is therefore a typical design center for many of the application circuits.

Note 5: Clamp transistors are included on the IC to prevent the input voltages from swinging below ground more than approximately −0.3 V_{DS}. The negative input currents which may result from large signal overdrive with capacitance input coupling need to be externally limited to values of approximately 1 mA. Negative input currents in excess of 4 mA will cause the output voltage to drop to a low voltage. This maximum current applies to any one of the input terminals. If more than one of the input terminals are simultaneously driven negative smaller maximum currents are allowed. Common-mode current biasing can be used to prevent negative input voltages; see for example, the "Differentiator Circuit" in the applications section.

Note 6: These specs apply for −55°C ≤ T_A ≤ +125°C, unless otherwise stated.

Typical Applications (Continued)

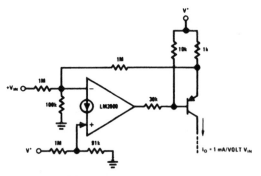

**Voltage-Controlled Current Source
(Transconductance Amplifier)**

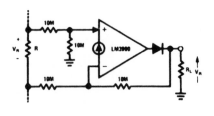

**Ground-Referencing a
Differential Input Signal**

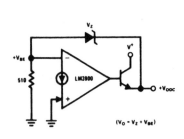

Voltage Regulator

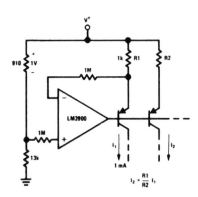

Fixed Current Sources

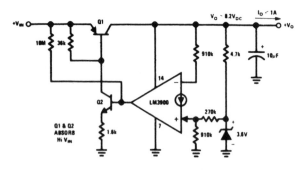

Hi V$_{IN}$, Lo (V$_{IN}$ − V$_O$) Self-Regulator

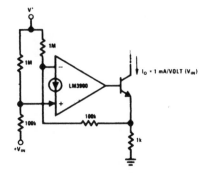

**Voltage-Controlled Current Sink
(Transconductance Amplifier)**

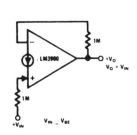

Buffer Amplifier

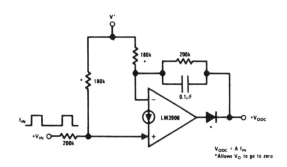

Tachometer

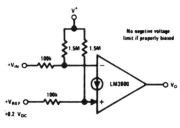

Low-Voltage Comparator

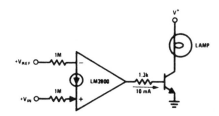

Power Comparator

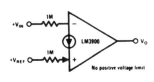

Comparator

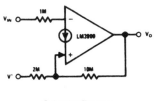

Schmitt-Trigger

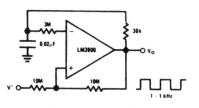

Square-Wave Oscillator

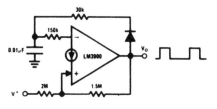

Pulse Generator

Typical Applications (Continued)

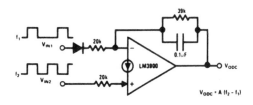

Frequency Differencing Tachometer

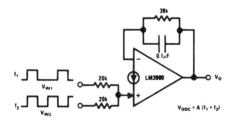

Frequency Averaging Tachometer

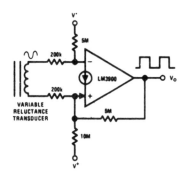

Squaring Amplifier (W/Hysteresis)

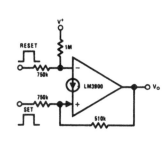

Bi-Stable Multivibrator

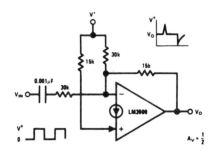

Differentiator (Common-Mode Biasing Keeps Input at +V$_{BE}$)

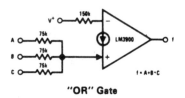

"OR" Gate

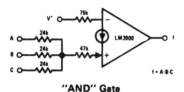

"AND" Gate

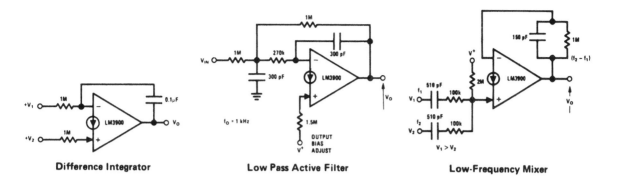

Difference Integrator

Low Pass Active Filter

Low-Frequency Mixer

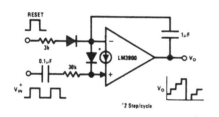

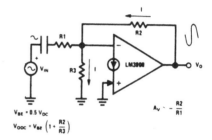

Staircase Generator

V_{BE} Biasing

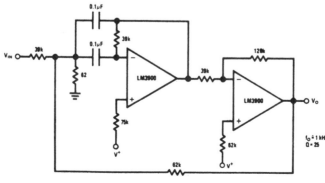

Bandpass Active Filter

Typical Applications (Continued)

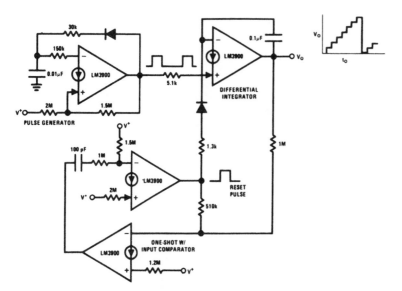

Free-Running Staircase Generator/Pulse Counter

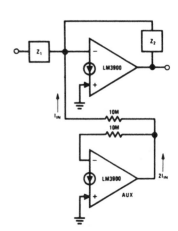

Supplying I_{IN} with Aux. Amp
(to Allow Hi-Z Feedback Networks)

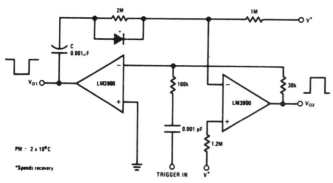

One-Shot Multivibrator

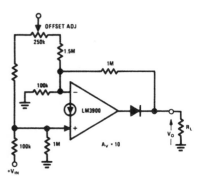

Non-Inverting DC Gain to (0,0)

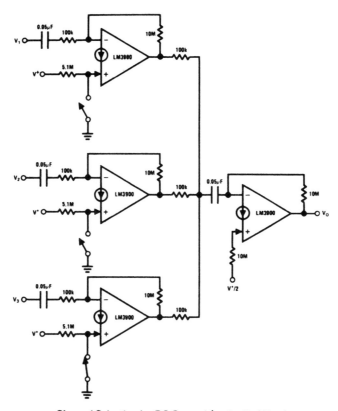

Channel Selection by DC Control (or Audio Mixer)

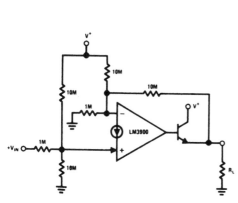

Power Amplifier

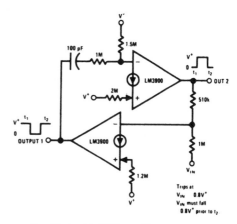

One-Shot with DC Input Comparator

Typical Applications (Continued)

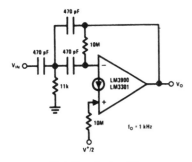

High Pass Active Filter

$f_0 = 1 \text{ kHz}$

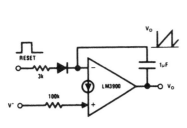

Sawtooth Generator

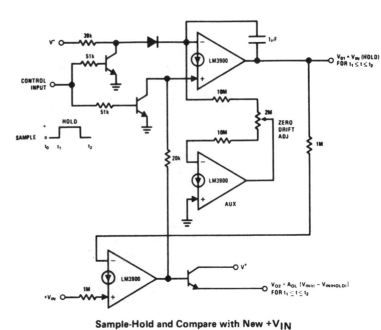

Sample-Hold and Compare with New +V$_{IN}$

$V_{O1} = V_{IN} \text{ (HOLD)}$
FOR $t_1 \leq t \leq t_2$

$V_{O2} = A_{OL} (V_{IN(t)} - V_{IN(HOLD)})$
FOR $t_1 \leq t \leq t_2$

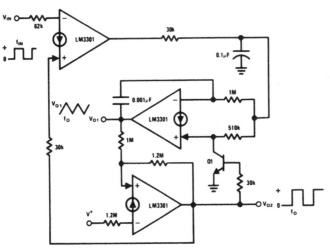

Phase-locked Loop

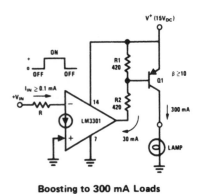

Boosting to 300 mA Loads

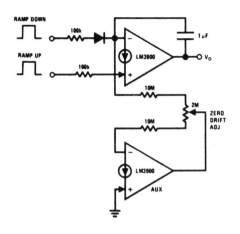

Low-Drift Ramp and Hold Circuit

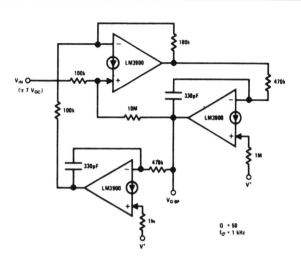

**Bi-Quad Active Filter
(2nd Degree State-Variable Network)**

Split-Supply Applications $(V^+ = +15 \; V_{DC} \; \& \; V^- = -15 \; V_{DC})$

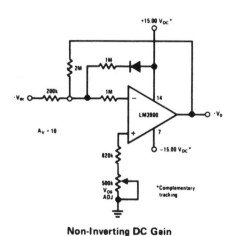

Non-Inverting DC Gain

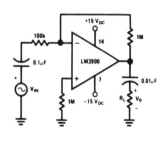

AC Amplifier

Appendix C μA7800 Series 3-Terminal Positive Voltage Regulators

Linear Products

Description

The μA78M00 series of 3-Terminal Medium Current Positive Voltage Regulators is constructed using the Fairchild Planar epitaxial process. These regulators employ internal current-limiting, thermal-shutdown and safe-area compensation making them essentially indestructible. If adequate heat sinking is provided, they can deliver in excess of 500 mA output current. They are intended as fixed voltage regulators in a wide range of applications including local or on-card regulation for elimination of noise and distribution problems associated with single point regulation. In addition to use as fixed voltage regulators, these devices can be used with external components to obtain adjustable output voltages and currents.

- **OUTPUT CURRENT IN EXCESS OF 0.5 A**
- **NO EXTERNAL COMPONENTS**
- **INTERNAL THERMAL-OVERLOAD PROTECTION**
- **INTERNAL SHORT-CIRCUIT CURRENT LIMITING**
- **OUTPUT TRANSISTOR SAFE-AREA COMPENSATION**
- **AVAILABLE IN JEDEC TO-220 AND TO-39 PACKAGES**
- **OUTPUT VOLTAGES OF 5 V, 6 V, 8 V, 12 V, 15 V, AND 24 V**
- **MILITARY AND COMMERCIAL TEMPERATURE RANGE**

Absolute Maximum Ratings

Input Voltage	
(5 V through 15 V)	35 V
(20 V, 24 V)	40 V
Internal Power Dissipation	Internally Limited
Storage Temperature Range	
TO-39	−65°C to + 150°C
TO-220	−55°C to + 150°C
Operating Junction Temperature Range	
μA78M00	−55°C to + 150°C
μA78M00C	0°C to + 125°C
Pin Temperatures	
(Soldering, 60 s time limit)	
TO-39	300°C
(Soldering, 10 s time limit)	
TO-220	230°C

Connection Diagram
TO-39 Package

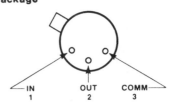

(Top View)

Order Information

Type	Package	Code	Part No.
μA78M05	Metal	FC	μA78M05HM
μA78M06	Metal	FC	μA78M06HM
μA78M08	Metal	FC	μA78M08HM
μA78M12	Metal	FC	μA78M12HM
μA78M15	Metal	FC	μA78M15HM
μA78M24	Metal	FC	μA78M24HM
μA78M05C	Metal	FC	μA78M05HC
μA78M06C	Metal	FC	μA78M06HC
μA78M08C	Metal	FC	μA78M08HC
μA78M12C	Metal	FC	μA78M12HC
μA78M15C	Metal	FC	μA78M15HC

Connection Diagram
TO-220 Package

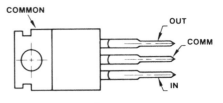

(Side View)

Order Information

Type	Package	Code	Part No.
μA78M05C	Molded Power Pack	GH	μA78M05UC
μA78M06C	Molded Power Pack	GH	μA78M06UC
μA78M08C	Molded Power Pack	GH	μA78M08UC
μA78M12C	Molded Power Pack	GH	μA78M12UC
μA78M15C	Molded Power Pack	GH	μA78M15UC
μA78M24C	Molded Power Pack	GH	μA78M24UC

(Courtesy of Fairchild Camera & Instrument Corporation, Linear Division, © copyright 1982.)

Equivalent Circuit

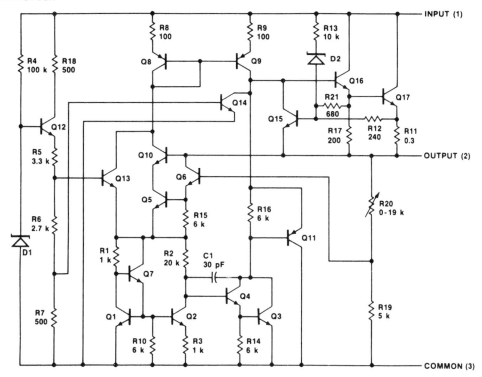

μA7805
Electrical Characteristics V_{IN} = 10 V, I_{OUT} = 500 mA, −55°C ≤ T_J ≤ 150°C, C_{IN} = 0.33 μF, C_{OUT} = 0.1 μF, unless otherwise specified.

Characteristic		Condition (Note)		Min	Typ	Max	Unit
Output Voltage		T_J = 25°C		4.8	5.0	5.2	V
Line Regulation		T_J = 25°C	7 V ≤ V_{IN} ≤ 25 V		3	50	mV
			8 V ≤ V_{IN} ≤ 12 V		1	25	mV
Load Regulation		T_J = 25°C	5 mA ≤ I_{OUT} ≤ 1.5 A		15	100	mV
			250 mA ≤ I_{OUT} ≤ 750 mA		5	25	mV
Output Voltage		8.0 V ≤ V_{IN} ≤ 20 V 5 mA ≤ I_{OUT} ≤ 1.0 A P ≤ 15 W		4.65		5.35	V
Quiescent Current		T_J = 25°C			4.2	6.0	mA
Quiescent Current Change	with line	8 V ≤ V_{IN} ≤ 25 V				0.8	mA
	with load	5 mA ≤ I_{OUT} ≤ 1.0 A				0.5	mA
Output Noise Voltage		T_A = 25°C, 10 Hz ≤ f ≤ 100 kHz			8	40	μV / V_{OUT}
Ripple Rejection		f = 120 Hz, 8 V ≤ V_{IN} ≤ 18 V		68	78		dB
Dropout Voltage		I_{OUT} = 1.0 A, T_J = 25°C			2.0	2.5	V
Output Resistance		f = 1 kHz			17		mΩ
Short-Circuit Current		T_J = 25°C, V_{IN} = 35 V			0.75	1.2	A
Peak Output Current		T_J = 25°C		1.3	2.2	3.3	A
Average Temperature Coefficient of Output Voltage		I_{OUT} = 5 mA	−55° C ≤ T_J ≤ +25°C			0.4	mV/°C/
			+25°C ≤ T_J ≤ +150°C			0.3	V_{OUT}

μA7805C
Electrical Characteristics V_{IN} = 10 V, I_{OUT} = 500 mA, 0°C ≤ T_J ≤ 125°C, C_{IN} = 0.33 μF, C_{OUT} = 0.1 μF, unless otherwise specified.

Characteristic		Condition (Note)		Min	Typ	Max	Unit
Output Voltage		T_J = 25°C		4.8	5.0	5.2	V
Line Regulation		T_J = 25°C	7 V ≤ V_{IN} ≤ 25 V		3	100	mV
			8 V ≤ V_{IN} ≤ 12 V		1	50	mV
Load Regulation		T_J = 25°C	5 mA ≤ I_{OUT} ≤ 1.5 A		15	100	mV
			250 mA ≤ I_{OUT} ≤ 750 mA		5	50	mV
Output Voltage		7 V ≤ V_{IN} ≤ 20 V 5 mA ≤ I_{OUT} ≤ 1.0 A P ≤ 15 W		4.75		5.25	V
Quiescent Current		T_J = 25°C			4.2	8.0	mA
Quiescent Current Change	with line	7 V ≤ V_{IN} ≤ 25 V				1.3	mA
	with load	5 mA ≤ I_{OUT} ≤ 1.0 A				0.5	mA
Output Noise Voltage		T_A = 25°C, 10 Hz ≤ f ≤ 100 kHz			40		μV
Ripple Rejection		f = 120 Hz, 8 V ≤ V_{IN} ≤ 18 V		62	78		dB
Dropout Voltage		I_{OUT} = 1.0 A, T_J = 25°C			2.0		V
Output Resistance		f = 1 kHz			17		mΩ
Short-Circuit Current		T_J = 25°C, V_{IN} = 35 V			750		mA
Peak Output Current		T_J = 25°C			2.2		A
Average Temperature Coefficient of Output Voltage		I_{OUT} = 5 mA, 0°C ≤ T_J ≤ 125°C			1.1		mV/°C

μA7806C

Electrical Characteristics V_{IN} = 11 V, I_{OUT} = 500 mA, 0°C $\leq T_J \leq$ 125°C, C_{IN} = 0.33 μF, C_{OUT} = 0.1 μF, unless otherwise specified.

Characteristic	Condition (Note)		Min	Typ	Max	Unit
Output Voltage	T_J = 25°C		5.75	6.0	6.25	V
Line Regulation	T_J = 25°C	8 V $\leq V_{IN} \leq$ 25 V		5	120	mV
		9 V $\leq V_{IN} \leq$ 13 V		1.5	60	mV
Load Regulation	T_J = 25°C	5 mA $\leq I_{OUT} \leq$ 1.5 A		14	120	mV
		250 mA $\leq I_{OUT} \leq$ 750 mA		4	60	mV
Output Voltage	8 V $\leq V_{IN} \leq$ 21 V 5 mA $\leq I_{OUT} \leq$ 1.0 A P $\leq$ 15 W		5.7		6.3	V
Quiescent Current	T_J = 25°C			4.3	8.0	mA
Quiescent Current Change	with line	8 V $\leq V_{IN} \leq$ 25 V			1.3	mA
	with load	5 mA $\leq I_{OUT} \leq$ 1.0 A			0.5	mA
Output Noise Voltage	T_A = 25°C, 10 Hz $\leq$ f $\leq$ 100 kHz			45		μV
Ripple Rejection	f = 120 Hz, 9 V $\leq V_{IN} \leq$ 19 V		59	75		dB
Dropout Voltage	I_{OUT} = 1.0 A, T_J = 25°C			2.0		V
Output Resistance	f = 1 kHz			19		mΩ
Short-Circuit Current	T_J = 25°C, V_{IN} = 35 V			550		mA
Peak Output Current	T_J = 25°C			2.2		A
Average Temperature Coefficient of Output Voltage	I_{OUT} = 5 mA, 0°C $\leq T_J \leq$ 125°C			0.8		mV / °C

Note

1. For all tables, all characteristics except noise voltage and ripple rejection ratio are measured using pulse techniques ($t_w \leq$ 10 ms, duty cycle $\leq$ 5%). Output voltage changes due to changes in internal temperature must be taken into account separately.

μA7815

Electrical Characteristics V_{IN} = 23 V, I_{OUT} = 500 mA, $-55°C \leq T_J \leq 150°C$, C_{IN} = 0.33 μF, C_{OUT} = 0.1 μF, unless otherwise specified

Characteristic		Condition (Note)		Min	Typ	Max	Unit
Output Voltage		$T_J = 25°C$		14.4	15.0	15.6	V
Line Regulation		$T_J = 25°C$	$17.5\ V \leq V_{IN} \leq 30\ V$		11	150	mV
			$20\ V \leq V_{IN} \leq 26\ V$		3	75	mV
Load Regulation		$T_J = 25°C$	$5\ mA \leq I_{OUT} \leq 1.5\ A$		12	150	mV
			$250\ mA \leq I_{OUT} \leq 750\ mA$		4	75	mV
Output Voltage		$18.5\ V \leq V_{IN} \leq 30\ V$ $5\ mA \leq I_{OUT} \leq 1.0\ A$ $P \leq 15\ W$		14.25		15.75	V
Quiescent Current		$T_J = 25°C$			4.4	6.0	mA
Quiescent Current Change	with line	$18.5\ V \leq V_{IN} \leq 30\ V$				0.8	mA
	with load	$5\ mA \leq I_{OUT} \leq 1.0\ A$				0.5	mA
Output Noise Voltage		$T_A = 25°C$, $10\ Hz \leq f \leq 100\ kHz$			8	40	μV / V_{OUT}
Ripple Rejection		f = 120 Hz, $18.5\ V \leq V_{IN} \leq 28.5\ V$		60	70		dB
Dropout Voltage		$I_{OUT} = 1.0\ A$, $T_J = 25°C$			2.0	2.5	V
Output Resistance		f = 1 kHz			19		mΩ
Short-Circuit Current		$T_J = 25°C$, $V_{IN} = 35\ V$			0.75		A
Peak Output Current		$T_J = 25°C$		1.3	2.2	3.3	A
Average Temperature Coefficient of		$I_{OUT} = 5\ mA$	$-55°\ C \leq T_J \leq +25°C$			0.4	mV / °C /
Output Voltage			$+25°C \leq T_J \leq +150°C$			0.3	V_{OUT}

μA7815C

Electrical Characteristics V_{IN} = 23 V, I_{OUT} = 500 mA, $0°C \leq T_J \leq 125°C$, C_{IN} = 0.33 μF, C_{OUT} = 0.1 μF, unless otherwise specified.

Characteristic		Condition (Note)		Min	Typ	Max	Unit
Output Voltage		$T_J = 25°C$		14.4	15.0	15.6	V
Line Regulation		$T_J = 25°C$	$17.5\ V \leq V_{IN} \leq 30\ V$		11	300	mV
			$20\ V \leq V_{IN} \leq 26\ V$		3	150	mV
Load Regulation		$T_J = 25°C$	$5\ mA \leq I_{OUT} \leq 1.5\ A$		12	300	mV
			$250\ mA \leq I_{OUT} \leq 750\ mA$		4	150	mV
Output Voltage		$17.5\ V \leq V_{IN} \leq 30\ V$ $5\ mA \leq I_{OUT} \leq 1.0\ A$ $P \leq 15\ W$		14.25		15.75	V
Quiescent Current		$T_J = 25°C$			4.4	8.0	mA
Quiescent Current Change	with line	$17.5\ V \leq V_{IN} \leq 30\ V$				1.0	mA
	with load	$5\ mA \leq I_{OUT} \leq 1.0\ A$				0.5	mA
Output Noise Voltage		$T_A = 25°C$, $10\ Hz \leq f \leq 100\ kHz$			90		μV
Ripple Rejection		f = 120 Hz, $18.5\ V \leq V_{IN} \leq 28.5\ V$		54	70		dB
Dropout Voltage		$I_{OUT} = 1.0\ A$, $T_J = 25°C$			2.0		V
Output Resistance		f = 1 kHz			19		mΩ
Short-Circuit Current		$T_J = 25°C$, $V_{IN} = 35\ V$			230		A
Peak Output Current		$T_J = 25°C$			2.1		A
Average Temperature Coefficient of Output Voltage		$I_{OUT} = 5\ mA$, $0°C \leq T_J \leq 125°C$			1.0		mV / °C

Typical Performance Curves

Worst Case Power Dissipation Versus Ambient Temperature (TO-3)

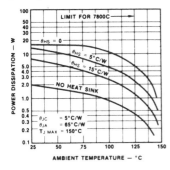

Worst Case Power Dissipation Versus Ambient Temperature (TO-220)

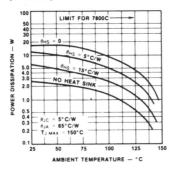

Dropout Voltage as a Function of Junction Temperature

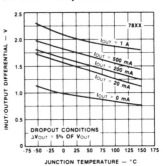

Ripple Rejection as a Function of Frequency

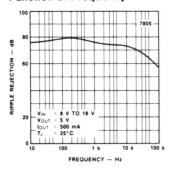

Dropout Characteristics

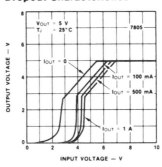

Line Transient Response

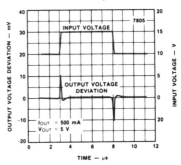

Peak Output Current as a Function of Input/Output Differential Voltage

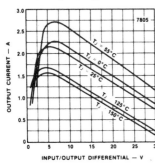

Load Transient Response

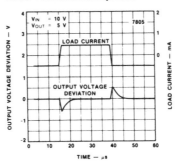

Quiescent Current as a Function of Input Voltage

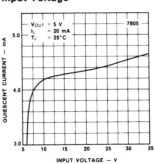

Typical Performance Curves (Cont.)

**Output Voltage as a
Function of
Junction Temperature**

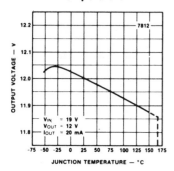

Current Limiting Characteristics

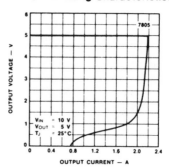

**Output Impedance as a
Function of Frequency**

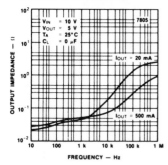

**Quiescent Current as a
Function of Temperature**

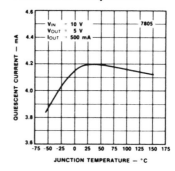

Note
The other μA7800 series devices have similar curves.

DC Parameter Test Circuit

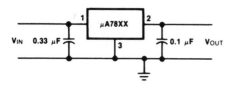

Design Considerations
The μA7800 fixed voltage regulator series has
thermal-overload protection from excessive power
dissipation, internal short circuit protection which
limits the regulator's maximum current, and output
transistor safe area-compensation for reducing the
output current as the voltage across the pass
transistor is increased.

Although the internal power dissipation is limited, the
junction temperature must be kept below the maximum
specified temperature (150°C for 7800, 125°C for
7800C) in order to meet data sheet specifications. To
calculate the maximum junction temperature or heat
sink required, the following thermal resistance values
should be used:

Package	Typ θ_{JC} °C/W	Max θ_{JC} °C/W	Typ θ_{JA} °C/W	Max θ_{JA} °C/W
TO-3	3.5	5.5	40	45
TO-220	3.0	5.0	60	65

$$P_{D(MAX)} = \frac{T_{J(Max)} - T_A}{\theta_{JC} + \theta_{CA}} \text{ or } \frac{T_{J(Max)} - T_A}{\theta_{JA}}$$
(Without heat sink)

$$\theta_{CA} = \theta_{CS} + \theta_{SA}$$

solving for T_J: $T_J = T_A + P_D (\theta_{JC} + \theta_{CA})$
or $T_A + P_D\theta_{JA}$ (Without heat sink)

where T_J = Junction Temperature
T_A = Ambient Temperature
P_D = Power Dissipation
θ_{JC} = Junction-to-case-thermal resistance
θ_{CA} = Case-to-ambient thermal resistance
θ_{CS} = Case-to-heat sink to thermal resistance
θ_{SA} = Heat sink-to-ambient thermal resistance
θ_{JA} = Junction-to-ambient thermal resistance

Typical Applications

Fixed Output Regulator

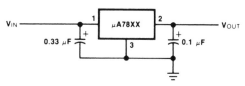

Notes
1. To specify an output voltage, substitute voltage value for "XX."
2. Bypass capacitors are recommended for optimum stability and transient response, and should be located as close as possible to the regulator.

High Input Voltage Circuits

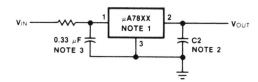

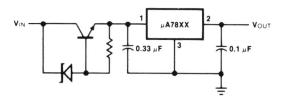

Positive and Negative Regulator

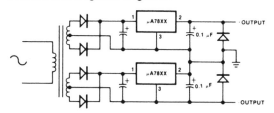

High Current Voltage Regulator

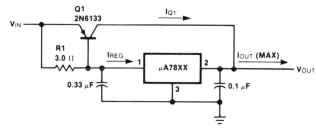

$$\beta(Q1) \geq \frac{I_{OUT(Max)}}{I_{REG(Max)}}$$

$$R1 = \frac{0.9}{I_{REG}} = \frac{\beta(Q1) \, V_{BE(Q1)}}{I_{REG(Max)} \, (\beta + 1) \, - I_{OUT(Max)}}$$

Dual Supply
Operational Amplifier Supply ($\pm$ 15 V @ 1.0 A)

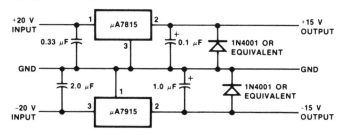

**High Output Current,
Short Circuit Protected**

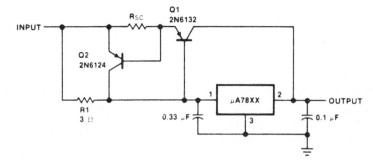

$$R_{SC} = \frac{0.8}{I_{SC}}$$

$$R1 = \frac{\beta \, V_{BE(Q1)}}{I_{REG(Max)} \, (\beta + 1) \; - \; I_{OUT(Max)}}$$

Positive and Negative Regulator

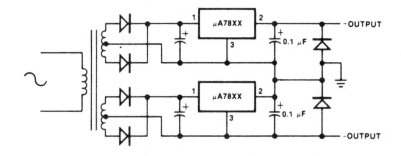

µA78S40 Universal Switching Regulator Subsystem

Description

The µA78S40 is a Monolithic Regulator Subsystem consisting of all the active building blocks necessary for switching regulator systems. The device consists of a temperature-compensated voltage reference, a duty-cycle controllable oscillator with an active current limit circuit, an error amplifier, high-current, high-voltage output switch, a power diode and an uncommitted operational amplifier. The device can drive external npn or pnp transistors when currents in excess of 1.5 A or voltages in excess of 40 V are required. The device can be used for step-down, step-up or inverting switching regulators as well as for series pass regulators. It features wide supply voltage range, low standby power dissipation, high efficiency and low drift. It is useful for any stand-alone, low part count switching system and works extremely well in battery operated systems.

- **STEP-UP, STEP DOWN OR INVERTING SWITCHING REGULATORS**
- **OUTPUT ADJUSTABLE FROM 1.3 to 40 V**
- **PEAK CURRENTS TO 1.5 A WITHOUT EXTERNAL TRANSISTORS**
- **OPERATION FROM 2.5 to 40 V INPUT**
- **LOW STANDBY CURRENT DRAIN**
- **80 dB LINE AND LOAD REGULATION**
- **HIGH GAIN, HIGH CURRENT, INDEPENDENT OP AMP**
- **PULSE WIDTH MODULATION WITH NO DOUBLE PULSING**

Connection Diagram
16-Pin DIP

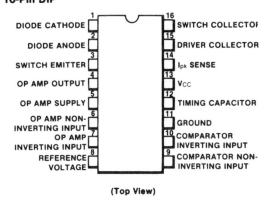

Pin	Left	Pin	Right
1	DIODE CATHODE	16	SWITCH COLLECTOR
2	DIODE ANODE	15	DRIVER COLLECTOR
3	SWITCH EMITTER	14	I_{pk} SENSE
4	OP AMP OUTPUT	13	V_{CC}
5	OP AMP SUPPLY	12	TIMING CAPACITOR
6	OP AMP NON-INVERTING INPUT	11	GROUND
7	OP AMP INVERTING INPUT	10	COMPARATOR INVERTING INPUT
8	REFERENCE VOLTAGE	9	COMPARATOR NON-INVERTING INPUT

(Top View)

Order Information

Type	Package	Code	Part No.
µA78S40	Ceramic DIP	6B	µA78S40DM
µA78S40	Ceramic DIP	6B	µA78S40DC
µA78S40	Molded DIP	9B	µA78S40PC

Block Diagram

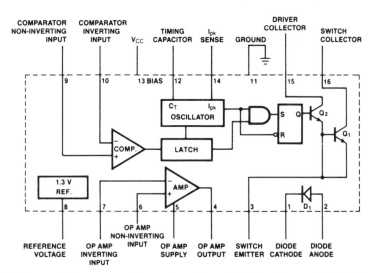

(Courtesy of Fairchild Camera & Instrument Corporation, Linear Division, © copyright 1982.)

Absolute Maximum Ratings

Input Voltage from V+ to V−	40 V
Input Voltage from V+	
Op Amp to V−	40 V
Common Mode Input Range	
(Error Amplifier and Op Amp)	−0.3 to V+
Differential Input Voltage (Note 1)	± 30 V
Output-Short Circuit Duration	
(Op Amp)	continuous
Current from V_{REF}	10 mA
Voltage from Switch	
Collectors to GND	40 V
Voltage from Switch	
Emitters to GND	40 V
Voltage from Switch	
Collectors to Emitter	40 V
Voltage from Power Diode	
to GND	40 V
Reverse Power Diode	
Voltage	40 V
Current through Power Switch	1.5 A
Current through Power Diode	1.5 A
Internal Power Dissipation	
(Note 2)	
Molded DIP	1500 mW
Ceramic DIP	1000 mW
Storage Temperature Range	−65°C to +150°C
Operating Temperature Range	
Military (µA78S40M)	−55°C to 125°C
Commercial (µA78S40C)	0°C to 70°C
Pin Temperature	
Ceramic DIP (Soldering, 60 s)	300°C
Molded DIP (Soldering, 10 s)	260°C

Notes

1. For supply voltages less than 30 V, the absolute maximum voltage is equal to the supply voltage.
2. Ratings apply to 25°C ambient, derate ceramic DIP at 8 mW/°C and plastic DIP at 14 mW/°C.

Functional Description

The µA78S40 is a variable frequency, variable duty cycle device. The initial switching frequency is set by the timing capacitor. The initial duty cycle is 6:1. This switching frequency and duty cycle can be modified by two mechanisms—the current limit circuitry (I_{pk} sense) and the comparator.

The comparator modifies the OFF time. When the output voltage is correct, the comparator output is in the HIGH state and has no effect on the circuit operation. If the output voltage is too high then the comparator output goes LOW. In the LOW state the comparator inhibits the turn on of the output stage switching transistors. As long as the comparator is LOW the system is in OFF time. As the output current rises the OFF time decreases. As the output current nears its maximum the OFF time approaches its minimum value. The comparator can inhibit several ON cycles, one ON cycle or any portion of an ON cycle. Once the ON cycle has begun the comparator cannot inhibit until the beginning of the next ON cycle.

The current limit modifies the ON time. The current limit is activated when a 300 mV potential appears between pin 13 (V_{CC}) and pin 14 (I pk). This potential is intended to result when designed for peak current flows through R_{SC}. When the peak current is reached the current limit is turned on. The current limit circuitry provides for a quick end to ON time and the immediate start of OFF time. Generally the oscillator is free running but the current limit action tends to reset the timing cycle.

Increasing load results in more current limited ON time and less OFF time. The switching frequency increases with load current.

V_D is the forward voltage drop across the internal power diode. It is listed on the data sheet as 1.25 V typical, 1.5 V maximum. If an external diode is used, then its own forward voltage drop must be used for V_D.

V_S is the voltage across the switch element (output transistors Q1 and Q2) when the switch is closed or on. This is listed on the data sheet as output saturation voltage.

Output saturation voltage 1 — defined as the switching element voltage for Q2 and Q1 in the Darlington configuration with collectors tied together. On the data sheet this applies to *Figure 1*, the step down mode.

Output saturation voltage 2 — switching element voltage for just Q1 used as a transistor switch. This applies to *Figure 2* of the data sheet, the step-up mode.

For the inverting mode, *Figure 3*, the saturation voltage of the external transistor should be used for V_S.

Electrical Characteristics $V_{IN} = 5.0$ V, $V_{Op\ Amp} = 5.0$ V, T_A = Operating temperature range, unless otherwise specified.

Characteristic	Condition		Min	Typ	Max	Unit
General Characteristics						
Supply Voltage			2.5		40	V
Supply Current (Op Amp Disconnected)	$V_{IN} = 5.0$ V $V_{IN} = 40$ V			1.8 2.3	3.5 5.0	mA mA
Supply Current Op Amp Connected	$V_{IN} = 5.0$ $V_{IN} = 40$ V				4.0 5.5	mA mA
Reference Section						
Reference Voltage	$I_{REF} = 1.0$ mA	$0 < T_A < 70°C\ \mu A78S40C$	1.180	1.245	1.310	V
		$-55°C < T_A < 125°C$ $\mu A78S40M$				
Reference Voltage Line Regulation	$V_{IN} = 3.0$ V to $V_{IN} = 40$ V, $I_{REF} = 1.0$ mA, $T_A = 25°C$			0.04	0.2	mV / V
Reference Voltage Load Regulation	$I_{REF} = 1.0$ mA to $I_{REF} = 10$ mA, $T_A = 25°C$			0.2	0.5	mV / mA
Oscillator Section						
Charging Current	$V_{IN} = 5.0$ V, $T_A = 25°C$		20		50	μA
Charging Current	$V_{IN} = 40$ V, $T_A = 25°C$		20		70	μA
Discharge Current	$V_{IN} = 5.0$ V, $T_A = 25°C$		150		250	μA
Discharge Current	$V_{IN} = 40$ V, $T_A = 25°C$		150		350	μA
Oscillator Voltage Swing	$V_{IN} = 5$ V, $T_A = 25°C$			0.5		V
t_{on} / t_{off}				6.0		$\mu s / \mu s$
Current Limit Section						
Current Limit Sense Voltage	$T_A = 25°C$		250		350	mV
Output Switch Section						
Output Saturation Voltage 1	$I_{SW} = 1.0$ A, *Figure 1*			1.1	1.3	V
Output Saturation Voltage 2	$I_{SW} = 1.0$ A, *Figure 2*			0.45	0.7	V
Output Transistor h_{FE}	$I_C = 1.0$ A, $V_{CE} = 5.0$ V, $T_A = 25°C$			70		
Output Leakage Current	$V_{OUT} = 40$ V, $T_A = 25°C$			10		nA
Power Diode						
Forward Voltage Drop	$I_D = 1.0$ A			1.25	1.5	V
Diode Leakage Current	$V_D = 40$ V, $T_A = 25°C$			10		nA
Comparator						
Input Offset Voltage	$V_{CM} = V_{REF}$			1.5	15	mV
Input Bias Current	$V_{CM} = V_{REF}$			35	200	nA
Input Offset Current	$V_{CM} = V_{REF}$			5.0	75	nA
Common Mode Voltage Range	$T_A = 25°C$		0		V+ -2	V
Power Supply Rejection Ratio	$V_{IN} = 3.0$ V to 40 V, $T_A = 25°C$		70	96		dB

Electrical Characteristics $V_{IN} = 5.0$ V, $V_{Op\ Amp} = 5.0$ V, T_A = Operating temperature range, unless otherwise specified.

Characteristic	Condition	Min	Typ	Max	Unit
Output Operational Amplifier					
Input Offset Voltage	$V_{CM} = 2.5$ V		4.0	15	mV
Input Bias Current	$V_{CM} = 2.5$ V		30	200	nA
Input Offset Current	$V_{CM} = 2.5$ V		5.0	75	nA
Voltage Gain +	$R_L = 2.0$ k to GND; $V_O = 1.0$ to 2.5 V, $T_A = 25°C$	25 k	250 k		V/V
Voltage Gain −	$R_L = 2.0$ k to V+ Op Amp; $V_O = 1.0$ to 2.5 V, $T_A = 25°C$	25 k	250 k		V/V
Common Mode Voltage Range	$T_A = 25°C$	0		V+ −2	V
Common Mode Rejection Ratio	$V_{CM} = 0$ to 3.0 V, $T_A = 25°C$	76	100		dB
Power Supply Rejection Ratio	V+ Op Amp = 3.0 to 40 V, $T_A = 25°C$	76	100		dB
Output Source Current	$T_A = 25°C$	75	150		mA
Output Sink Current	$T_A = 25°C$	10	35		mA
Slew Rate	$T_A = 25°C$		0.6		V/μs
Output LOW Voltage	$I_L = −5.0$ mA, $T_A = 25°C$			1.0	V
Output HIGH Voltage	$I_L = 50$ mA, $T_A = 25°C$	V+OP Amp −3.0 V			V

Design Formulas

Characteristic	Step Down	Step Up	Inverting	Unit
I_{pk}	$2\ I_{OUT(Max)}$	$2\ I_{OUT(Max)} \bullet \dfrac{V_{OUT} + V_D - V_S}{V_{IN} - V_S}$	$2\ I_{OUT(Max)} \bullet \dfrac{V_{IN} + |V_{OUT}| + V_D - V_S}{V_{IN} - V_S}$	A
R_{SC}	$0.33/I_{pk}$	$0.33\ I_{pk}$	$0.33\ I_{pk}$	Ω
$\dfrac{t_{on}}{t_{off}}$	$\dfrac{V_{OUT} + V_D}{V_{IN} - V_S - V_{OUT}}$	$\dfrac{V_{OUT} + V_D - V_{IN}}{V_{IN} - V_S}$	$\dfrac{|V_{OUT}| + V_D}{V_{IN} - V_S}$	
L	$\dfrac{V_{OUT} + V_D}{I_{pk}} \bullet t_{off}$	$\dfrac{V_{OUT} + V_D - V_{IN}}{I_{pk}} \bullet t_{off}$	$\dfrac{|V_{OUT}| + V_D}{I_{pk}} \bullet t_{off}$	μH
t_{off}	$\dfrac{I_{pk} \bullet L}{V_{OUT} + V_D}$	$\dfrac{I_{pk} \bullet L}{V_{OUT} + V_D - V_{IN}}$	$\dfrac{I_{pk} \bullet L}{|V_{OUT}| + V_D}$	μs
C_T (μF)	$45 \times 10^{-5}\ t_{off}(\mu s)$	$45 \times 10^{-5}\ t_{off}(\mu s)$	$45 \times 10^{-5}\ t_{off}(\mu s)$	μF
C_O	$\dfrac{I_{pk} \bullet (t_{on} + t_{off})}{8\ V_{ripple}}$	$\dfrac{(I_{pk} - I_{OUT})^2 \bullet t_{off}}{2\ I_{pk} \bullet v_{ripple}}$	$\dfrac{(I_{pk} - I_{OUT})^2 \bullet t_{off}}{2\ I_{pk} \bullet v_{ripple}}$	μF
Efficiency	$\dfrac{V_{IN} - V_S + V_D}{V_{IN}} \bullet \dfrac{V_{OUT}}{V_{OUT} + V_D}$	$\dfrac{V_{IN} - V_S}{V_{IN}} \bullet \dfrac{V_{OUT}}{V_{OUT} + V_D - V_S}$	$\dfrac{V_{IN} - V_S}{V_{IN}} \bullet \dfrac{|V_{OUT}|}{V_{OUT} + V_D}$	
$I_{IN(Avg)}$ (Max load Condition)	$\dfrac{I_{pk}}{2} \bullet \dfrac{V_{OUT} + V_D}{V_{IN} - V_S + V_D}$	$\dfrac{I_{pk}}{2}$	$\dfrac{I_{pk}}{2} \bullet \dfrac{|V_{OUT}| + V_D}{V_{IN} + |V_{OUT}| + V_D - V_S}$	A

Typical Step-Down Performance
$T_A = 25°C$

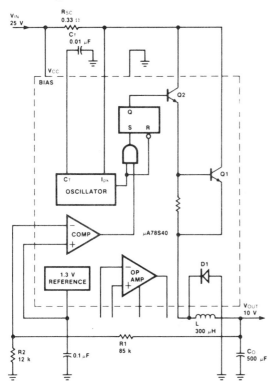

Typical Step-Up Operational Performance
$T_A = 25°C$

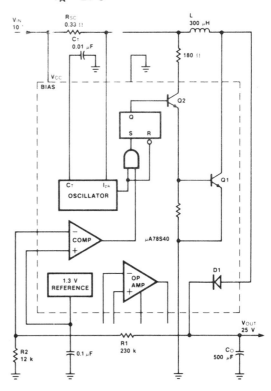

Characteristic	Condition	Typical Value
Output Voltage	I_{OUT} = 200 mA	10 V
Line Regulation	$20 \leq V_{IN} \leq 30$ V	1.5 mV
Load Regulation	5 mA $\leq I_{OUT}$	
	$I_{OUT} \leq 300$ mA	3.0 mV
Max Output Current	V_{OUT} = 9.5 V	500 mA
Output Ripple	I_{OUT} = 200 mA	50 mV
Efficiency	I_{OUT} = 200 mA	74%
Standby Current	I_{OUT} = 200 mA	2.8 mA

Characteristic	Condition	Typical Value
Output Voltage	I_{OUT} = 50 mA	25 V
Line Regulation	5 V $\leq V_{IN} \leq 15$ V	4.0 mV
Load Regulation	5 mA $\leq I_{OUT}$	
	$I_{OUT} \leq 100$ mA	2.0 mV
Max Output Current	V_{OUT} = 23.75 V	160 mA
Output Ripple	I_{OUT} = 50 mA	30 mV
Efficiency	I_{OUT} = 50 mA	79%
Standby Current	I_{OUT} = 50 mA	2.6 mA

Notes

1. For $I_{OUT} \geq 200$ mA use external diode to limit on chip power dissipation.

2. It is recommended that the internal reference (pin 8) be bypassed by a 0.1 μF capacitor directly to (pin 11) the ground point of the $\mu A78S40$.

Typical Inversion Operational Performance
$T_A = 25°C$

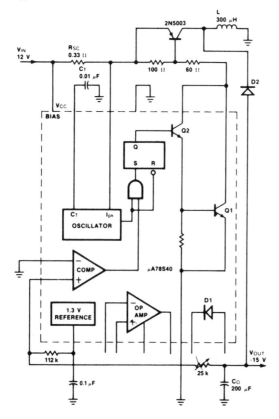

Typical Performance Curves

C_T as a Function of t_{off}

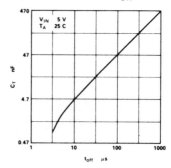

V_{REF} as a Function of T_J

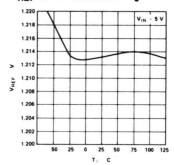

$I_{discharge}$ as a Function of V_{IN}

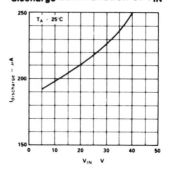

Characteristic	Condition	Typical Value
Output Voltage	$I_{OUT} = 100$ mA	−15 V
Line Regulation	$8\ V \leq V_{IN} \leq 18\ V$	5.0 mV
Load Regulation	$5\ mA \leq I_{OUT}$	
	$I_{OUT} \leq 150$ mA	3.0 mV
Max Output Current	$V_{OUT} = 14.25$ V	160 mA
Output Ripple	$I_{OUT} = 100$ mA	20 mV
Efficiency	$I_{OUT} = 100$ mA	70%
Standby Current	$I_{OUT} = 100$ mA	2.3 mA

Typical Performance Curves (Cont.)

V_{Sense} as a Function of V_{IN}

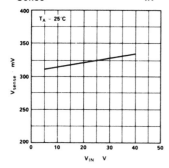

Typical Pulse Width Modulator Application

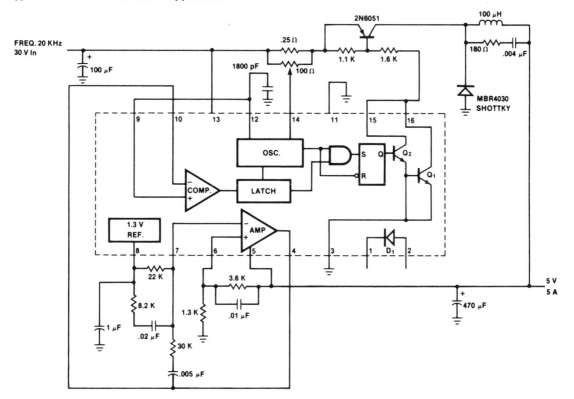

Appendix E XR-2211 FSK Demodulator/ Tone Decoder

Features

- Wide Frequency Range — 0.01Hz to 300kHz
- Wide Supply Voltage Range — 4.5V to 20V
- DTL/TTL/ECL Logic Compatibility
- FSK Demodulation with Carrier-Detector
- Wide Dynamic Range — 2mV to 3V$_{RMS}$
- Adjustable Tracking Range — ±1% to ±80%
- Excellent Temperature Stability — 20ppm/°C Typical

Applications

- FSK Demodulation
- Data Synchronization
- Tone Decoding
- FM Detection
- Carrier Detection

Description

The XR-2211 is a monolithic phase-locked loop (PLL) system especially designed for data communications. It is particularly well suited for FSK modem applications, and operates over a wide frequency range of 0.01Hz to 300kHz. It can accommodate analog signals between 2mV and 3V, and can interface with conventional DTL, TTL and ECL logic families. The circuit consists of a basic PLL for tracking an input signal frequency within the passband, a quadrature phase detector which provides carrier detection, and an FSK voltage comparator which provides FSK demodulation. External components are used to independently set carrier frequency, bandwidth, and output delay.

Schematic Diagram

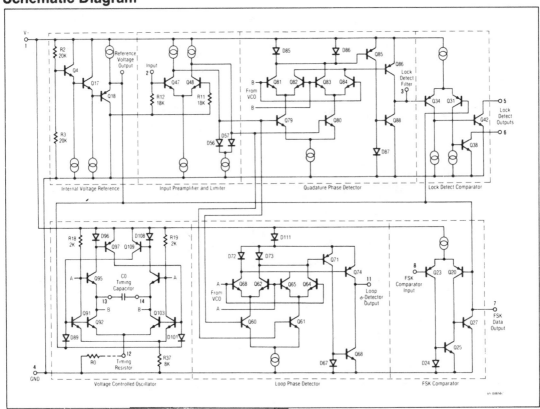

(Courtesy of Raytheon Company, Semiconductor Division, © copyright.)

Connection Information

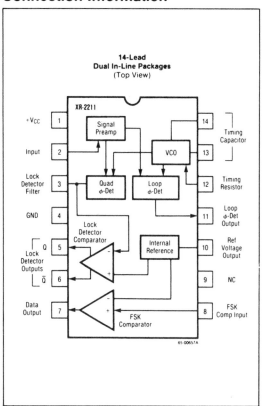

Functional Block Diagram

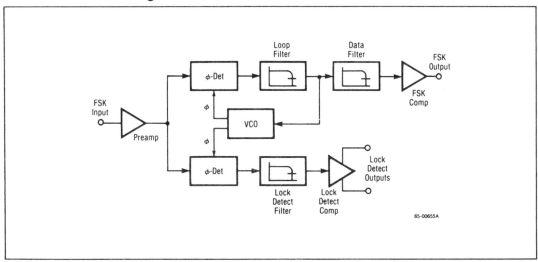

Thermal Characteristics

	14-Lead Plastic DIP	14-Lead Ceramic DIP
Max. Junction Temp.	125°C	175°C
Max. P_D T_A < 50°C	468mW	1042mW
Therm. Res. θ_{JC}	—	50°C/W
Therm. Res. θ_{JA}	160°C/W	120°C/W
For T_A > 50°C Derate at	6.25mW per °C	8.33mW per °C

Ordering Information

Part Number	Package	Operating Temperature Range
XR-2211M	Ceramic	–55°C to +125°C
XR-2211N	Ceramic	–40°C to +85°C
XR-2211P	Plastic	–40°C to +85°C
XR-2211CN	Ceramic	0°C to +75°C
XR-2211CP	Plastic	0°C to +75°C

Absolute Maximum Ratings

Supply Voltage .... +20V
Input Signal Level .... 3V_{RMS}
Storage Temperature
   Range .... –65°C to +150°C

Operating Temperature Range
   XR-2211CN/CP .... 0°C to +75°C
   XR-2211N/P .... –40°C to +85°C
   XR-2211M .... –55°C to +125°C
Lead Soldering
   Temperature (60 Sec) .... 300°C

Electrical Characteristics

Test Conditions V+ = +12V, T_A = +25°C, R0 = 30kΩ, C0 = 0.033μF.
See Figure 1 for component designations.

Parameters	Conditions	XR-2211/M			XR-2211C			Units
		Min	Typ	Max	Min	Typ	Max	
General								
Supply Voltage		4.5		20	4.5		20	V
Supply Current	R0 ≥ 10kΩ		4.0	9.0		5.0	11.0	mA
Oscillator								
Frequency Accuracy	Deviation from f_0 = 1/R0C0		±1.0	±3.0		±1.0		%
Frequency Stability Temperature Coeffient	R1 = ∞		±20	±50		±20		ppm/°C
Power Supply Rejection	V+ = 12 ±1V		0.05	0.5		0.05		%/V
	V+ = 5 ±0.5V		0.2			0.2		%/V
Upper Frequency Limit	R0 = 8.2kΩ, C0 = 400pF	100	300			300		kHz
Lowest Practical Operating Frequency	R0 = 2MΩ C0 = 50 μF		0.01			0.01		Hz
Timing Resistor, R0 Operating Range		5.0		2000	5.0		2000	kΩ
Recommended Range		15		100	15		100	kΩ

Electrical Characteristics (Cont'd)

Test Conditions V^+ = +12V, T_A = +25°C, R0 = 30kΩ, C0 = 0.033μF.
See Figure 1 for component designations.

Parameters	Conditions	XR-2211/M			XR-2211C			Units
		Min	Typ	Max	Min	Typ	Max	
Loop Phase Detector								
Peak Output Current	Meas. at Pin 11	±150	±200	±300	±100	±200	±300	μA
Output Offset Current			±1.0			±2.0		μA
Output Impedance			1.0			1.0		MΩ
Maximum Swing	Ref. to Pin 10	±4.0	±5.0		±4.0	±5.0		V
Quadrature Phase Detector								
Peak Output Current	Meas. at Pin 3	100	150			150		μA
Output Impedance			1.0			1.0		MΩ
Maximum Swing			11			11		Vp-p
Input Preamp								
Input Impedance	Meas. at Pin 2		20			20		kΩ
Input Signal Voltage Required to Cause Limiting			2.0	10		2.0		mV$_{RMS}$
Voltage Comparator								
Input Impedance	Meas. at Pins 3 & 8		2.0			2.0		MΩ
Input Bias Current			100			100		nA
Voltage Gain	R_L = 5.1kΩ	55	70		55	70		dB
Output Voltage Low	I_C = 3mA		300			300		mV
Output Leakage Current	V_O = 12V		0.01			0.01		μA
Internal Reference								
Voltage Level	Meas. at Pin 10	4.9	5.3	5.7	4.75	5.3	5.85	V
Output Impedance			100			100		Ω

Description of Circuit Controls

Signal Input (Pin 2)

Signal is ac coupled to this terminal. The internal impedance at pin 2 is 20kΩ. Recommended input signal level is in the range of 10mV$_{RMS}$ to 3V$_{RMS}$.

Quadrature Phase Detector Output (Pin 3)

This is the high-impedance output of quadrature phase detector, and is internally connected to the input of lock-detect voltage-comparator. In tone-detection applications, pin 3 is connected to ground through a parallel combination of R_D and C_D (see Figure 1) to elimate the chatter at lock-detect outputs. If this tone-detect section is not used, pin 3 can be left open circuited.

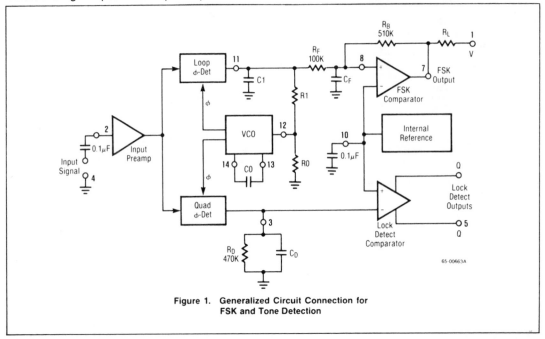

Figure 1.   Generalized Circuit Connection for FSK and Tone Detection

Lock-Detect Output, Q (Pin 5)

The output at pin 5 is at "high" state when the PLL is out of lock and goes to "low" or conducting state when the PLL is locked. It is an open-collector type output and requires a pull-up resistor, R_L, to V+ for proper operation. At "low" state, it can sink up to 5mA of load current.

Lock-Detect Complement, $\overline{Q}$ (Pin 6)

The output at pin 6 is the logic complement of the lock-detect output at pin 5. This output is also an open-collector type stage which can sink 5mA of load current at low or "on" state.

FSK Data Output (Pin 7)

This output is an open-collector logic stage which requires a pull-up resistor, R_L, to to V+ for proper operation. It can sink 5mA of load

current. When decoding FSK signals, FSK data output is at "high" or off state for low input frequency; and at "low" or on state for high input frequency. 1f no input signal is present, the logic state at pin 7 is indeterminate.

FSK Comparator Input (Pin 8)

This is the high-impedance input to the FSK voltage comparator. Normally, an FSK post-detection or data filter is connected between this terminal and the PLL phase-detector output (pin 11). This data filter is formed by R_F and C_F of Figure 1. The threshold voltage of the comparator is set by the internal reference voltage, V_R, available at pin 10.

Reference Voltage, V_R (Pin 10)

This pin is internally biased at the reference voltage level, V_R; $V_R = $ V+/2 – 650mV. The dc voltage level at this pin forms an internal reference for the voltage levels at pins 3, 8, 11 and 12. Pin 10 must be bypassed to ground with a 0.1μF capacitor, for proper operation of the circuit.

Loop Phase Detector Output (Pin 11)
This terminal provides a high-impedance output for the loop phase-detector. The PLL loop filter is formed by R1 and C1 connected to pin 11 (see Figure 1). With no input signal, or with no phase-error within the PLL, the dc level at pin 11 is very nearly equal to V_R. The peak voltage swing available at the phase detector output is equal to $\pm V_R$.

VCO Control Input (Pin 12)
VCO free-running frequency is determined by external timing resistor, R0, connected from this terminal to ground. The VCO free-running frequency, f_O, is:

$$f_O = \frac{1}{R0C0} \text{ Hz}$$

where C0 is the timing capacitor across pins 13 and 14. For optimum temperature stability, R0 must be in the range of 10kΩ to 100kΩ (see Typical Electrical Characteristics).

This terminal is a low-impedance point, and is internally biased at a dc level equal to V_R. The maximum timing current drawn from pin 12 must be limited to $\leq$ 3mA for proper operation of the circuit.

VCO Timing Capacitor (Pins 13 and 14)
VCO frequency is inversely proportional to the external timing capacitor, C0, connected across these terminals. C0 must non-polar, and in the range of 200pF to 10μF.

VCO Frequency Adjustment
VCO can be fine-tuned by connecting a potentiometer, R_X, in series with R0 at pin 12 (see Figure 2).

VCO Free-Running Frequency, f_O
The XR-2211 does not have a separate VCO output terminal. Instead, the VCO outputs are internally connected to the phase-detector sections of the circuit. However, for set-up or adjustment purposes, VCO free-running frequency can be measured at pin 3 (with C_D disconnected), with no input and with pin 2 shorted to pin 10.

Design Equations

See Figure 1 for Definitions of Components.

1. VCO Center Frequency, f_O:
$$f_O = 1/R0C0 \text{ Hz}$$

2. Internal Reference Voltage, V_R (measured at pin 10):
$$V_R = V^+/2 - 650\text{mV}$$

3. Loop Lowpass Filter Time Constant, τ:
$$\tau = R1C1$$

4. Loop Damping, ζ:
$$\zeta = 1/4 \sqrt{\frac{C0}{C1}}$$

5. Loop Tracking Bandwidth, $\pm\Delta f/f_O$:
$$\Delta f/f_O = R0/R1$$

$\Delta f/f_0 = R0/R1$

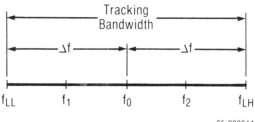

65-00664A

6. FSK Data Filter Time Constant, τ_F:
$$\tau_F = R_F C_F$$

7. Loop Phase Detector Conversion Gain, K_ϕ: (K_ϕ is the differential dc voltage across pins 10 and 11, per unit of phase error at phase-detector input):
$$K_\phi = -2V_R/\pi \text{ Volts/Radian}$$

8. VCO Conversion Gain, K0 is the amount of change in VCO frequency, per unit of dc voltage change at pin 11):
$$K0 = -1/V_R \text{ C0R1 Hz/Volt}$$

9. Total Loop Gain, K_T:
$$K_T = 2\pi K_\phi K0 = 4/C0R1 \text{ Rad/Sec/Volt}$$

10. Peak Phase-Detector Current, I_A:
$$I_A = V_R \text{ (Volts)}/25 \text{ mA}$$

Linear FM Detection

The XR-2211 can be used as a linear FM detector for a wide range of analog communications and telemetry applications. The recommended circuit connection for the application is shown in Figure 5. The demodulated output is taken from the loop phase detector output (pin 11), through a post detection filter made up of R_F and C_F, and an external buffer amplifier. This buffer amplifier is necessary because of the high impedance output at pin 11. Normally, a non-inverting unity gain op amp can be used as a buffer amplifier, as shown in Figure 5.

The FM detector gain, i.e., the output voltage change per unit of FM deviation, can be given as:

$$V_{OUT} = R1 \ V_R/100 \ R0 \ \text{Volts/\% deviation}$$

where V_R is the internal reference voltage. ($V_R = V^+/2 - 650\text{mV}$). For the choice of external components R1, R0, C_D, C1 and C_F, see section on Design Equations.

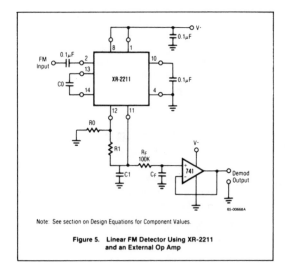

Note: See section on Design Equations for Component Values.

Figure 5.   Linear FM Detector Using XR-2211 and an External Op Amp

Appendix F NE/SE 565 Phase Locked Loop

DESCRIPTION

The SE/NE565 Phase-Locked Loop (PLL) is a self-contained, adaptable filter and demodulator for the frequency range from 0.001Hz to 500kHz. The circuit comprises a voltage-controlled oscillator of exceptional stability and linearity, a phase comparator, an amplifier and a low-pass filter as shown in the block diagram. The center frequency of the PLL is determined by the free-running frequency of the VCO; this frequency can be adjusted externally with a resistor or a capacitor. The low-pass filter, which determines the capture characteristics of the loop, is formed by an internal resistor and an external capacitor.

APPLICATIONS

- Frequency shift keying
- Modems
- Telemetry receivers
- Tone decoders
- SCA receivers
- Wideband FM discriminators
- Data synchronizers
- Tracking filters
- Signal restoration
- Frequency multiplication & division

FEATURES

- Highly stable center frequency (200ppm/°C typ.)
- Wide operating voltage range (±6 to ±12 volts)
- Highly linear demodulated output (0.2% typ.)
- Center frequency programming by means of a resistor or capacitor, voltage or current
- TTL and DTL compatible square-wave output; loop can be opened to insert digital frequency divider
- Highly linear triangle wave output
- Reference output for connection of comparator in frequency discriminator
- Bandwidth adjustable from < ±1% to > ±60%
- Frequency adjustable over 10 to 1 range with same capacitor

PIN CONFIGURATIONS

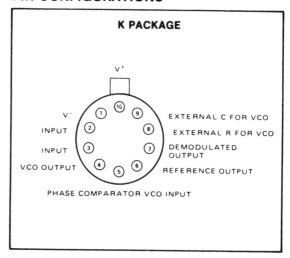

BLOCK DIAGRAM

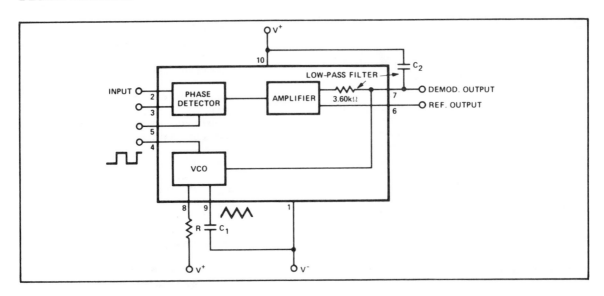

EQUIVALENT SCHEMATIC

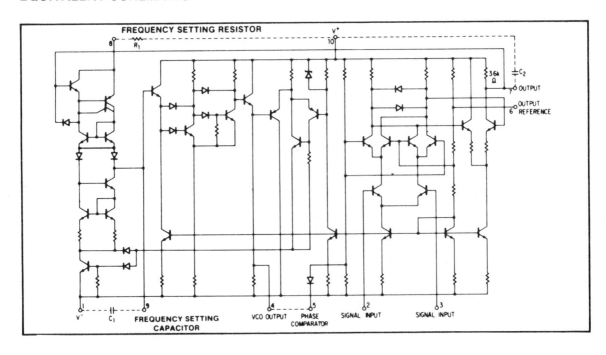

ABSOLUTE MAXIMUM RATINGS $T_A = 25°C$ unless otherwise specified.

PARAMETER	RATING	UNIT
Maximum operating voltage	26	V
Input voltage	3	Vp-p
Storage temperature	–65 to +150	°C
Operating temperature range		
NE565	0 to +70	°C
SE565	–55 to +125	°C
Power dissipation	300	mW

ELECTRICAL CHARACTERISTICS $T_A = 25°C$, $V_{CC} = ±6V$ unless otherwise specified.

PARAMETER	TEST CONDITIONS	SE565			NE565			UNIT
		Min	Typ	Max	Min	Typ	Max	
SUPPLY REQUIREMENTS								
Supply voltage		12	±12		±6		±12	V
Supply current			8	12.5		8	12.5	mA
INPUT CHARACTERISTICS								
Input impedance[1]		7	10		5	10		kΩ
Input level required for	$f_o = 50kHz$, ±10%	10	1		10	1		mVrms
tracking	frequency deviation							
VCO CHARACTERISTICS								
Center frequency								
Maximum value	$C_1 = 2.7pF$	300	500			500		kHz
Distribution[2]	Distribution taken about							
	$f_o = 50kHz$, $R_1 = 5.0kΩ$, $C_1 = 1200pF$	–10	0	+10	–30	0	+30	%
Drift with temperature	$f_o = 50kHz$		200			300		ppm/°C
Drift with supply voltage	$f_o = 50kHz$, $V_{CC} = ±6$ to ±7 volts		0.1	1.0		0.2	1.5	%/V
Triangle wave								
Output voltage level		1.9	0		1.9	0		V
Amplitude			2.4	3		2.4	3	Vp-p
Linearity			0.2			0.5		%
Square wave								
Logical "1" output voltage	$f_o = 50kHz$	+4.9	+5.2		+4.9	+5.2		V
Logical "0" output voltage	$f_o = 50kHz$		–0.2	+0.2		–0.2	+0.2	V
Duty cycle	$f_o = 50kHz$	45	50	55	40	50	60	%
Rise time			20	100		20		ns
Fall time			50	200		50		ns
Output current (sink)		0.6	1		0 6	1		mA
Output current (source)		5	10		5	10		mA
DEMODULATED OUTPUT CHARACTERISTICS								
Output voltage level	Measured at pin 7	4.25	4.5	4.75	4.0	4.5	5.0	V
Maximum voltage swing[3]			2			2		Vp-p
Output voltage swing	±10% frequency deviation	250	300		200	300		mVp-p
Total harmonic distortion			0.2	0.75		0.4	1.5	%
Output impedance[4]			3.6			3.6		kΩ
Offset voltage (V6-V7)			30	100		50	200	mV
Offset voltage vs temperature (drift)			50			100		μV/°C
AM rejection		30	40			40		dB

NOTES

1. Both input terminals (pins 2 and 3) must receive identical dc bias. This bias may range
from 0 volts to –4 volts.
2. The external resistance for frequency adjustment (R1) must have a value between 2kΩ
and 20kΩ.
3. Output voltage swings negative as input frequency increases.
4. Output not buffered.

NE/SE 567 Tone Decoder/ Phase Locked Loop

DESCRIPTION

The SE/NE567 tone and frequency decoder is a highly stable phase-locked loop with synchronous AM lock detection and power output circuitry. Its primary function is to drive a load whenever a sustained frequency within its detection band is present at the self-biased input. The bandwidth center frequency, and output delay are independently determined by means of four external components.

FEATURES

- Wide frequency range (.01Hz to 500kHz)
- High stability of center frequency
- Independently controllable bandwidth (up to 14 percent)
- High out-band signal and noise rejection
- Logic-compatible output with 100mA current sinking capability
- Inherent immunity to false signals
- Frequency adjustment over a 20 to 1 range with an external resistor
- Military processing available

APPLICATIONS

- Touch Tone® decoding
- Carrier current remote controls
- Ultrasonic controls (remote TV, etc.)
- Communications paging
- Frequency monitoring and control
- Wireless intercom
- Precision oscillator

PIN CONFIGURATIONS

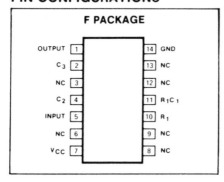

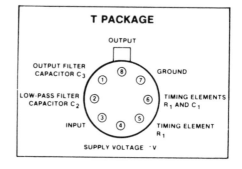

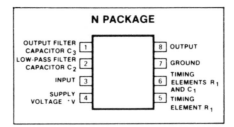

(Courtesy of Signetics, a subsidiary of U.S. Philips Corporation, © copyright 1982.)

BLOCK DIAGRAM

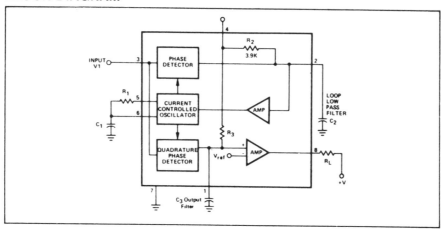

ABSOLUTE MAXIMUM RATINGS

PARAMETER	RATING	UNIT
Operating temperature		
NE567	0 to +70	°C
SE567	–55 to +125	°C
Operating voltage	10	V
Positive voltage at input	$0.5 + V_S$	V
Negative voltage at input	–10	Vdc
Output voltage (collector	15	Vdc
of output transistor)		
Storage temperature	–65 to +150	°C
Power dissipation	300	mW

EQUIVALENT SCHEMATIC

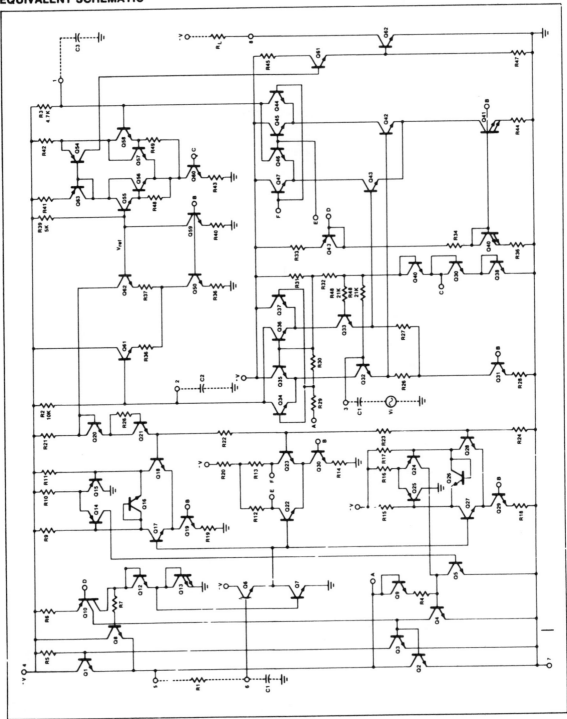

TYPICAL PERFORMANCE CHARACTERISTICS (Cont'd)

PARAMETER	TEST CONDITIONS	SE567			NE567			UNIT
		Min	Typ	Max	Min	Typ	Max	
CENTER FREQUENCY[1]								
Highest center frequency (f_o)		100	500		100	500		kHz
Center frequency stability[2]	−55 to +125°C		35±140			35±140		ppm/°C
	0 to +70°C		35±60			35±60		ppm/°C
Center frequency shift with supply voltage	f_o = 100kHz		0.5	1		0.7	2	%/V
DETECTION BANDWIDTH								
Largest detection bandwidth	f_o = 100kHz	12	14	16	10	14	18	% of f_o
Largest detection bandwidth skew			1	2		2	3	% of f_o
Largest detection bandwidth— variation with temperature	V_i = 300mVrms		±0.1			±0.1		%/°C
Largest detection bandwidth— variation with supply voltage	V_i = 300mVrms		±2			±2		%/V
INPUT								
Input resistance			20			20		kΩ
Smallest detectable input voltage (V_i)	I_L = 100mA, $f_i = f_o$		20	25		20	25	mVrms
Largest no-output input voltage	I_L = 100mA, $f_i = f_o$	10	15		10	15		mVrms
Greatest simultaneous outband signal to inband signal ratio			+6			+6		dB
Minimum input signal to wideband noise ratio	B_n = 140kHz		−6			−6		dB
OUTPUT								
Fastest on-off cycling rate			f_o/20			f_o/20		
"1" output leakage current			0.01	25		0.01	25	A
"0" output voltage	I_L = 30mA		0.2	0.4		0.2	0.4	V
	I_L = 100mA		0.6	1.0		0.6	1.0	V
Output fall time[3]	R_L = 50Ω		30			30		ns
Output rise time[3]	R_L = 50Ω		150			150		ns
GENERAL								
Operating voltage range		4.75		9.0	4.75		9.0	V
Supply current quiescent			6	8		7	10	mA
Supply current—activated	R_L = 20kΩ		11	13		12	15	mA
Quiescent power dissipation			30			35		mW

NOTES

1. Frequency determining resistor R_1 should be between 1 and 20kΩ.
2. Applicable over 4.75 to 5.75 volts. See graphs for more detailed information.
3. Pin 8 to Pin 1 feedback R_L network selected to eliminate pulsing during turn-on and turn-off.

LM135/LM235/LM335,LM135A/ LM235A/LM335A

General Description

The LM135 series are precision, easily-calibrated, integrated circuit temperature sensors. Operating as a 2-terminal zener, the LM135 has a breakdown voltage directly proportional to absolute temperature at +10 mV/°K. With less than 1Ω dynamic impedance the device operates over a current range of 400 μA to 5 mA with virtually no change in performance. When calibrated at 25°C the LM135 has typically less than 1°C error over a 100°C temperature range. Unlike other sensors the LM135 has a linear output.

Applications for the LM135 include almost any type of temperature sensing over a −55°C to +150°C temperature range. The low impedance and linear output make interfacing to readout or control circuitry especially easy.

The LM135 operates over a −55°C to +150°C temperature range while the LM235 operates over a −40°C to +125°C temperature range. The LM335 operates from −40°C to +100°C. The LM135/LM235/LM335 are available packaged in hermetic TO-46 transistor packages while the LM335 is also available in plastic TO-92 packages.

Features

- Directly calibrated in °Kelvin
- 1°C initial accuracy available
- Operates from 400 μA to 5 mA
- Less than 1Ω dynamic impedance
- Easily calibrated
- Wide operating temperature range
- 200°C overrange
- Low cost

Schematic Diagram

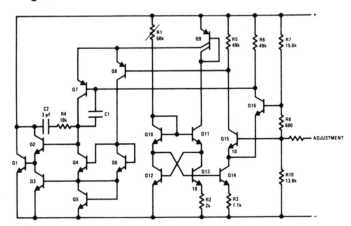

Typical Applications

Basic Temperature Sensor

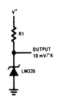

Calibrated Sensor

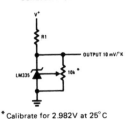

*Calibrate for 2.982V at 25°C

Wide Operating Supply

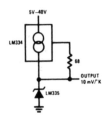

Absolute Maximum Ratings

Reverse Current	15 mA
Forward Current	10 mA
Storage Temperature	
TO-46 Package	-60°C to $+180^{\circ}$C
TO-92 Package	-60°C to $+150^{\circ}$C
Specified Operating Temperature Range	

	Continuous	Intermittent (Note 2)
LM135, LM135A	-55°C to $+150^{\circ}$C	150°C to 200°C
LM235, LM235A	-40°C to $+125^{\circ}$C	125°C to 150°C
LM335, LM335A	-40°C to $+100^{\circ}$C	100°C to 125°C
Lead Temperature (Soldering, 10 seconds)		300°C

Temperature Accuracy LM135/LM235, LM135A/LM235A (Note 1)

PARAMETER	CONDITIONS	LM135A/LM235A			LM135/LM235			UNITS
		MIN	TYP	MAX	MIN	TYP	MAX	
Operating Output Voltage	$T_C = 25^{\circ}$C, $I_R = 1$ mA	2.97	2.98	2.99	2.95	2.98	3.01	V
Uncalibrated Temperature Error	$T_C = 25^{\circ}$C, $I_R = 1$ mA		0.5	1		1	3	$^{\circ}$C
Uncalibrated Temperature Error	$T_{MIN} < T_C < T_{MAX}$, $I_R = 1$ mA		1.3	2.7		2	5	$^{\circ}$C
Temperature Error with 25°C Calibration	$T_{MIN} < T_C < T_{MAX}$, $I_R = 1$ mA		0.3	1		0.5	1.5	$^{\circ}$C
Calibrated Error at Extended Temperatures	$T_C = T_{MAX}$ (Intermittent)		2			2		$^{\circ}$C
Non-Linearity	$I_R = 1$ mA		0.3	0.5		0.3	1	$^{\circ}$C

Temperature Accuracy LM335, LM335A (Note 1)

PARAMETER	CONDITIONS	LM335A			LM335			UNITS
		MIN	TYP	MAX	MIN	TYP	MAX	
Operating Output Voltage	$T_C = 25^{\circ}$C, $I_R = 1$ mA	2.95	2.98	3.01	2.92	2.98	3.04	V
Uncalibrated Temperature Error	$T_C = 25^{\circ}$C, $I_R = 1$ mA		1	3		2	6	$^{\circ}$C
Uncalibrated Temperature Error	$T_{MIN} < T_C < T_{MAX}$, $I_R = 1$ mA		2	5		4	9	$^{\circ}$C
Temperature Error with 25°C Calibration	$T_{MIN} < T_C < T_{MAX}$, $I_R = 1$ mA		0.5	1		1	2	$^{\circ}$C
Calibrated Error at Extended Temperatures	$T_C = T_{MAX}$ (Intermittent)		2			2		$^{\circ}$C
Non-Linearity	$I_R = 1$ mA		0.3	1.5		0.3	1.5	$^{\circ}$C

Electrical Characteristics (Note 1)

PARAMETER	CONDITIONS	LM135/LM235 LM135A/LM235A			LM335 LM335A			UNITS
		MIN	TYP	MAX	MIN	TYP	MAX	
Operating Output Voltage Change with Current	$400\,\mu$A $< I_R <$ 5 mA At Constant Temperature		2.5	10		3	14	mV
Dynamic Impedance	$I_R = 1$ mA		0.5			0.6		Ω
Output Voltage Temperature Drift			+10			+10		mV/$^{\circ}$C
Time Constant	Still Air		80			80		sec
	100 ft/Min Air		10			10		sec
	Stirred Oil		1			1		sec
Time Stability	$T_C = 125^{\circ}$C		0.2			0.2		$^{\circ}$C/khr

Note 1: Accuracy measurements are made in a well-stirred oil bath. For other conditions, self heating must be considered.

Note 2: Continuous operation at these temperatures for 10,000 hours for H package and 5,000 hours for Z package may decrease life expectancy of the device.

Application Hints

CALIBRATING THE LM135

Included on the LM135 chip is an easy method of calibrating the device for higher accuracies. A pot connected across the LM135 with the arm tied to the adjustment terminal allows a 1-point calibration of the sensor that corrects for inaccuracy over the full temperature range.

This single point calibration works because the output of the LM135 is proportional to absolute temperature with the extrapolated output of sensor going to 0V output at $0^\circ K$ ($-273.15^\circ C$). Errors in output voltage versus temperature are only slope (or scale factor) so a slope calibration at one temperature corrects at all temperatures.

The output of the device (calibrated or uncalibrated) can be expressed as:

$$V_{OUT_T} = V_{OUT_{T_0}} \times \frac{T}{T_0}$$

where T is the unknown temperature and T_0 is a reference temperature, both expressed in degrees Kelvin. By calibrating the output to read correctly at one temperature the output at all temperatures is correct. Nominally the output is calibrated at $10\,mV/^\circ K$.

To insure good sensing accuracy several precautions must be taken. Like any temperature sensing device, self heating can reduce accuracy. The LM135 should be operated at the lowest current suitable for the application. Sufficient current, of course, must be available to drive both the sensor and the calibration pot at the maximum operating temperature.

If the sensor is used in an ambient where the thermal resistance is constant, self heating errors can be calibrated out. This is possible if the device is run with a temperature stable current. Heating will then be proportional to zener voltage and therefore temperature. This makes the self heating error proportional to absolute temperature the same as scale factor errors.

WATERPROOFING SENSORS

Meltable inner core heat shrinkable tubing such as manufactured by Raychem can be used to make low-cost waterproof sensors. The LM335 is inserted into the tubing about 1/2" from the end and the tubing heated above the melting point of the core. The unfilled 1/2" end melts and provides a seal over the device.

Typical Applications

Simple Temperature Control

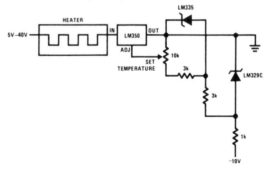

Simple Temperature Controller

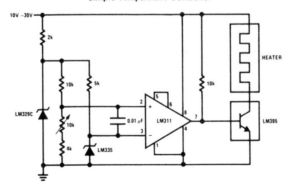

Fast Charger for Nickel-Cadmium Batteries

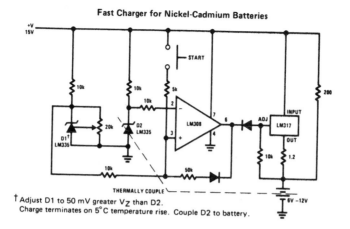

† Adjust D1 to 50 mV greater V_Z than D2.
Charge terminates on 5°C temperature rise. Couple D2 to battery.

Air Flow Detector*

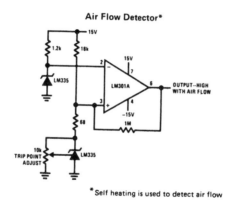

*Self heating is used to detect air flow

Isolated Temperature Sensor

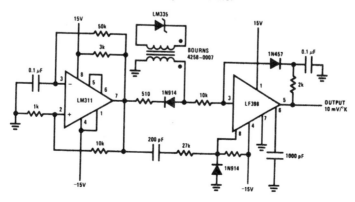

Ground Referred Fahrenheit Thermometer

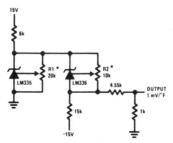

*Adjust R2 for 2.554V across LM336.
 Adjust R1 for correct output.

Centigrade Thermometer

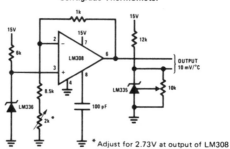

*Adjust for 2.73V at output of LM308

Fahrenheit Thermometer

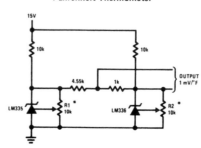

*To calibrate adjust R2 for 2.554V across LM336.
 Adjust R1 for correct output.

Minimum Temperature Sensing

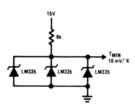

Average Temperature Sensing

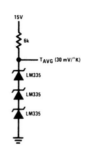

Remote Temperature Sensing

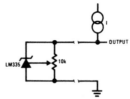

Wire length for 1° C error due to wire drop

AWG	I_R = 1 mA FEET	I_R = 0.5 mA FEET
14	4000	8000
16	2500	5000
18	1600	3200
20	1000	2000
22	625	1250
24	400	800

Definition of Terms

Operating Output Voltage: The voltage appearing across the positive and negative terminals of the device at specified conditions of operating temperature and current.

Uncalibrated Temperature Error: The error between the operating output voltage at 10 mV/$^\circ$K and case temperature at specified conditions of current and case temperature.

Calibrated Temperature Error: The error between operating output voltage and case temperature at 10 mV/$^\circ$K over a temperature range at a specified operating current with the 25°C error adjusted to zero.

Connection Diagrams

TO-92
Plastic Package

BOTTOM VIEW

TO-46
Metal Can Package*

BOTTOM VIEW

*Case is connected to negative pin

Physical Dimensions inches (millimeters)

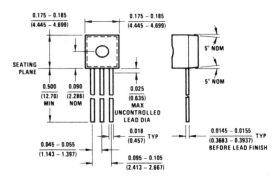

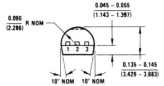

TO-92 Plastic Package (Z)
Order Number LM335Z
or LM335AZ
NS Package Number Z03A

TO-46 Metal Can Package (H)
Order Number LM135H,
LM235H, LM335H, LM135AH,
LM235AH or LM335AH
NS Package Number H03H

National Semiconductor Corporation
2900 Semiconductor Drive
Santa Clara, California 95051
Tel.: (408)737-5000
TWX: (910)339-9240

National Semiconductor GmbH
Eisenheimerstrasse 61/II
8000 München 21
West Germany
Tel.: (089)576091
Telex: 05-22772

NS International Inc., Japan
Miyake Building
1-9 Yotsuya, Shinjuku-ku 160
Tokyo, Japan
Tel.: (03)355-3711
TWX: 232-2015 NSCJ-J

National Semiconductor (Hong Kong) Ltd.
8th Floor,
Cheung Kong Electronic Bldg
4 Hing Yip Street
Kwun Tong
Kowloon, Hong Kong
Tel : 3-899235
Telex: 43866 NSEHK HX
Cable: NATSEMI

NS Electronics Do Brasil
Avda Brigadeiro Faria Lima 844
11 Andar Conjunto 1104
Jardim Paulistano
Sao Paulo, Brasil
Telex
1121008 CABINE SAO PAULO

NS Electronics Pty. Ltd.
Cnr. Stud Rd. & Mtn. Highway
Bayswater, Victoria 3153
Australia
Tel.: 03-729-6333
Telex: 32096

Answers to Selected Questions and Problems

Questions and Problems requiring essay answers are not included in this answer section.

CHAPTER 2

2.5.   0.5 V/μs
2.7.   300
2.9.   100 mV
2.11. 8000
2.13. 120 mW
2.15. 333 Ω

CHAPTER 3

3.1.   (1) b,c,h; (2) i; (3) c; (4) b,g,h,i,k,l;
       (5) g,k,l; (6) a,c,f,; (7) b,h; (8) e;
       (9) j; (10) d
3.3.   24 V

CHAPTER 4

4.11. 25%
4.19. 15 V

CHAPTER 5

5.5.   (1) a,c,e; (2) d,f,g; (3) b; (4) a; (5) b,c;
       (6) f; (7) d; (8) g; (9) b

CHAPTER 6

6.13. 10 s
6.15. select C = 0.01 μF, then R_1 = 10 kΩ
6.17. select C = 0.1 μF, R_1 = 25 kΩ; R_{sh}
       = 250 kΩ; R_2 = 22.7 kΩ; $V_{out(peak)}$
       = 750 mV
6.19. a. $V_{out(sqr)}$ = ± 16 V; b. $V_{out(tri)}$ = ± 8 V;
       c. f_{op} = 12.5 Hz

CHAPTER 7

7.3.   select C = 1 μF, R_2 = 10 kΩ, R_3 =
       20 kΩ; then R_1 = 1600 Ω; A_{dB} = 9.5
       dB; breakpoint = 6.5 dB

7.5. a. scaling: select $C_1 = 0.05\ \mu\text{F}$; then $R_1 = R_2 = 22.5\ \text{k}\Omega$; $C_2 = 0.1\ \mu\text{F}$; $A_{dB} = 0\ \text{dB}$; breakpoint $= -3\ \text{dB}$
b. equal components: select $C_1 = C_2 = 0.05\ \mu\text{F}$; then $R_1 = R_2 = 31.8\ \text{k}\Omega$; $A_{dB} = 4\ \text{dB}$; breakpoint $= 1\ \text{dB}$

7.7. a. wideband filter; b. assume equal components, $A_v = 1.59$; c. $A_{dB} = 4$ dB; d. $BW = 2700\ \text{Hz}$; e. second order

7.9. $A_v = 4.22$; $A_{dB} = 12.5\ \text{dB}$

7.11. $f_{high} = 2200\ \text{Hz}$; $Q = 2.5$

7.13. $f_{low} = 53\ \text{Hz}$; $f_{high} = 67\ \text{Hz}$

CHAPTER 8

8.5. $0.56\ \mu\text{s}$

8.11. $V_{uth} = 2.45\ \text{V}$; $V_{lth} = 1.55\ \text{V}$

8.13. $V_H = 0.90\ \text{V}$

8.15. $C = 25\ \mu\text{F}$

CHAPTER 9

9.7. 1024

9.9. a. 2.5 V; b. 1.75 V; c. 3.75 V; d. 4.5 V; e. 63.75 V

9.11. 0.25 V

9.15. a. 0101010110; b. 3.42 ms; c. 0.0098%

9.17. 100 ms

CHAPTER 10

10.5. refer to Figures 10–6 and 10–11

10.7. refer to Figure 10–20

10.9. 30 MHz

10.11. 6 mV

10.13. 100%

CHAPTER 11

11.13. $242\ \mu\text{s}$

11.15. $27.3\ \mu\text{F}$

11.19. $110\ \mu\text{s}$

11.21. a. 1.04 ms; b. 0.3475 ms; c. 1.3875 ms; d. 742 Hz

11.23. a. 34000 s; b. 17000 s; c. 51000 s

11.25. select $C = 470\ \mu\text{F}$; then $R = 395\ \text{M}\Omega$

11.27. a. 92%; b. 8 %

CHAPTER 12

12.5. Channel 3 = 61.25 MHz; Channel 4 = 67.25 MHz; $L_1 = 0.14\ \mu\text{H}$ (tunable); $C_2 = 39\ \text{pF}$; $C_7 = 12\ \text{pF}$ (tunable)

12.15. 91 nF

12.17. 220 ms

Index